Recent Advances in Mechatronics

Tomas Brezina and Ryszard Jablonski (Eds.)

Recent Advances in Mechatronics

2008-2009

Springer

Prof. Tomas Brezina
Brno University of Technology
Faculty of Mechanical Engineering
Institute of Automation and Computer Science
Technická 2896/2
616 69 Brno
Czech Republic

Prof. Ryszard Jablonski
Warsaw University of Technology
Faculty of Mechatronics
Institute of Metrology and Biomedical Engineering
Sw. A. Boboli Street 8
02-525 Warsaw
Poland

ISBN 978-3-642-42497-7 ISBN 978-3-642-05022-0 (eBook)

DOI 10.1007/978-3-642-05022-0

Softcover re-print of the Hardcover 1st edition 2010

Typesetting: Data supplied by the authors

Production & Cover Design: Scientific Publishing Services Pvt. Ltd., Chennai, India

Printed in acid-free paper

9 8 7 6 5 4 3 2 1

springer.com

Preface

This book comprises the best contributions presented at the 8th International Conference "Mechatronics 2009", organized by Brno Technical University, Faculty of Mechanical Engineering, held on November 18–20, 2009, in Luhačovice, Czech Republic.

For the first time, this conference took place in 1994 in the Czech Republic and since then it has been organized alternately in the Czech Republic as "Mechatronics, Robotics and Biomechanics", and in Poland as "Mechatronics". Until 2005 it was held annually, since that time every second year. This year we used the name "Mechatronics" for the Czech conference for the first time and decided to continue with the Polish conference numbering. Each of the conferences provided a gathering place for academicians and researchers focused on different topics, allowing them to exchange ideas and to inspire each other mainly by specific forms and areas of use of spatial and functional integration.

When choosing the papers to be published in this volume, as is our tradition, we looked for originality and quality within the thematic scope of mechatronics, understood as synergic combination of suitable technologies with application of the advanced simulation tools, aimed at reduction of complexity by spatial and functional integration. Hence, the conference topics include Modelling and Simulation, Metrology & Diagnostics, Sensorics & Photonics, Control & Robotics, MEMS Design & Mechatronic Products, Production Machines and Biomechanics.

We express our thanks to all of the authors for their contribution to this book.

Tomáš Březina
Conference Chairman
Brno University of Technology

Preface

Contents

Modelling and Simulation

Metrology and Diagnostics, Sensorics and Photonics

Control and Robotics

MEMS Design and Mechatronic Products

Production Machines

Biomechanics

Elastic Constants of Austenitic and Martensitic Phases of NiTi Shape Memory Alloy

P. Šesták, M. Černý, and J. Pokluda

Brno University of Technology, Faculty of Mechanical Engineering, Institute of Physical Engineering, Technická 2896/2, Brno, Czech Republic
sestak@fme.vutbr.cz

Abstract. NiTi shape memory alloys start to be widely used in mechatronic systems. In this article, theoretical elastic constants of austenitic and martensitic phases of perfect NiTi crystals and martensitic crystals containing twins in compound twinning mode are presented as computed by using first principles methods. The comparison of elastic constants of the twinned NiTi martensite with those for perfect crystals helps us to understand the transition from elastic to pseudoplastic behavior of NiTi alloys. The results indicate that the elastic response is not influenced by the presence of the twins.

1 Introduction

The NiTi shape memory alloy (SMA) has been discovered in 1963 [1] and, since that time, this material has been used in mechatronic (actuators), medicine (stents, bone implants) [2] and other branches due to their pronounced shape memory effect (SME). This effect is caused by transformation from the martensitic to the austenitic phase and vice versa (see Fig.1) and can be started by an external deformation or a temperature change. This particularly means that, after a deformation-induced shape change in the martensitic condition, the SMA returns to its original geometrical shape when being warmed up to the austenitic state. Such a behavior is facilitated by a reversible creation and vanishing of selected twining variants in the domain-like martensitic microstructure. There are several possible types of phase transformations depending on a particular alloy composition. An extensive overview of a current state of the art can be found in the paper by Otsuka and Ren [3]. There are also some papers investigating this alloy using the first principles (ab-initio) calculations [4-7].

The elastic response corresponds to the near-equilibrium state and, in the case of SMA, the transition from elastic to pseudoplastic behavior is of a great practical importance. The elastic response of materials is characterized by elastic constants c_{ij}. However, these constants for NiTi martensite have been unknown until the end of 2008 when the theoretical (ab-initio) data of these constants were published [5, 10].

It is generally known that the shape memory effect is based on twinning during the pseudoplastic deformation of the NiTi martensite. In general, there are three types of twinnig modes: Type-I, Type-II and compound [3]. Since all the previous

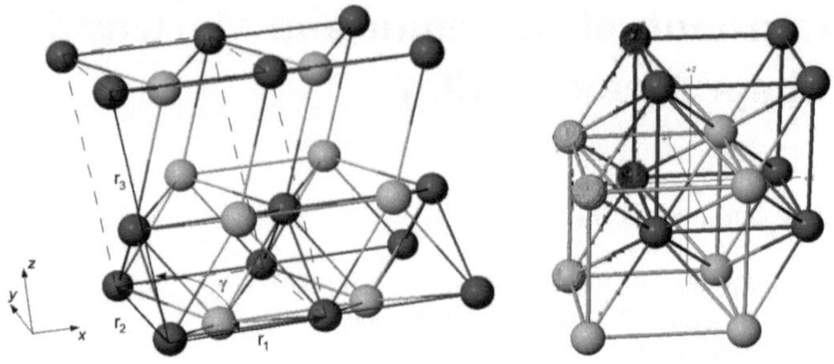

Fig. 1. Martensitic (monoclinic structure B19') and austenitic (cubic structure B2) phase of NiTi shape memory alloy.

theoretical results on c_{ij} [5, 10] were computed for perfect crystals, the influence of twins on elastic characteristics remains still unknown. This influence can be assessed only when the data of elastic characteristics are available for both twinned and perfect NiTi martensite crystals. Indeed, the experimental determination of elastic characteristics of the perfect structure is impossible due to fact that its preparation is beyond the capability of contemporary technologies. Thus, the theoretical simulation represents the only way how to investigate this influence.

The aim of this article is to compute elastic constants of twinned and untwinned martensitic structure as well as those of the austenitic one. Previously published ab-initio results revealed that the B33 orthorhombic martensitic structure possesses a lower energy than the B19' structure usually considered as the ground – state structure. However, the B19' structure is stabilized by residual stresses remaining after the cooling [8, 9]. For that reason, this structure is also studied in this work.

2 The First Principles Calculations

The total energy E_{tot} and the stress tensor τ_i (in the Voigt notation) of the studied system have been computed by the Abinit program code [11]. Abinit is an efficient tool for electronic structure calculations developed by the team of Prof. Xavier Gonze at the Université Catholique de Louvain, which is distributed under GNU General Public Licence. Another additional package including pseudopotentials together with its generators, manuals, tutorials, examples, etc. are available in [12].

The calculations were performed using GGA PAW pseudopotentials and the cutoff energy was set to 270 eV. The solution was considered to be self-consistent when the energy difference of three consequent iterations became smaller than 1.0 μeV.

3 Computation of Elastic Constants

The elastic constants can be computed from the dependence of the total energy E_{tot} on applied deformations (ground state calculations - GS) using the relation

$$c_{ij} = \frac{1}{V_0} \frac{\partial^2 E_{tot}}{\partial \varepsilon_i \partial \varepsilon_j},$$

where ε_i correspond to applied strains, and V_0 is equilibrium volume. The elastic constants c_{ij} can be also computed from the stress – strain dependence as

$$c_{ij} = \frac{d\tau_i}{d\varepsilon_j}.$$

Some elastic constants were obtained in this way but most of them were computed by means of the Linear Response Function method (RF) implemented in the Abinit program code [13]. This approach enables us to obtain elastic constants during a single program run. The elastic constants of a super-cell containing twins have been calculated from the stress-strain dependence.

4 Construction of the Super-Cell

The simulation cell was build as a supper-cell composed of eight primitive cells (of two different bases). The first base corresponds to a standard B19` martensite and the other one represents a tilted base of B19` martensite. The tilted base was created by giving the translation vector $\boldsymbol{r}_3$ a tilt that leads to an increase of the γ angle – see the scheme in Fig. 2.

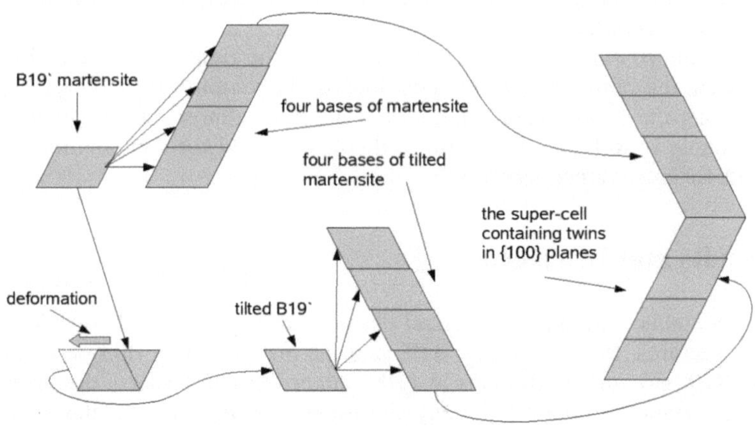

Fig. 2. The process of building the computational super-cell containing {100} twins.

Such a simulation cell is shown in Fig. 3 on the left. However, this cell could not be used for computations of elastic constant c_{ij} yet, because the values of the stress tensor and forces acting on individual atoms at the twin interface were still too high. For this reason, the translation vectors describing the primitive cell and

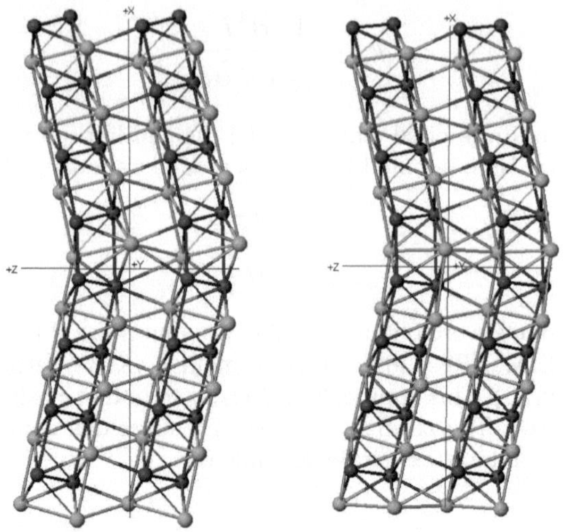

Fig. 3. The super-cell containing twins in {100} planes before optimization of ionic position at the interface (on the left) and after the optimization (on the right)

the ionic positions at the twin interface have been optimized using a relaxation procedure that guarantees the stress values lower than 500 MPa and the atomic forces below 10^{-1} eV/Å. It is very difficult to relax the stresses and forces to lower values because the cell contains an interface between two different variants of B19' martensite and the optimization process must be partially constrained to preserve the twinned structure.

The optimized simulation cell is displayed on the right hand side of Fig. 3. As can be seen, the optimized atomic positions in the vicinity of the interface are arranged along the {100} plane, making the interface almost flat in agreement with data available in Ref. [7]. The optimized cell was used for computation of elastic constants for the twinned structure.

5 Results and Discussion

Table 1 contains computed theoretical elastic constants c_{ij} for all considered martensitic structures; the monoclinic B19` and the orthorhombic B33 perfect crystals and the B19` structure with twins in {100} plane. As can be seen, the investigated twinning variant does not exhibit any significant influence on the elastic constants c_{ij}. Indeed, the c_{ij}-values for the twinned martensite lie well within the range of those for both B19' and B33 perfect crystals.

It should be emphasized that relevant experimental data of the Young modulus E for the B19' structure lie within the range of 90 – 120 GPa [14] which is in agreement with our previous Young's moduli calculations performed for the untwinned B19' structure [4]. This also implies that the twinning has no substantial influence on elastic properties of the NiTi martensite.

Table 1. Theoretical elastic constants for B19' and B33 perfect crystals computed using the Abinit [10] and VASP [5] program codes along with the present results for the super-cell containing (100) twins.

MARTENSITE	c_{11}	c_{22}	c_{33}	c_{12}	c_{13}	c_{23}	c_{44}	c_{55}	c_{66}
Abinit - B19' (RF)	188	231	245	122	89	108	77	45	90
VASP - B19' (GS)	200	241	223	125	99	129	77	21	76
Abinit – B33 (RF)	166	255	268	137	75	98	81	36	108
VASP – B33 (GS)	191	231	247	134	96	137	91	6	83
present (GS)	201	228	224	126	109	127	74	45	72

The theoretical results of elastic constants for the austenitic B2 structure are displayed in Table 2 along with available experimental results. As can be seen, our theoretical data are in a good agreement with those experimentally measured. This confirms a reasonable validity of theoretical ab-initio approaches used in our analysis.

Table 2. The theoretical and experimental data on elastic constants cij for B2 structure

AUSTENITE	c_{11}	c_{12}	c_{44}
Abinit - B2	190	136	50
experiment – B2	180	150	40

6 Conclusion

The presented theoretical data on elastic constants c_{ij} of austenite and martensite structures of NiTi are in a good agreement with available experimental or other theoretical data. The presence of twins in the martensite does not change its elastic response.

Acknowledgement. This research was supported by the Ministry of Education, Youth and Sport of the Czech Republic in the frame of MSM 0021630518 and 2E08017 projects.

References

[1] Buehler, W.J., Gilfrich, J.V., Wiley, R.C.: Effect of low-temperature phase changes on the mechanical properties of alloys near composition TiNi. Journal of Applied Physics 34, 1475–1477 (1963)

[2] Duerig, T., Pelton, A., Stöckel, D.: An overview of nitinol medical applications. Materials Science & Engineering A 273, 149–160 (1999)

[3] Otsuka, K., Ren, X.: Physical metallurgy of Ni-Ti - based shape memory alloys. Progress in Materials Science 50, 511–678 (2005)

[4] Šesták, P., Černý, M., Pokluda, J.: "Elastic properties of B19' structure of NiTi alloy under uniaxial and hydrostatic loading from first principles". Strength of Materials 40, 12–15 (2008)

[5] Wagner, M.F.-X., Windl, W.: Lattice stability, elastic constants and macroscopic moduli of NiTi martensites from first principles. Acta Materialia 56, 6232–6245 (2008)

[6] Wagner, M.F.-X., Windl, W.: Elastic anisotropy of Ni4Ti3 from first principles. Scripta Materialia 60, 207–210 (2009)

[7] Waitz, T., Spišák, D., Hafner, J., Karnthaler, H.P.: Size-dependent martensitic transformation path causing atomic-scale twinning of nanocrystalline niti shape memory alloys. Europhysics Letters 71, 98–103 (2005)

[8] Huang, X., Ackland, G.J., Rabe, K.M.: Crystal structures and shape-memory behaviour of NiTi. Nature Materials 2, 307–311 (2003)

[9] Zhao, J., Meng, F.L., Zheng, W.T., Li, A., Jiang, Q.: Theoretical investigation of atomic-scale (001) twinned martensite in the NiTi alloy. Materials Letters 62, 964–966 (2008)

[10] Šesták, P., Černý, M., Pokluda, J.: The elastic constants of austenitic and martensitic phases of NiTi shape memory alloy. Materials Science and Technology, 120–124 (2008)

[11] Gonze, X., Beuken, J.-M., Caracas, R., Detraux, F., Fuchs, M., Rignanese, G.-M., Sindic, L., Verstraete, M., Zerah, G., Jollet, F., Torrent, M., Roy, A., Mikami, M., Ghosez, Ph., Raty, J.-Y., Allan, D.C.: First-principles computation of material properties: the Abinit software project. Computational Materials Science 25, 478–492 (2002)

[12] Information, `http://www.abinit.org`

[13] Gonze, X.: First-principles responses of solids to atomic displacements and homogeneous electric fields: Implementation of a conjugate-gradient algorithm. Phys. Rev. B 55, 10337–10354 (1997)

[14] Rajagopalan, S., Little, A.L., Bourke, M.A.M., Vaidyanathan, R.: Elastic modulus of shape-memory NiTi from in situ neutron diffraction during macroscopic loading, instrumented indentation, and extenzometry. Applied Physics Letters 86, 81901 (2005)

Simulation Modeling of Mechatronic Drive Systems with Chaotic Behavior

L. Houfek, M. Houfek, and C. Kratochvíl

Brno University of Technology, Faculty of Mechanical Engineering,
Institute of Solid Mechanics, Mechatronics and Biomechanics,
Technicka 2896/2, Brno, Czech Republic
houfek@fme.vutbr.cz

Abstract. The paper is focused on analysis of dynamic properties of controlled drive systems. It describes the possible ways of stability analysis. Paper is also focused on bifurcation of steady states and possible occurence of chaotic behavior.

1 Introduction

Stability analysis cannot be omitted when examining the dynamic properties of controlled drive systems. In case of nonlinear systems and its models one can also expect occurrence of chaotic movements. The approach towards the analysis of its occurrence possibilities will be different when analyzing models with one or a few degrees of freedom or models of real technical systems. [1], [2] Those problems are addressed in the contribution.

2 Occurrence of Chaos in Dissipative Systems and Its Modelling

Dissipative dynamic system can be characterized as systems whose behaviour with increasing time asymptotically approaches steady states if there is no energy added from the outside. Such system description is possible with relatively simple nonlinear equations of motion. For certain values of parameters of those equations the solution does not converge towards expected values, but chaotically oscillates. Strong dependency on small changes of initial conditions occurs as well. When analyzing such phenomena its mathematical essence can be connected with existence of "strange attractor" in phase plane. Possible creation of chaos can be seen in repeated bifurcation of solution, with so called cumulation point behind which the strange attractor is generated. Phase diagram of system solution then transfers from stable set of trajectories towards new, unstable and chaotic set. Creating the global trajectory diagrams is of essential importance. When succesfull, the asymptotic behavior of systems model is described.[3], [4].

3 Global Behavior of Simple Model of Drive System

Let's assume that mathematical model of simple system can be described by non-linear equation:

$$I\ddot{\phi}+b_T\dot{\phi}+k_T\phi+f(\phi)=0 \tag{1}$$

Nonlinear function of displacement is considered in form of $f(\phi)=k_{3T}\phi^3$. Using well known rearrangements the equation (1) can be transformed into more suitable form:

$$\ddot{\phi}+2\kappa\dot{\phi}+\alpha\phi+\beta\phi^3=0 \tag{2}$$

where $\kappa=\frac{b_T}{2I}$, $\alpha=\frac{k_T}{I}\equiv\omega_0^2$ is the natural frequency of undamped model and $\beta=\frac{k_{3T}}{I}$ is the damping.

Now we can observe the changes in steady state of the model and movements around those states when changing parameters of equation (2):

1. lets search for changes of steady states of undampled model when changing parameter α. For values of $\alpha>0$ the system has one steadystate stable position (center). For values of $\alpha<0$ the original steady state breaks up into three new states, two of them stable (centers) and one unstable. The critical bifurcation value is therefore obviously $\alpha=0$.
2. if the value of $\alpha>0$ and value of $\beta<0$, then the original state changes into new one, represented by three steady states, this time two unstable saddles and one stable center. The critical bifurcation value is $\beta=0$.
3. in the dumped model case the state is similar. Original steady state ($\alpha>0$, $\beta>0$), see, characterized by stable focal point changes for $\alpha>0$ and $\beta<0$ again into three steady state, one stable focal point and two unstable saddles. In the case of $\alpha<0$ and $\beta>0$ we obtain two stable focal points and one unstable saddle, see T1,F. Critical bifurcation values are $\alpha=0$ and $\beta=0$, while $\alpha\neq\beta$.

Above shown bifurcations are known as bifurcations of I. type and can (mainly when combined with fluctuation of initial conditions) evoke chaotic movements, which are usually dumped or transferred into different steady states. It's physical interpretation is obvious – classical flexible links with stiff and soft characteristics.

Bifurcation of type II. (Hopf) can occur in the case of change of parameters of models complex conjugate eigenvalues:

$$A_{1.2} = \gamma \pm \Omega \tag{3}$$

In this case the transfer of steady state from left portion of Gauss plane can occur (for $\gamma < 0$) towards the right side (for $\gamma > 0$), that means that steady state stable focal point is changed into unstable focal point and new steady state periodical movement can occur, characterized by limit cycle. The critical bifurcation value is then $\gamma = 0$. It is questionable how such situation can be physically real for particular values (for model according to Eq. (2)), as $\gamma = -\chi$ and $\Omega = \left(\chi^2 + \omega_0^2\right)^{1/2}$, with no energy added from outside environment.

4 Analysis of Real Drive System Model Properties

During the analysis of stability properties of models of real drive systems we often deal with difficulties coming from its structure. Partial results, found by analysis of models with few DOF, see eq. (1) and (2), enable to determine certain areas of design parameters values ensuring the reduction of possible chaotic areas, but with increasing complexity of the model the situation becomes immeasurable. However, there is alternative solution, which comes from the properties of integration formulae, used in current programs for dynamic system analysis. Those formula are sufficiently powerful to enable the detailed evaluation of substitution points when observing the response of analyzed system in phase plane and therefore to reach its full phase portrait. Let's make this case clear on following example.

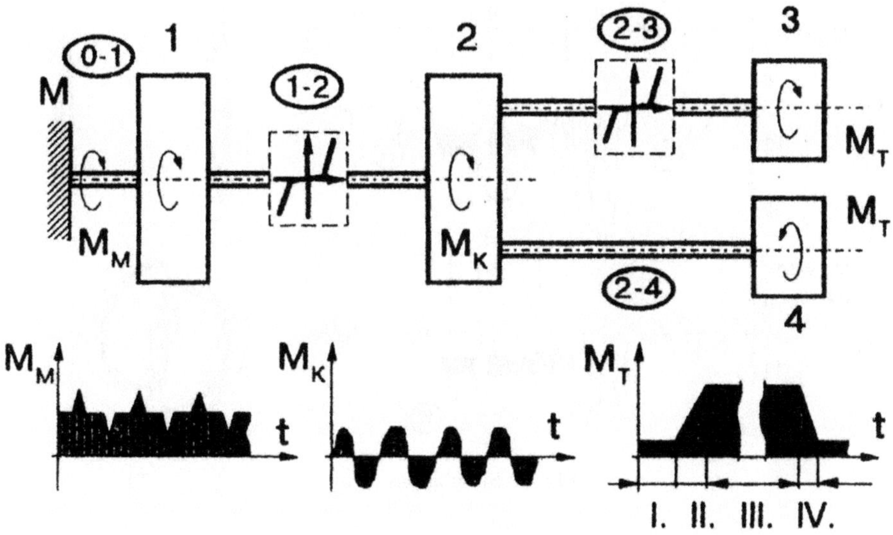

Fig. 1. Model of real drive system

Fig. 1. shows the model of real drive system and model of its revolutions control with complex working state which contains idle run (phase I and end of phase IV), transition states (phase II and beginning of phase IV) and working (operational) state (phase II). Fig. 2 then shows the courses of restoring torques in particular flexible links of the model depending on those phases.

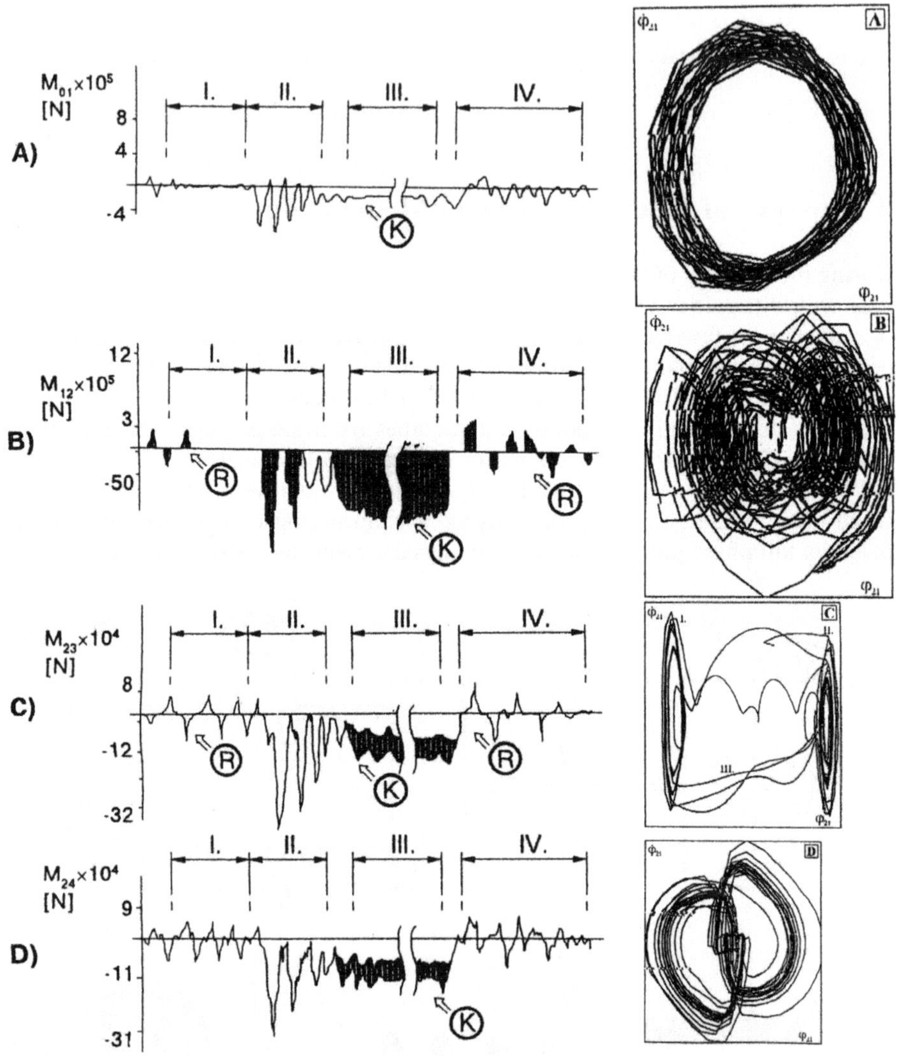

Fig. 2. Simulations results

During the restoring torque $M_{12}(t)$ "disconnection" of the system can occur during idle operation, as well as impacts occurrence and for certain values of gaping the repeated presence of chaotic motion can occur – see Fig. 2b. To clarify this phenomena a number of computer simulations with different value of gap was performed. Phase diagrams in the link (1 - 2) are shown on Fig. 2 right side. Steady states change from relatively static course for very small gap (A) denoted as u_0 towards typically chaotic state (B) for gap value of $2.5u_0$. When further increasing the gap value the parasitic movements occur in the limit of gap value $5u_0$, see (C), finally reaching relatively static state (D) for relatively high gap of $10u_0$. While given states can be considered as attractors, the states among those levels were unstable and corresponded more to complex periodic movements rather than chaotic ones.

Based on given analysis the attributes of chaotic motion can be characterized as follows:

- sensitivity of responses to changes o selected parameters, or initial conditions,
- increasing complexity of regular movements when changing certain parameter (including known motion "period doubling",
- wide Fourier spectrum of system responses (excited by the input with one or only a few frequencies) when in chaotic state and
- introduction of transiting non-periodic oscillating movement which sequentially relaxes towards complex but regular multifrequency motion.

5 Conclusion

Chaos became phenomenon in variety of engineering problems in last years. Therefore we focused on it also in analysis of drive systems. Based on performed analysis we can state following recommendations:

- when evaluating the properties and behavior of dynamic system it is useful to define such parameters of models, which can influence the occurrence of parasitic motion including chaotic one (fluctuation of initial conditions, links gaps, control parameters),
- to observe the evolution of responses in phase planes based on changes of selected parameters and to identify typical chaos effects,
- if such effect occur then evaluate Fourier spectrum of responses. Chaotic motion corresponds to broadband spectra, even when exciting spectra is narrowband.

With respecting given recommendations it is not difficult to identify the areas of possible occurrence of chaos in technical systems using mathematical modelling. However, we do not want to disvalue the analytical approaches with above described alternative approach.

Acknowledgement. Published results were acquired with the support of the research plan of Ministry of Education, Youth and Sports, nr. MSM 0021630518 – Simulation modeling of mechatronics systems and Grant agency of Czech Republic, grant nr. 101/08/0282 – Mechatronic drive systems with nonlinear couplings.

References

[1] Nayfeh, A.H., Balachandran, B.: Applied Nonlinear Dynamics. John Wiley & sons, Inc., New York (1995)

[2] Kamm, L.J.: Understanding Electro-Machanical Engineering (An Introduction to Mechatronics). IEEE Press, Inc., New York (1996)

[3] Kratochvíl, C., Heriban, P., Houfek, L., Houfek, M.: Mechatronické pohonové soustavy, VUT Brno (2008) ISBN 978-80-214-3790-6

[4] Kratochvíl et al: Bifurkace a chaos v technických soustavách a jejich modelování, ÚT AV ČR, Praha (2008) ISBN 978-80-214-3720-3

[5] Moon, F.C.: Chaotic vibrations. John Wiley & sons, Inc., New York (1987)

[6] Procházka, F., Kratochvíl, C.: Úvod do matematického modelování pohonových soustav, CERN, s.r.o., Brno (2002) ISBN 80-7204-256-4

[7] Kratochvíl, C., Procházka, F., Pulkrábek, J.: Pohonové soustavy v mechatronických objektech. In: Int. Conf. Computional Mechanisc 2005, Hrad Nečtiny (2005)

Experimental Research of Chaos and Its Visualization

C. Kratochvil, L. Houfek, and M. Houfek

Brno University of Technology, Faculty of Mechanical Engineering,
Institute of Solid Mechanics, Mechatronics and Biomechanics,
Technicka 2896/2, Brno, Czech Republic

Abstract. Chaos theory as scientific discipline is being developed since the sixties of the last century. Most of the publications are focused on theoretical aspects of this phenomena and the research in case of technical applications is usually using model systems with small number of DOF. In this paper we present the results of simulation studies of chaotic phenomena obtained using so called chaos module on models of nonlinear dynamic systems. Persistence storage oscilloscope is used to visualize obtained results.

1 Introduction

Chaotic behavior of dynamic systems is usually characterized as unpredictable and transitive. However, if we take a look at its visualization using e.g. fractal geometry [1], [4], [5], there are certain laws and order accompanied with the chaos. If we want to understand the relations within chaos, it is useful to study it from different perspectives. One of possible approaches towards study of chaos in real systems is the use of electronic equipment called chaos module.

2 Chaos Module Characteristic

Chaos module is electronic device developed by Yamakawa's Lab & FLSI for modeling and analysis of chaotic states of discrete nonlinear dynamic systems using storage oscilloscope and computers with PSpice program with respect to the changes of selected parameters of dynamic systems [2]. The device uses chaos chip connection enabling activation of chaos module electronic circuit. The device was designed in a way that minimum number of external equipment is required. For example in the simplest wiring it only needs clock signal (rectangular voltage generator), two channels storage oscilloscope and system to be measured. For higher precision measurements one can add external resistors, precise power sources, voltmeters and potentiometers (this device was made on Department of Power Electrical and Electronic Engineering, Faculty of Electrical Engineering and Communication, Brno University of Technology). Internal structure of chaos chip circuit is shown on figure 1. [3].

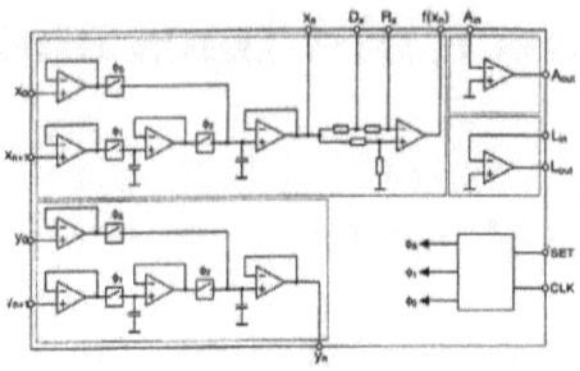

Fig. 1. Internal structure of chaos chip

On Fig 1 one can see the basic structural elements of the chaos chip circuit: delay circuit, summator, inverter and timing circuit. We will further focus on possibility to visualize chaotic states using bifurcation diagrams and using chaotic attractors via storage oscilloscope.

3 Realization of Chaos Chip Wiring into Measuring System

Two variants were implemented using the chaos chip system:

- Modeling of bifurcation diagrams (on 1-D system)
- Modeling of chaotic attractors (on 2-D system)

The block diagrams and some of the results of the experiments are shown in following paragraphs.

3.1 Implementation of 1-D Nonlinear Dynamic System

Figure 2 shows the block diagram of 1-D system with chaos chip. The equation describing such circuit is of form:

$$x_{n+1} = \alpha \cdot \beta \cdot f_{(x_n)}$$

for n = 0,1,2,3,…, where α and β are the gains.

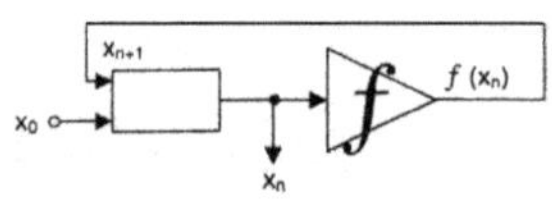

Fig. 2. Block diagram of 1-D system

The goal of this arrangement is to model bifurcation diagrams [3], [4]. As generally known, bifurcation diagrams show, how change of single parameter of the circuit can change behavior of the whole system. The values of parameter that is changed are on horizontal axis from left to right, the state of observed system x_n is on vertical axis.

Circuit diagram of 1-D system is shown on Figure 3. Prior to computational iteration process the SET terminal must be set to positive value and all parameters of nonlinear system must be set, that is R2, R3, U1, U2, gains α and β and initial

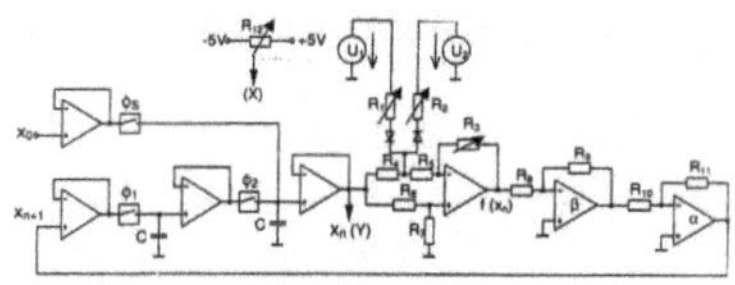

Fig. 3. Circuit diagram of 1-D system

value of iteration x_0. The parameter, whose influence on complete system behavior we observe (e.g. R1) is connected with resistor R12. Its output is the voltage that follows parameter value change. On oscilloscope we connect this voltage to horizontal axis X. Output observed variable (system state) is connected to vertical axis Y.

After setting all the values we bring negative voltage to SET terminal and start computational iteration process. After setting required ranges of X and Y inputs we start to record observed variables in connected storage oscilloscope. At the same time we very slowly change bifurcation parameter (R1 in our case) in given range. This way we obtain on screen bifurcation diagram we are searching for. Examples of bifurcation diagrams calculated using chaos chip for various bifurcation parameters are shown on figures 4, 5 and 6.

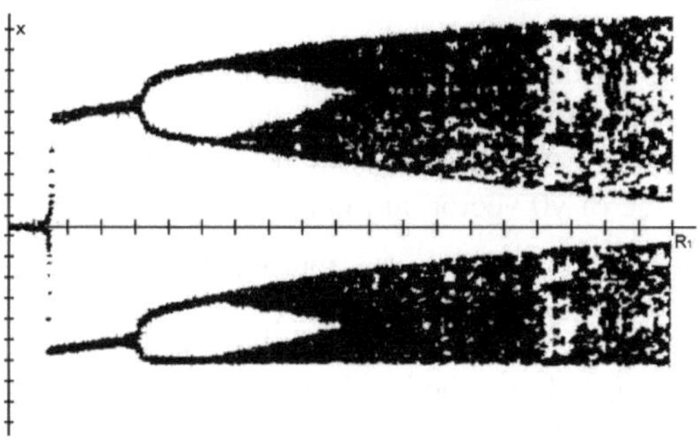

Fig. 4. Bifurcation diagram of the system with R1 parameter

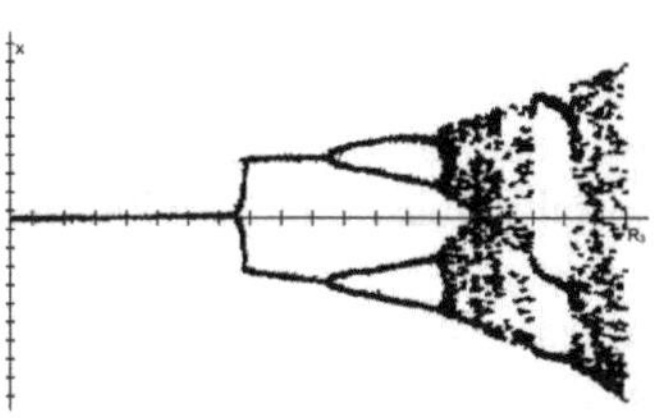

Fig. 5. Bifurcation diagram of the system with R3 parameter

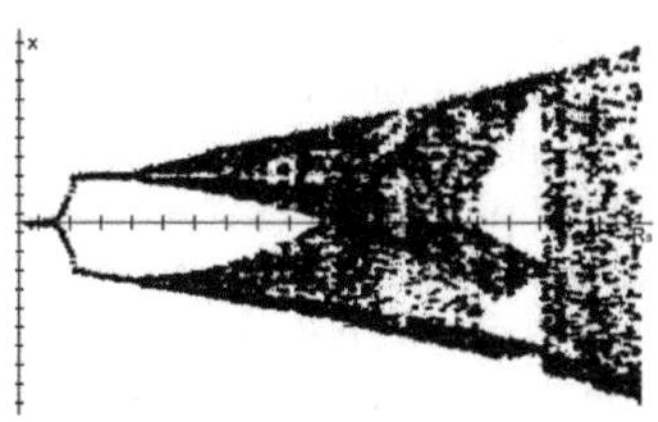

Fig. 6. Bifurcation diagram of the system with R3 parameter but with different gain α compared to case in Fig. 5

3.2 Implementation of 2-D Nonlinear Dynamic System

Figure 7 shows the block diagram of 2-D system with chaos chip. The equations describing such circuit are of form:

$$x_{n+1} = f_{(x_n)} - \alpha \cdot y_n \qquad x_{n+1} = x_n$$

for n = 0,1,2,3,…, where α and β are the gains.

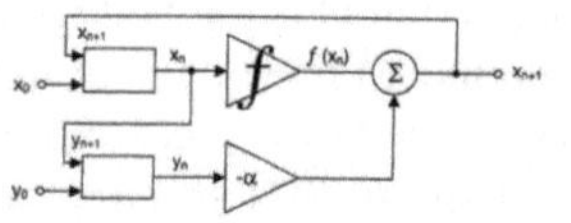

Fig. 7. Block diagram of 2-D system

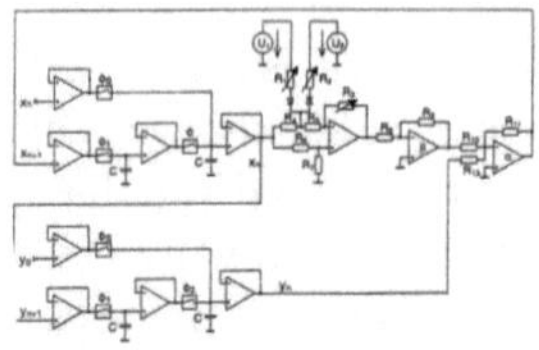

Fig. 8. Circuit diagram of 2-D system

The goal of this arrangement is to model chaotic attractors. Let's note that we consider attractor as sets of responses gained by the state vector of dynamic system during sufficiently long time period from initialization in t_0 time. Attractors in its simplest form are so called fixed points or limit cycles towards which the trajectories of the system are "attracted".

Circuit diagram of 2-D system is shown on Figure 8.

Setting the circuit parameters and initial conditions prior to iterative computation is done in the same way as in 1-D system. Moreover, apart from initial vector x0 there is initial value of y0 vector and gains α and β of particular signals can be set independently for vectors x and y. Output x_n is connected to X axis while y_n output is connected to Y axis. After bringing negative voltage to SET terminal the screen of oscilloscope shows the image which however does not have to be the attractor we are searching for. It strongly depends on setting all parameters of the circuit and setting the initial values of iteration process. Most commonly it is required after starting iteration process to continuously change circuit parameters to put system into chaotic state and therefore to obtain particular attractor. The parameters close to the unstable state must be changed gently, as with even very small change of one or more parameters in "undesired direction" the system immediately gets into stable state. In such a case the iteration process must be stopped, its parameters set again together with initial values, computation must be restarted and "tuned".

Figures below show selected results of numerical experiments. All numerical values within these figures are final, meaning written in the moment of attractor appearance. Only the initial values of iteration correspond to the data regarding the iteration process, as the circuit parameters were "tuned" during the process. Attractor shown on Figure 9 is of particular interest. In this case it was very difficult to stabilize the attractor and obtain input parameters. Moreover, we were unable to

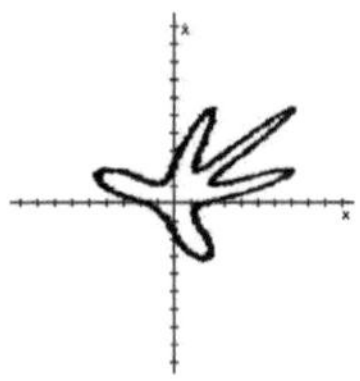

Fig. 9. Immediate attractor corresponding to "just before chaos" state

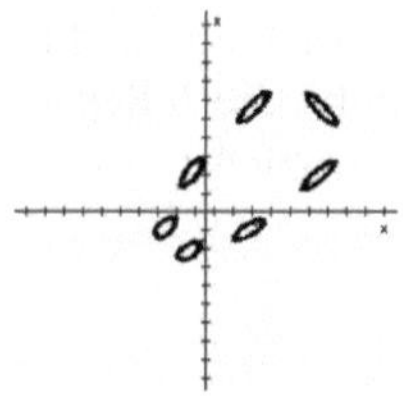

Fig. 10. Dynamic system state prior to attractor formation

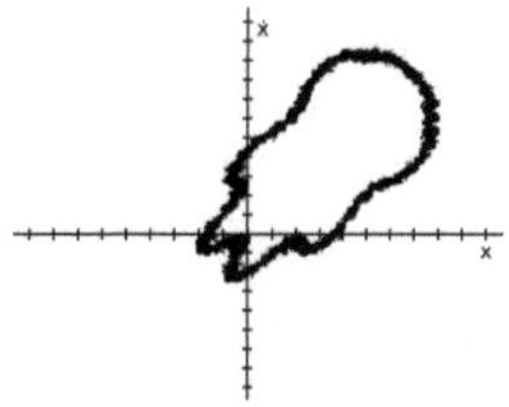

Fig. 11. Stabilized attractor of dynamic system

obtain this attractor repeatedly. This case documents extreme sensitivity of unstable states of system on small changes in initial conditions.

Another interesting case is the portrait of the system with exhibited seven pseudo-steady limit cycles running around partial steady state (practically unstable), see Figure 10. Even very small change of U2 voltage (from -1.12 V to -1.15V) led to creation of stable attractor that was captured repeatedly – see Figure 11.

The paper deals with the behavior of discrete non-linear systems with the aim to obtain its behavior in chaotic states. Chaos module and its circuit using chaos chip were used. Depending on changes of selected parameters of experimental 1-D and 2-D systems we tried to present both bifurcation diagrams and chaotic attractors. Obtained results confirm that [3], [7], [8]:

- It is possible to visualize chaotic states of dynamic systems on storage oscilloscope screen, that means in common laboratory conditions,
- After reaching critical values of bifurcation parameters there really appear expected phenomena preceding chaos, such as period doubling, creation of state with quadruperiod, chaos realization and consequent relaxation states creation (see figures 4, 5 and 6),
- It is possible to present even complex states of system on storage oscilloscope. As an example we can mention bifurcation diagram on figure 6, that corresponds to the system setting on figure 5 and that exhibit strong change in systems behavior after small change of β gain (from $\beta = -1.00$ on figure 5 to $\beta = -0.33$ on figure 6),
- We also proved that even on simple device, such as storage oscilloscope (even if of high quality) it is possible to observe and stabilize complex chaotic attractors, commonly obtained by computers. At the same time the extreme sensitivity of dynamic systems behavior on small changes of its parameters is confirmed.

The research of bifurcation and chaotic behavior in electronic and electromechanical system, mainly in system used in mechatronics applications, is of importance not only as an example of analysis of nonlinear systems behavior in extreme conditions, but is of importance with respect to development of diagnostic methods and with respect of selection, setting and optimization of control structures.

Acknowledgement. Published results were acquired using the subsidisation of the Ministry of Education, Youth and Sports of the Czech Republic, research plan MSM0021630518 "Simulation modelling of mechatronics systems" and GAČR 101/08/0282.

References

[1] Yamakowa, T., Miki, T., Uchyno, E.: Chaotic Chip for Analyzing Nonlinear Discrete Dynamical Network System. In: Proc. Of the 2th Inter. Conf. On Fuzzy Logic&Neural Network, Iizuka, Japan, pp. 563–566 (1992)

[2] Honzák, A.: Komplexní nelineární dynamický systém se změnou parametrů, Diplomová práce UVEE, FEKT VUT v Brně (2001)

[3] Kratochvíl, C., Koláčný, J.: a kol: Bifurkace a chaos v technických soustavách a jejich modelování. ISBN: 978-80-214-3720-3, 108p. ÚT AVČR, Brno (2008)

[4] Nayfeh, A.H., Balachandran, B.: Applied Nonlinear Dynamics. J. Wiley & Sons, Inc., New York (1995)

[5] de Silva, C.W.: Mechatronics (An Integrated Approach). CRC Press, Boca Raton (2005)

[6] Macul, J.: Dynamic systems

[7] Houfek, L., et al.: Bifucation and chaos in Drive Systems. Enginering Mechanics 15(6) (2008)

[8] Byrtus, M.: Qvalitative Analysis of Nonlinear Gear Drive Vibration Consed by Internal Kinematics and Parametric Excitation. Enginering Mechanics 15(6) (2008)

Discrete-Difference Filter in Vehicle Dynamics Analysis

P. Porteš, M. Laurinec, and O. Blaťák

Brno University of Technology, Faculty of Mechanical Engineering,
Institute of Automotive Engineering, Technická 2896/2, Brno, Czech Republic

Abstract. This article presents possible benefits of using derivative-free filter to estimate vehicle dynamics states based on measured signals where the complexity of nonlinear dynamics limits the use of Extended Kalman Filter commonly used for nonlinear filtering. The filtration process was applied to real data obtained from testing manoeuvre. Bicycle model of a vehicle was used for state description and lateral dynamics investigation. Filter implementation in Matlab-Simulink software environment was used for analysis and comparison with earlier results published in [2].

1 Introduction

Modelling of vehicle dynamics for dynamic states analysis during real road tests using only erroneous measured quantities provide insufficient accuracy. State-space mathematical description of the dynamic system, such as vehicle, integrated with measured signals becomes a useful and sufficiently precise tool for state estimation.

The main purpose of our project is to extend preceding work published in [2], where the linear Kalman filter was applied to estimate vehicle states during an avoiding manoeuvre, whereof measured data were obtained.

In this project we focused our attention on more complex mathematical description of vehicle dynamics which makes the conventional Kalman filter unusable. New filter capable to estimate states of even nonlinear systems will be therefore presented. After its mathematical derivation a software implementation in Matlab/Simulink is shown, finally followed by a graphic confrontation of obtained results with results from linear Kalman filter to show an improvement of the new filter.

2 Nonlinear Systems and Discrete-Difference Filter

A usually used method for state estimation is the Kalman filter derived in 1960 by R.E. Kalman [1]. This approach is applicable in case of linear transformation. But by considering nonlinearities the classic Kalman Filter becomes unavailable.

A state-space model of a dynamic system, generally nonlinear, is given by

$$\begin{aligned} \mathbf{x}_{k+1} &= \mathbf{f}(\mathbf{x}_k, \mathbf{u}_k, \mathbf{v}_k) \qquad & \mathbf{v}_k &\sim \mathrm{N}(\overline{v}_k, \mathbf{Q}_k) \\ \mathbf{y}_k &= \mathbf{g}(\mathbf{x}_k, \mathbf{w}_k) \qquad & \mathbf{w}_k &\sim \mathrm{N}(\overline{v}_k, \mathbf{R}_k) \end{aligned} \tag{1}$$

The first part represents an evolution of the state $\mathbf{x}_k$ over time depending on an input $\mathbf{u}_k$ disturbed by a Gaussian random variable $\mathbf{v}_k$ called *process noise*. The second part express a relation between the state and the measurement also disturbed with a Gaussian random variable $\mathbf{w}_k$ called *measurement noise.*

The algorithm presented in [3] cover detailed derivation of presented method in the proceeding section. This filter uses the first two terms from Stirling's interpolation formula [4] and the Cholesky factorization of the covariance matrices:

$$\mathbf{R} = \mathbf{S}_w \mathbf{S}_w^T \qquad \mathbf{Q} = \mathbf{S}_v \mathbf{S}_v^T \qquad \hat{\mathbf{P}} = \hat{\mathbf{S}}_x \hat{\mathbf{S}}_x^T \qquad \overline{\mathbf{P}} = \overline{\mathbf{S}}_x \overline{\mathbf{S}}_x^T \tag{2}$$

The whole algorithm works with the factored covariance matrices, therefore also the resulting a priori and a posteriori covariance will be in factored form.

A priori update

The a priori state estimate and its factored covariance matrix is

$$\overline{\mathbf{x}}_{k+1} = \mathbf{f}(\hat{\mathbf{x}}_k, \mathbf{u}_k, \overline{\mathbf{v}}_k) \qquad \overline{\mathbf{S}}_{x,k+1} = \left[\mathbf{S}_{x\hat{x},k} \quad \mathbf{S}_{xv,k}\right], \text{ where} \tag{3}$$

$$\begin{aligned} \mathbf{S}_{x\hat{x},k}(i,j) &= \left[f_i(\hat{\mathbf{x}}_k + h\hat{\mathbf{s}}_{x,j}, \mathbf{u}_k, \overline{\mathbf{v}}_k) - f_i(\hat{\mathbf{x}}_k - h\hat{\mathbf{s}}_{x,j}, \mathbf{u}_k, \overline{\mathbf{v}}_k)\right] / 2h \\ \mathbf{S}_{xv,k}(i,j) &= \left[f_i(\hat{\mathbf{x}}_k, \mathbf{u}_k, \overline{\mathbf{v}}_k + h\mathbf{s}_{v,j}) - f_i(\hat{\mathbf{x}}_k, \mathbf{u}_k, \overline{\mathbf{v}}_k - h\mathbf{s}_{v,j})\right] / 2h \end{aligned} . \tag{4}$$

The symbols $\hat{\mathbf{s}}_{x,j}$ and $\mathbf{s}_{v,j}$ represents jth column of $\hat{\mathbf{S}}_x$ and $\mathbf{S}_v$ respectively.

A posteriori update

The a priori measurement estimate and the corresponding factored covariance matrix is

$$\overline{\mathbf{y}}_k = \mathbf{g}(\overline{\mathbf{x}}_k, \overline{\mathbf{w}}_k) \qquad \mathbf{S}_{y,k} = \left[\mathbf{S}_{y\overline{x},k} \quad \mathbf{S}_{yw,k}\right], \tag{5}$$

where

$$\mathbf{S}_{y\bar{x},k}(i,j) = \left[g_i(\bar{\mathbf{x}}_k + h\bar{\mathbf{s}}_{x,j}, \bar{\mathbf{w}}_k) - g_i(\bar{\mathbf{x}}_k - h\bar{\mathbf{s}}_{x,j}\bar{\mathbf{w}}_k)\right]/2h$$
$$\mathbf{S}_{yw,k}(i,j) = \left[g_i(\bar{\mathbf{x}}_k, \bar{\mathbf{w}}_k + h\mathbf{s}_{w,j}) - g_i(\bar{\mathbf{x}}_k, \bar{\mathbf{w}}_k - h\mathbf{s}_{w,j})\right]/2h \tag{6}$$

The a posteriori state estimate and the appropriate factored covariance are then given by

$$\hat{\mathbf{x}}_k = \bar{\mathbf{x}}_k + \mathbf{K}_k(\mathbf{y}_k - \bar{\mathbf{y}}_k) \qquad \hat{\mathbf{S}}_k = \left[\bar{\mathbf{S}}_{x,k} - \mathbf{K}_k\mathbf{S}_{yx,k} \quad \mathbf{K}_k\mathbf{S}_{yw,k}\right], \tag{7}$$

where $\mathbf{K}_k$ represents the Kalman gain and it is given by

$$\mathbf{K}_k = \bar{\mathbf{S}}_{x,k}\mathbf{S}^T_{y\bar{x},k}\left[\mathbf{S}_{y,k}\mathbf{S}^T_{y,k}\right]^{-1}. \tag{8}$$

3 Measurement

For the application of derived filter, we selected testing manoeuvre of a passenger car according to ISO/WD 3888-2 [5] international standard (Fig. 1).

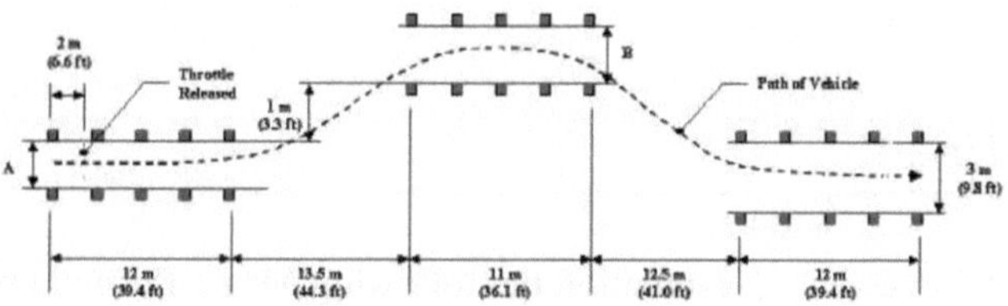

Fig. 1. ISO/WD 3888-2 manoeuvre.

The whole track was passed with relaxed accelerator pedal, i.e. at almost constant velocity. The experimental car was equipped with measuring instruments (V1, HS-CE) for velocity and slip angle measurement and with marking device for vehicle trajectory logging.

4 Filter Utilization

For all practical purposes, we created a mathematical vehicle model to which previously mentioned filter algorithms were applied. The following table shows chosen state variables and measured quantities for vehicle model design.

For a state-space mathematical description, we used equations of lateral vehicle dynamics described in [7] extended with relationships describing yaw angle and y-axis position. The measurement equations were derived according to measured quantities from Table 1 and their dependence on state variables with respect to sensor placement. The whole state-space model is as follows.

Table 1. State variables and measured quantities for filter implementation

State variables		Measured quantities	
Sideslip velocity:	V	Velocity from HS-CE and V1 sensors:	$\lvert \mathbf{v}_{HS-CE} \rvert, \lvert \mathbf{v}_{V1} \rvert$
Yaw rate:	r	Slip angle from V1 sensor:	α_{V1}
Yaw angle:	ε	Slip angle from HS-CE sensor:	α_{HS-CE}
y-axis position:	y_0	y-axis position from marking device:	y_{MD}

Table 2. State space model for filter implementation

State equations:	Measurement equations:
$\dot{V} = \frac{S_F + S_R}{m} - Ur$	$v_{V1,y} = x_{V1} r + V$
$\dot{r} = \frac{S_F l_F - S_R l_R}{J_z}$	$v_{HS-CE,y} = x_{HS-CE} r + V$
$\dot{\varepsilon} = r$	$y_{MD} = y_0$
$\dot{y}_0 = U\varepsilon + V$	

For the tire forces $S_{F(R)}$ evaluation the following Magic Formula was used [8], which was the main reason of Extended Kalman Filter inapplicableness.

$$S_{F(R)} = D \sin\left[C \arctan\left\{B\alpha_{F(R)} - E\left(B\alpha_{F(R)} - \arctan\left(B\alpha_{F(R)}\right)\right)\right\}\right] \quad (9)$$

Vehicle parameters l_F, l_R, J_Z, m in Table 2 stands for front axle distance from CoG, rear axle distance from CoG, mass moment of inertia and vehicle mass respectively and their values could be found in [6] together with x_{HS-CE}, x_{V1} representing sensors placement with respect to the vehicle CoG.

5 Results Comparison

Fig. 2 and Fig. 3 illustrate the filter estimates for yaw rate and vehicle trajectory respectively. To illustrate the difference between derivative-free filters and the linear Kalman filter performance results from [2] were added to the graphs. These data were obtained from the linear vehicle model, i.e. without any presence of the Magic formula, using the same state variables and measurement quantities as in discrete difference filters. The improvement in state estimation is obvious.

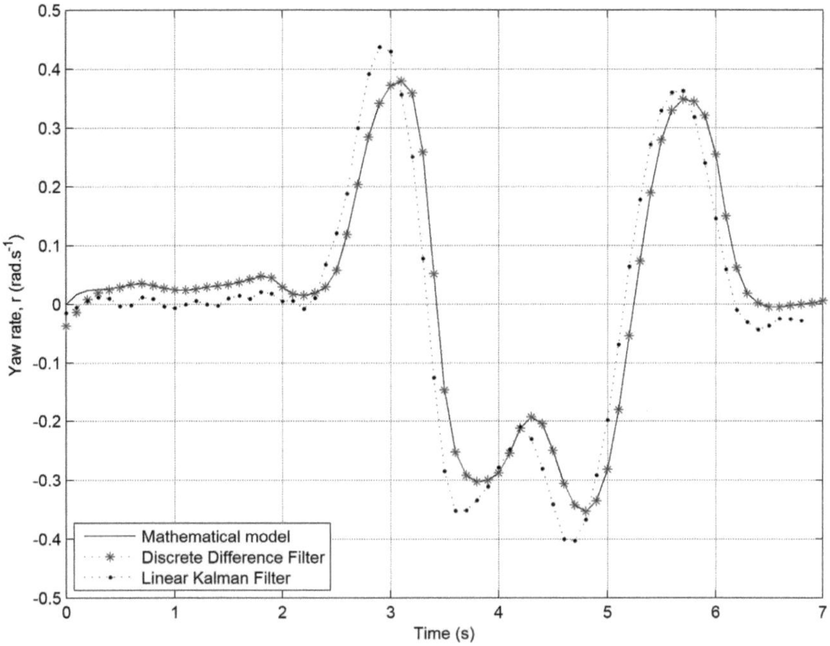

Fig. 2. Comparison of the yaw rate

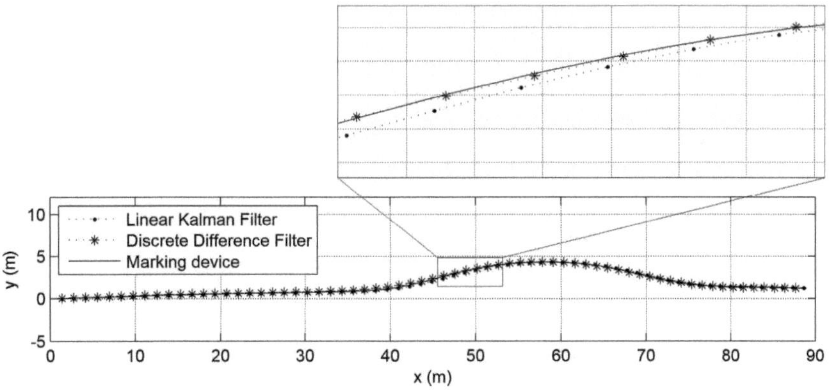

Fig. 3. Comparison of the estimated trajectory

References

[1] Kalman, R.E.: A new approach to linear filtering and prediction problems. Transaction of the ASME - Journal of basic engineering, 35–45 (1960)

[2] Porteš, P., Laurinec, M., Blaťák, O.: Analysis of Vehicle Dynamics using Kalman Filter. In: Simulation Modelling of Mechatronic Systems III, Brno University of Technology, Faculty on Mechanical Engineering, pp. 215–232 (2007) ISBN: 978-80-214-3559-9

[3] Nørgaard, M., Poulsen, N.K., Ravn, O.: Advances in Derivative-Free State Estimation for Nonlinear Systems. Revised Edition, IMM-Technical Report-1998-15 (2004)
[4] Froberg, C.E.: Introduction to Numerical Analysis, p. 433. Addison-Wesley, Reading (1970) ASIN B000NKJ5LC
[5] ISO/WD 3888-2, 1999(E) Passenger cars Test track for a severe lane-change manoeuvre, Part 2: Obstacle avoidance
[6] Kledus, R., Porteš, P., Vémola, A., Zelinka, A.: Messung von Fahrmanövern von Kraftfahrzeugen. In: 10. EVU Jahrestagung des Europäisches Vereins für Unfallforschung und Unfallanalyse e.v (EVU) Brno / Tschechische Republik, Institute of Forensic Engineering of Brno University of Technology, pp. 6–45 (2001)
[7] Vlk, F.: Dynamika motorových vozidel. Publisher VLK, Brno (2000)
[8] Bakker, E., Pacejka, H.B., Lidner, L.: A new Tire model with an Application in Vehicle Dynamics Studies. SAE 890087 (1989)
[9] Laurinec, M.: Extended and Derivative Free Kalman Filter. In: Advances in Automotive Engineering, vol. II. 1, pp. 135–279. Tribun EU, Brno (2008)

3D Slide Bearing Model for Virtual Engine

V. Píštěk, P. Novotný, and L. Drápal

Brno University of Technology, Faculty of Mechanical Engineering,
Institute of Automotive Engineering, Technicka 2896/2, Brno, Czech Republic
pistek.v@fme.vutbr.cz

Abstract. The paper focuses on the description of a 3D slide bearing model worked out as a virtual engine module. A complex computational model of a powertrain is assembled in multi-body systems. The slide bearing model makes a submodule of the virtual engine. The paper presents theoretical assumptions supplied with a numerical solution. The finite difference method with non-uniform integration step is introduced for the numerical solution. The results achieved using the slide bearing computational model help to develop modern diesel engines in the area of noise, vibrations and fatigue of the main parts.

1 Introduction

Present computational models of a slide bearing enable to describe a slide bearing behaviour in high details. These models are often very complicated and require long solution times even on condition that only one slide bearing model is being solved. However, the virtual engine sometime includes tents of slide bearings, therefore, all model features of slide bearings have to be carefully considered.

The loading capacity of a slide bearing model included in the virtual engine is considered in a radial direction and also incorporates pin tiltings, which means that radial forces and moments are included into the solution. For the solution of powertrain part dynamics elastic deformations can be neglected because integral values of pressure (forces and moments) for HD (hydrodynamic) and EHD (elastohydrodynamic) solution are approximately the same. This presumption is very important and it enables a simplification of the solution. On the other hand, the solution cannot be used for a detailed description of the slide bearing. Simultaneous solutions of tens of EHD slide bearing models seem to be extremely difficult and do not provide any fundamental benefits for general dynamics. Therefore, the virtual engine incorporates a compromise solution using the HD solution with elastic bearing shells and can be named (E)HD approach. A HD approach presumes basic premises [1].

Generally, oil temperature has a significant influence on slide bearing behaviour. Oil temperature is treated as a constant for whole oil film of the bearing. This presumption enables to include temperature influences after the hydrodynamic solution according to temperatures determined from similar engines.

2 Theoretical Assumptions

In general, if the equation of the motion and Continuity equation [1] are transformed for cylindrical forms of bearing oil gap together with restrictive conditions [1], the behaviour of oil pressure can be described by Reynolds differential equation. This frequently used equation is derivated for a bearing oil gap [1] or [2]. The oil film gap is defined as

$$h = R - r + e\cos(\varphi)\,, \tag{1}$$

where h is oil film gap, R is shell radius, r is pin radius, e is eccentricity and φ angle. Using dimensionless values [1] the dimensionless pressures can be defined as

$$\Pi_D = \frac{p_D \psi^2}{\eta \overline{\omega}} \quad \text{and} \quad \Pi_V = \frac{p_V \psi^2}{\eta \dot{\varepsilon}}\,, \tag{2}$$

where p is pressure and η is dynamic viscosity of oil. Π_D denotes dimensionless pressure for a tangential movement of the pin, Π_V is dimensionless pressure for a radial movement of the pin, $\overline{\omega}$ is effective angular velocity and $\dot{\varepsilon}$ is a derivative of dimensionless eccentricity with respect of time. Pin tilting angles can be introduced as

$$\gamma = \frac{tg\gamma^*}{tg\gamma_{\max}^*} \quad \text{and} \quad \delta = \frac{tg\delta^*}{tg\delta_{\max}^*}\,, \tag{3}$$

where γ is dimensionless pin tilting angle in the narrowest oil film gap and δ is a dimensionless tilting angle in the plane perpendicular to the plane of the narrowest oil film gap. γ^* denotes a real tilting angle in a plane of the narrowest oil film gap and γ^*_{max} denotes a maximal possible tilting angle in the plane of the narrowest oil film gap for given eccentricity. δ^* is a real tilting angle in the plane perpendicular to the plane of the narrowest oil film gap and δ^*_{max} is a maximal real tilting angle in the plane perpendicular to the plane of narrowest oil film gap for given eccentricity. Figure 1 presents the definition of pin tilting angles and the definition of general and maximal tilting angle in a plane of the narrowest oil film gap.

The final definition of the dimensionless oil film gap H depending on tilting angles is

$$H = H(\varphi, \varepsilon, \gamma, \delta) = (1 + \varepsilon\cos\varphi)(1 - \gamma Z\cos\varphi - \delta Z\sin\varphi) \tag{4}$$

and includes a dependency on two tilting angles. Z is dimensionless coordinate.

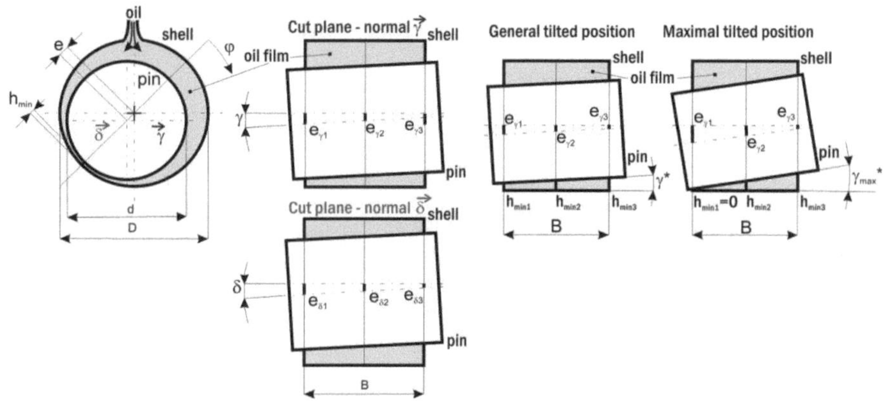

Fig. 1. Definition of tilting angles of pin and description of real tilting angles in plane of the narrowest oil film gap

If the dimensionless oil film gap is used for the Reynolds equations for tangential and radial movements of the pin, then the equation can be rewritten into two separate equations [1]. Likewise the dimensionless pressure is modified [1] and the equations from (2) and (3) are inputted into Reynolds dimensionless equations. The final forms of Reynolds equations are

$$\frac{\partial^2 \overline{\Pi}_D}{\partial \varphi^2} + \left(\frac{D}{B}\right)^2 \frac{\partial^2 \overline{\Pi}_D}{\partial Z^2} + a(\varphi, \varepsilon, Z, \gamma, \delta)\overline{\Pi}_D = b_D(\varphi, \varepsilon, Z, \gamma, \delta) \tag{5}$$

$$\frac{\partial^2 \overline{\Pi}_V}{\partial \varphi^2} + \left(\frac{D}{B}\right)^2 \frac{\partial^2 \overline{\Pi}_V}{\partial Z^2} + a(\varphi, \varepsilon, Z, \gamma, \delta)\overline{\Pi}_V = b_V(\varphi, \varepsilon, Z, \gamma, \delta) \tag{6}$$

The equation term $a(\varphi,\varepsilon,Z,\gamma,\delta)$ is defined as

$$a(\varphi, \varepsilon, Z, \gamma, \delta) = -\frac{3}{4} H^{-2}\left[2HH_{\varphi\varphi} + H_{\varphi}^2 + \left(\frac{D}{B}\right)^2 H_z^2\right] \tag{7}$$

and the equation terms $b_D(\varphi,\varepsilon,Z,\gamma,\delta)$ and $b_V(\varphi,\varepsilon,Z,\gamma,\delta)$ are defined as

$$b_V(\varphi, \varepsilon, Z, \gamma, \delta) = 6H^{-\frac{3}{2}} H_{\varphi} \quad \text{and} \quad b_D(\varphi, \varepsilon, Z, \gamma, \delta) = 12H^{-\frac{3}{2}} \cos\varphi \ . \tag{8}$$

Functions H_{φ}, H_Z and $H_{\varphi\varphi}$ are partial derivatives of the oil film gap and the definition of these function can be found in [3].

3 Numerical Solution

Equations (5) and (6) are solved numerically. The Finite Difference Method (FDM) is used for numeric solution. The FDM in basic form uses a constant integration

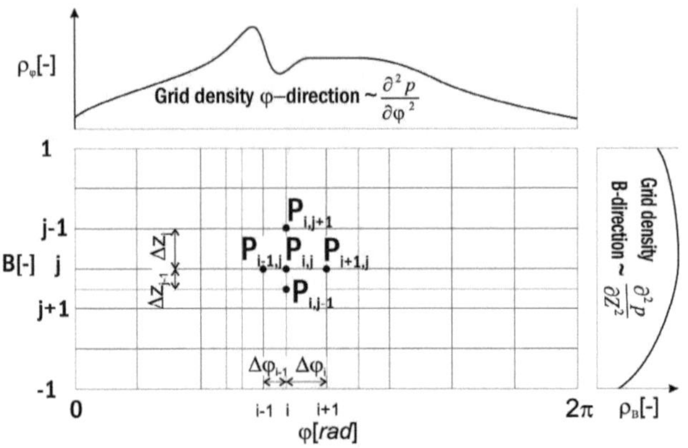

Fig. 2. Computational grid for FDM with variable integration step

step, however, this strategy can be disadvantageous because in case the pin eccentricities are very high, the oil film pressure becomes concentrated in small areas and it is necessary to use a very small integration step. This leads to higher computational models. Therefore, FDM using variable integration step combining with multigrid strategies is developed. The grid density is changed in dependency on prescribed conditions. Three point integration scheme is chosen for the solution because for small integration steps it is very fast. Figure 2 presents an example of computational grid for FDM with a variable integration step.

Resulted formula for iterative solution of dimensionless pressure at point *i,j* is defined as

$$\overline{\Pi}_{i,j_{D,V}} = \frac{\dfrac{\dfrac{1}{\Delta\varphi_i}\overline{\Pi}_{i+1,j} + \dfrac{1}{\Delta\varphi_{i-1}}\overline{\Pi}_{i-1,j}}{\Delta\varphi_j + \Delta\varphi_{j-1}} + 2\dfrac{D^2}{B^2}\dfrac{\dfrac{1}{\Delta Z_j}\overline{\Pi}_{i,j+1} + \dfrac{1}{\Delta Z_{j-1}}\overline{\Pi}_{i,j-1}}{\Delta Z_j + \Delta Z_{j-1}} - b_{D,V}}{\dfrac{2}{\Delta\varphi_j \Delta\varphi_{j-1}} + \dfrac{D^2}{B^2}\dfrac{2}{\Delta Z_j \Delta Z_{j-1}} - a} \quad . \tag{9}$$

The formula for the numerical solution (9) is different for tangential and radial pin movement only in the term b_D (for tangential pin movement) and b_V (for radial pin movement) respectively.

The solution approach with variable integration steps presumes sufficient density of a solution grid according to pressure differentiations with respect to the bearing angle and bearing width. This strategy enables solving problematic pressure zones in acceptable solution time.

Equation (9) is solved iteratively for the tangential pin movement as well as for the radial pin movement. Initial and boundary conditions are the same for both solutions. The first boundary condition describes

$$p_{\left(z=\pm\frac{B}{2}\right)} = 0 \Rightarrow \Pi_{(Z=\pm 1)} = 0 \quad . \tag{10}$$

The only initial condition describes

$$p_{(\varphi=0)} = 0 \Rightarrow \Pi_{(\varphi=0)} = 0 \quad \text{for the first iteration.} \tag{11}$$

This initial condition is used only for the first iteration and after that it is replaced by the cyclic boundary condition

$$p_{(\varphi=0)} = p_{(\varphi=2\pi)} \Rightarrow \Pi_{(\varphi=0)} = \Pi_{(\varphi=2\pi)} \; . \tag{12}$$

The cavitation condition is included during the numerical solution. This condition resets negative pressures to zero values.

$$p = 0 \in p < 0 \Rightarrow \Pi = 0 \in \Pi < 0 \, . \tag{13}$$

Equation (13) describes real physical processes and does not allow negative pressures in liquids (a cavitation).

The computed pressure distributions have to be transferred to equivalent force systems for a solution in MBS. The pressure on an elementary surface can be imagined as an elementary force *dF* on this elementary surface *dS*. Integral values are dimensionless reaction forces *F* and reaction moments *M* and can be found by an integration of pressure across the whole bearing surface with coordinates φ a z (see Figure 3).

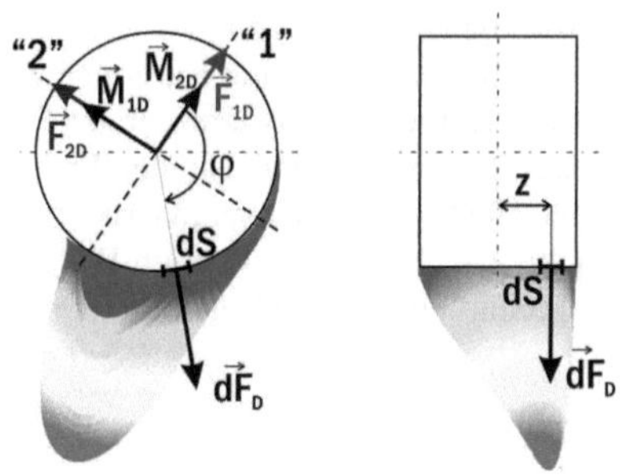

Fig. 3. Derivation of integral values

For example the elementary force for axis "*1*" and for tangential pin movement can be defined as follows

$$\begin{aligned} F_{1D} &= \int_S dF_{1D} = -\iint_S p_D \cos\varphi dS \\ &= -\int_{-\frac{B}{2}}^{\frac{B}{2}} \int_0^{2\pi} p_D \cos\varphi R d\varphi dz = -\frac{\eta\varpi}{\psi^2}\frac{DB}{4}\int_{-1}^{1}\int_0^{2\pi} \Pi_D \cos\varphi d\varphi dZ = \frac{\eta\varpi}{\psi^2}\frac{DB}{4}\Phi_{1D} \end{aligned} \; . \tag{14}$$

A simplification can be used, the pressure distribution for radial pin movement is symmetrical, therefore, the forces F_{2V} and moments M_{2V} equal zero. If all integration conditions are satisfied, the axis force components "*1*" and "*2*" and components of moments for axes can be found by numerical integrations. The details describing integrations of all terms can be found in [3].

Hydrodynamic databases include integral values Φ_{1D}, Φ_{2D}, Φ_{1V}, Θ_{1D}, Θ_{2D}, Θ_{1V} for chosen ratios *D/B* and pin tilting. Resulted forces and moments (F_{1D}, F_{2D}, F_{1V}, M_{1D}, M_{2D}, M_{1V}) inserted into MBS can be obtained by an inclusion of bearing sizes, bearing clearances, dynamic viscosity and pin kinematic values.

4 Conclusion

The results of slide bearing separate solution enable to give effective solution of powertrain dynamics and enable to use tens of slide bearing models in one complex model without significant increase of computational times.

Acknowledgments. The above activities have been supported by the grant provided by the GAČR (Grant Agency of the Czech Republic) reg. No. 101/09/1225, named "Interaction of elastic structures through thin layers of viscoelastic fluid". The authors would like to thank GAČR for the rendered assistance.

References

[1] Butenschön, H.J.: Das hydrodynamische, zylindrische Gleitlager endlicher Breite unter instationärer Belastung, PhD thesis. Universität Karlsruhe, Germany (1976)

[2] Rebbert, M.: Simulation der Kurbewellendynamik unter Berücksichtigung der hydrodynamischen Lagerung zur Lösung motorakustischer Fragen, PhD thesis. Rheinisch-Westfälischen Technischen Hochschule, Aachen, Germany (2000)

[3] Novotny, P.: Virtual Engine - A Tool for Powertrain Development, Inaugural dissertation. Brno University of Technology, Czech Republic (2009)

Powertrain Dynamics Solution Using Virtual Prototypes

D. Svída, P. Novotný, V. Píštěk, and R. Ambróz

Brno University of Technology, Faculty of Mechanical Engineering,
Institute of Automotive Engineering, Technicka 2896/2, Brno, Czech Republic
svida@fme.vutbr.cz

Abstract. The paper presents a computational approach to a powertrain dynamic solution using virtual prototypes. The main contribution is the fact that all models namely those of a cranktrain, a valvetrain, a gear timing mechanism and a fuel injection pump are solved simultaneously, using a complex computational model. Synchronous solutions can have a fundamental effect on results of powertrain dynamics solutions. Additionally, it might help to understand influences among powertrain parts. The virtual engine is assembled as well as numerically solved in Multi Body System.

1 Introduction

Modern powertrains are complex mechatronic systems improved by large and gradual development. Powertrain producers still increase engine parameters like engine performance together with a significant decrease of fuel consumption.

Generally, the increase of engine performance often leads to an increase of powertrain noise and vibrations. Noise and vibration problems can be resolved by experimental or computational methods. Experimental methods are often very expensive and this fact causes a fast development of modern computational methods. Modern computational methods can provide very exact results but only on condition that exact inputs are included. The inputs are often supplied by experimental methods. A portion of experimental methods continuously decreases, however, the experimental method still plays an important role in the powertrain development.

2 Computational Models

A complex computational model of the engine, in other words a virtual engine, is solved in time domain. This enables to incorporate different physical problems including various nonlinearities. The virtual engine is assembled as well as numerically solved in MBS (Multi Body System) ADAMS. ADAMS is a general code and enables an integration of user-defined models directly using ADAMS commands or using user written FORTRAN or C++ subroutines.

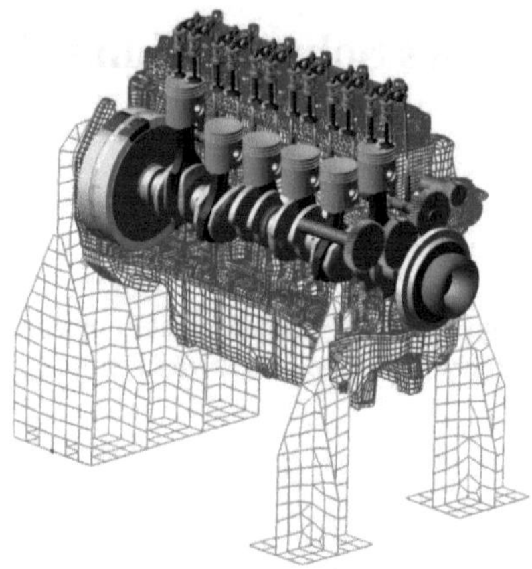

Fig. 1. Virtual engine including main subsystems

The virtual engine includes all significant components necessary for dynamics or fatigue analyses. The included modules are: a cranktrain, a valvetrain, a gear timing drive with fuel injection pump and a rubber damper. Figure 1 presents the virtual engine and its sub modules.

Flexible bodies represented by FE (Finite Element) models have decisive importance on powertrain dynamics simulations. The FE models of each components should be created with special effort. The uniform FE mesh is often preferred. Figure 2 shows FE models of main engine components.

A torsional vibration damper is an important component of some cranktrains. It can significantly increase fatigue of engine parts together with a decrease of noise and vibrations. Different designs of torsional vibration dampers can be used. The rubber damper is chosen for the target engine.

A rubber damper MBS model includes only global properties like torsional and axial stiffness and damping. The global stiffness values originate from a detailed solution of the three dimensional FE model in combinations with Matlab calculations.

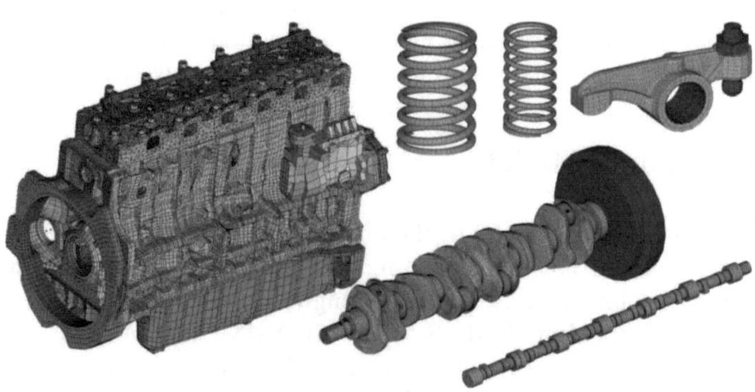

Fig. 2. FE models of main powertrain parts

The virtual engine includes a computational model of a gear timing drive. A model of a gear meshing includes variable stiffness of meshing. The variable stiffness of meshing enables to incorporate tooth meshing frequency as well as its harmonic components. The computational model of meshing helical gear also includes all resultant forces between teeth (radial, axial and tangential forces).

The variable meshing stiffness can be found by solution of a nonlinear contact FE model of the meshing gears. The resultant variable stiffness is included into the MBS computational model of meshing helical gears and into the gear timing drive computational model respectively. Damping values can be obtained from [1].

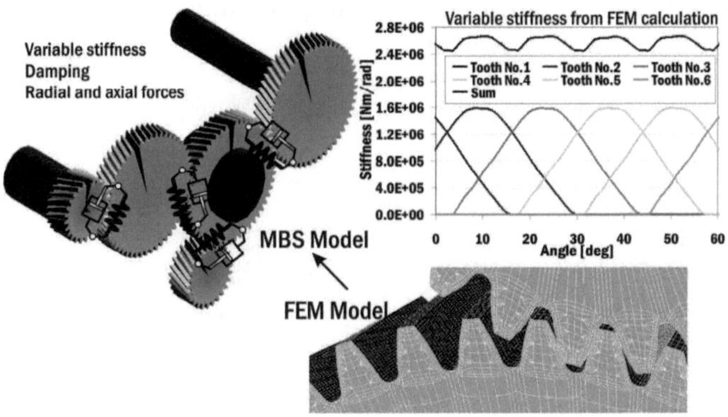

Fig. 3. Meshing gear FE model and gear timing drive MBS model with values of variable meshing stiffness for crankshaft and idle meshing gears

Concerning the powertrain dynamics, the injection pump can highly influence the dynamics of powertrain parts. In particular the valvetrain and the gear timing drive can be influenced by high peak torques of the injection pump.

In principal this type of injection pump includes injection pistons. The movement of the injection piston is controlled by a cam profile. Each cam interacts with a roller tappet. This interaction (cam – roller contact) produces time dependent torques on a pump shaft in each pump section. The injection pump torque values (see Figure 4) are determined with the help of [2].

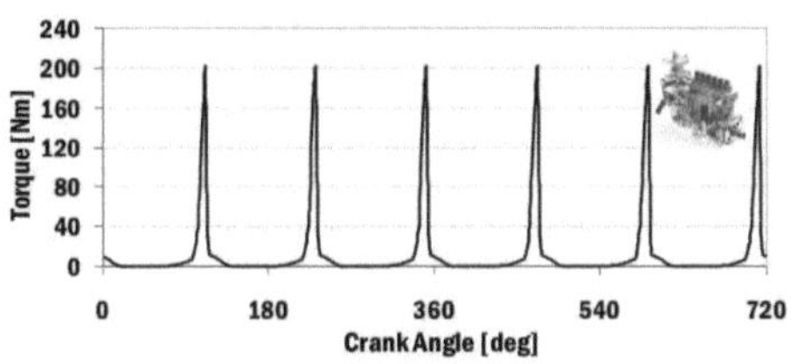

Fig. 4. Injection pump torque vs. crank angle

3 Powertrain Results

The determination of cranktrain torsional vibrations represents a fundamental step in a powertrain development. Cranktrain torsional vibrations influence torsional vibrations of each powertrain rotating component.

These vibrations can significantly influence forces in every single valvetrain or forces in a gear timing drive. Fatigue of powertrain components like a crankshaft or a camshaft is also affected. An experimental determination of torsional vibrations using, for example, laser vibration tools is an advance and it can help to validate computational models. Figure 5 shows computed and measured harmonic analysis results of cranktrain torsional vibrations of the powertrain with a rubber damper.

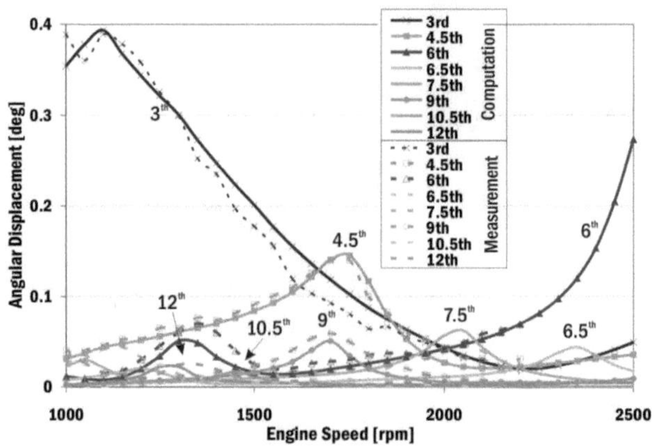

Fig. 5. Harmonic analysis of crankshaft pulley torsional vibrations of an inline six-cylinder engine with rubber torsional damper (a computation and a measurement)

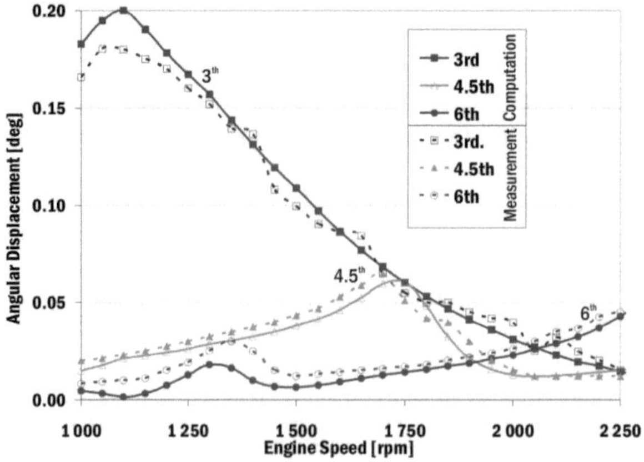

Fig. 6. Harmonic order analysis of camshaft end angular displacement (a computation and a measurement)

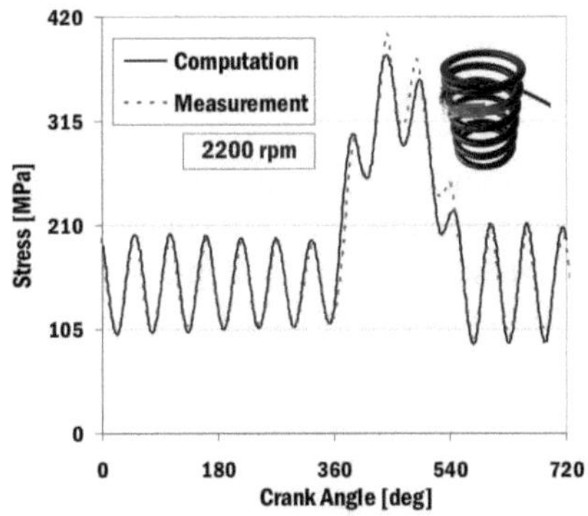

Fig. 7. Computed and measured stresses of the first cylinder intake rocker for engine speed 2200 rpm

The camshaft vibrations are influenced mainly by cranktrain torsional vibrations and also by all single valvetrain torques, a gear timing mechanism and an injection pump torque. Figure 6 presents harmonic analyses of measured and computed angular velocities of a camshaft end near to the camshaft bearing No. 1 and figure 7 presents computed and measured stresses of the first cylinder intake rocker.

Powertrain dynamic results are used for fatigue analyses of main powertrain components. The modal approach of a component stress determination used for fatigue analyses can be applied only to reduced FE bodies.Principles of a component stress determination consist in the combination of dimensionless modal stresses found by a computational modal analysis and scaled time variable generalized coordinates found by a powertrain dynamic solution. The combination of the dimensionless stresses and the generalized coordinates generates real stresses for each mode. A set of mode stresses gives a total stress for each node of FE body as a function of time. This time variable stresses are used for the component fatigue calculations.

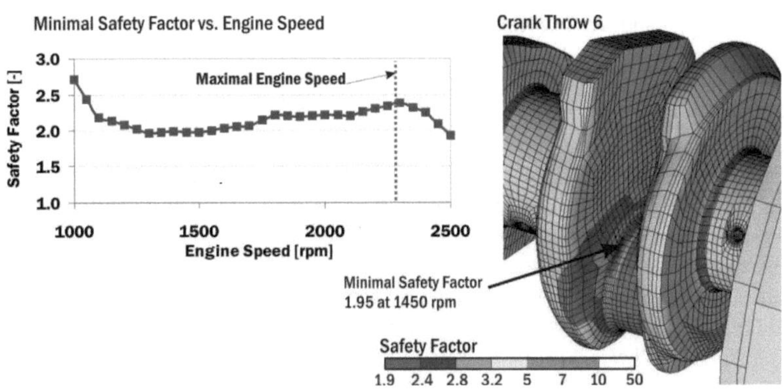

Fig. 8. Crankshaft fatigue analysis results – safety factor distributions on crankshaft surface for engine speed 1450 rpm and minimal safety factors vs. engine speed

Figure 8 presents fatigue analysis results of the crankshaft with rubber torsional damper for engine speed 1450 rpm.

Influences like material properties, surface roughness, component sizes, a stress gradient, means stress and so on are additional inputs for fatigue analyses. Fatigue analysis results can be characterized, for example, by safety factors. More information about fatigue analyses can be found in [3].

Figure 9 shows fatigue results of intake single valvetrain parts (a rocker, outer and inner valve springs) for engine speed 2200 rpm.

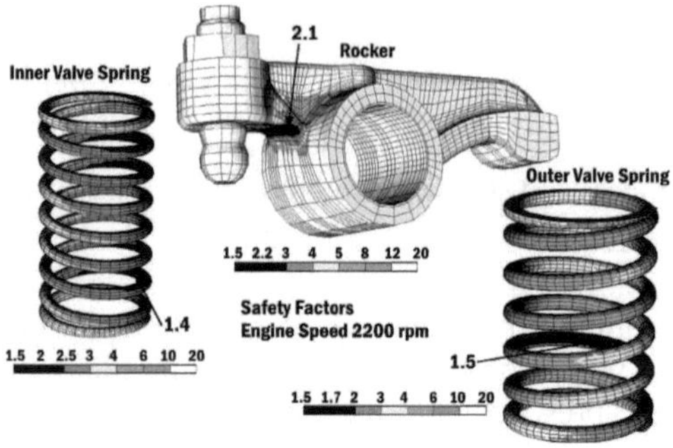

Fig. 9. Fatigue results of intake single valvetrain parts for engine speed 2200 rpm

4 Conclusion

The virtual engine results can help to understand the dynamic behaviour of a new powertrain and enable to speed-up the development process together with reductions of expensive prototypes. Therefore, the computational tools based on FEM, MBS, EHD or CFD principles together with experimental tests play an important role in the modern powertrain design.

Acknowledgments. Published results have been achieved with the financial help of the Ministry of Education of Czech Republic, project 1M6840770002 – Josef Božek Research Centre for Engine and Vehicle Technology II.

References

[1] Trampert, S.: Entwicklung eines realgeometrisch abwälzenden Zahnkontakt modells zur Simulation von Zahnrädertrieben in Verbrennungsmotoren. PhD thesis. Rheinisch-Westfälischen Technischen Hochschule, Aachen, Germany (2003)

[2] Siroky, M.: Interakce systémů hnacího ústrojí a palivového čerpadla vznětového motoru. Diploma thesis. Brno University of Technology, Czech Republic (1987)

[3] Hollmann, C.: Die Ubertragbarkeit von Schwingfestigkeits-Eigenschaften im Ortlichen Konzept, PhD thesis. Steyer, Austria (2004)

Description of Flow Intensities in Non-Homogeneous Materials

J. Malášek

Brno University of Technology, Faculty of Mechanical Engineering,
Institute of Automotive Engineering, Technická 2896/2, 616 69 Brno, Czech Republic
Tel.: +420 541 142 428; Phone/Fax: +420 541 142 425
malasek@fme.vutbr.cz

Abstract. The flowing and deformed areas in non-homogeneous materials described by means of vector fields of velocities identify the technological and energetic efficiencies of relevant processes, as for example mixing, compacting, storage, transport, flow, creep, deformation. Qualities of deformation processes can be described by kinetic intensity of transformation for example by step changes of relevant values acquired by means of mathematical image processing.

1 Introduction

Processing optimizations - for example by mixing, compacting, transport and storage of particular materials, powdery materials, dispersions and suspensions with high viscosities – are very complicated problems in various technologies.

When processing non-homogeneous materials by means of viscous flows and deformations, the common reality can be seen – a zone of processing arises as a cubic formation of shear curves and streamlines.

There is a possible way to find a solution by means of creation of actual mathematical models of flows and deformations.

The suggested way – unstable in time changing physical and constitutive characteristics of non-homogenous materials are empirical experience from experiments with at least discrete or in elementary subregions applicable descriptions of continuum mechanics.

Possible way to describe and to optimize processing of non-homogeneous materials with its viscous flows and deformations can be founded on the basis of mathematical image processing with value results of intensity of transformations - deformations and flows. These value results can be found by means of mathematical image processing as vector fields of velocities of flowing and deforming areas in non-homogeneous materials. These are the sufficient conditions to describe relevant parts of shear curves and streamlines.

2 Vector Fields of Velocities

Flows and deformations pictures can be described by means of scalar functions as a functional dependence of a brightness level in coordinates x, y, another variable is time t. These functions are called video functions in form $I = f(x, y, t)$. Digitalization is picture sampling of these functions in matrix of m x n elements and quantization of brightness level to every sample – pixel to one of the k intervals. Geometrical distortions which arise by location, position and shape of the camera lens must be eliminated by a geometrical transformation, which transforms point x, y into point x', y' and is defined by two component relations – polynomials of degree s, [1].

$$x' = \sum_{r=0}^{s} \sum_{q=0}^{s-r} a_{rk} . x^r . y^q \qquad y' = \sum_{r=0}^{s} \sum_{q=0}^{s-r} b_{rk} . x^r . y^q \tag{1}$$

These coefficients of transformation a_{rk} and b_{rk} are determined by the minimal square differences method on the basis of corresponding points x, y and x', y'.

The principle of searching for the vector field of velocities is following: In the time-successive pictures we search for the corresponding square element with dimensions Δx, Δy - its first position (x_i, y_i), in its new position $(x_i + dx_i, y_i + dy_i)$ by means of the summation of minimal square differences of the brightness level $I = f(x, y, t)$ of relevant pixels. Results of computing are the values dx_i, dy_i as the elementary shifts of non-homogeneous material along streamlines or along shear curves during the corresponding time t_i.

Coordinates of points from figure 1: $1[x_i, y_i]$,

$2[x_i + dx_i, y_i + dy_i]$, $3[x_i + dx_i + dx_{i+1}, y_i + dy_i + dy_{i+1}]$.

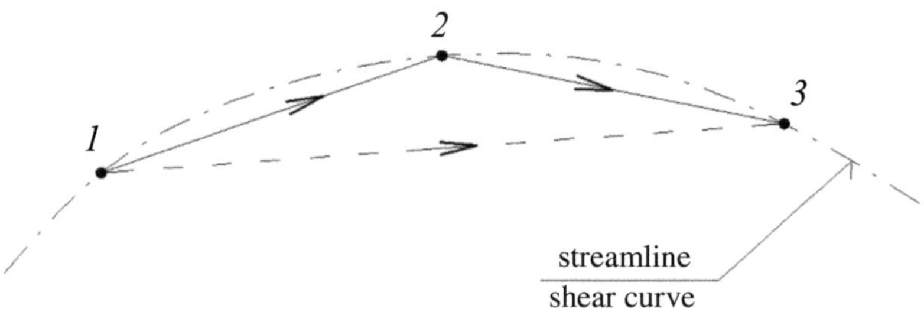

Fig. 1. Description of elementary deformation in non-homogeneous material

The direction of velocity $v_{1,3}$ - belongs to the point 2 - is identical with the direction $1-3$ in figure 1, it stands for a tangent-direction of a section of shear curve (or streamline) in the point 2.

Relevant velocity of movement in a selected point 2: [4]

$$\overrightarrow{v_{1,3}} \equiv \left[\frac{dx_i + dx_{i+1}}{\Delta t_i + \Delta t_{i+1}}, \frac{dy_i + dy_{i+1}}{\Delta t_i + \Delta t_{i+1}} \right] \tag{2}$$

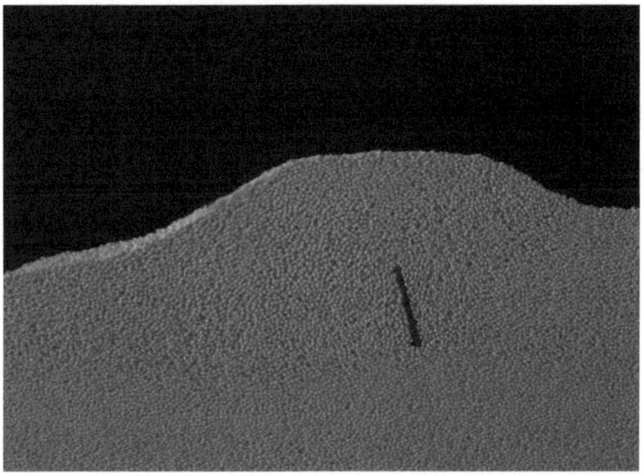

Fig. 2. Flowing – deformed non-homogeneous material along the moving blade

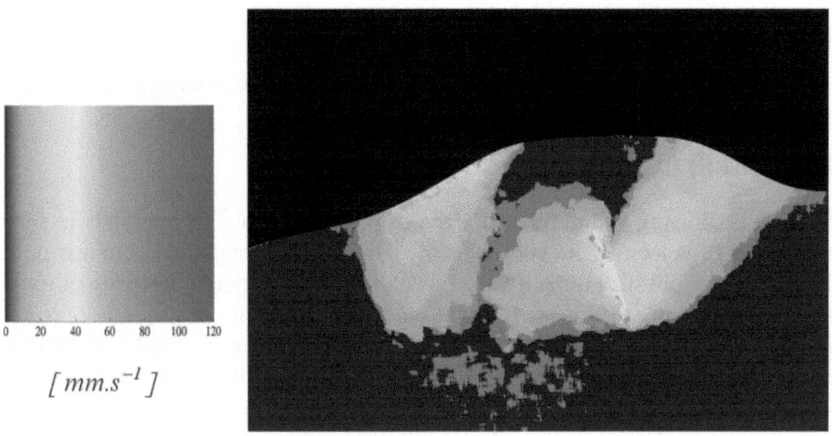

Fig. 3. Velocity scale, field of sizes of velocities of flows and deformations

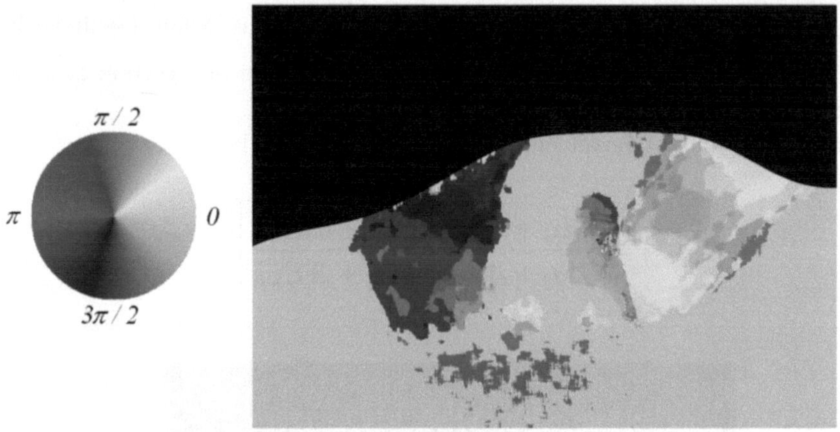

Fig. 4. Direction scale, field of directions of velocities of flows and deformations

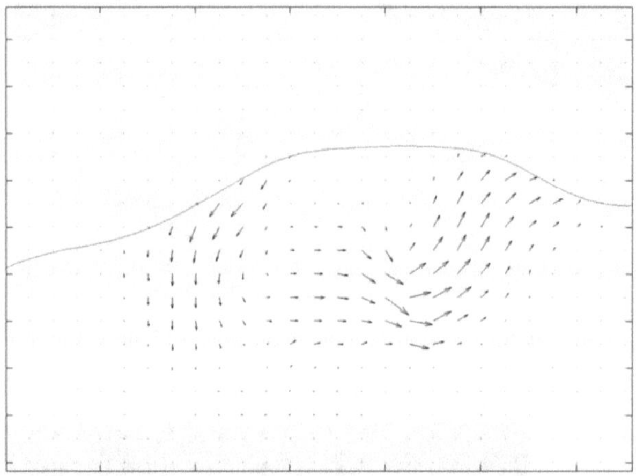

Fig. 5. Selected vectors of velocities of flows and deformations

3 Description of Flow Intensities

The pictures of flowing and deformed areas of non-homogeneous materials in form of vector fields of velocities are chosen as the carrier of information. From this point of view a reference physical quantity can be determined as a kinetic intensity of transformation. Kinetic intensity of transformation (transformation consists of flows and deformations) can describe qualities of mechanical processes in non-homogeneous materials. This physical quantity can be defined by means of the next equations:

Kinetic intensity of transformation in the plane:

$$I_{KS} = \sum_{i=1}^{l} \sum_{j=1}^{m} S_{ij} \cdot \left| \overrightarrow{v_{ij}} \right| \tag{3}$$

$$I_{KS} = \iint_S \left| \overrightarrow{v(x,y)} \right| dS \tag{4}$$

Kinetic intensity of transformation in the volume:

$$I_{KV} = \sum_{i=1}^{l} \sum_{j=1}^{m} \sum_{k=1}^{n} V_{ijk} \cdot \left| \overrightarrow{v_{ijk}} \right| \tag{5}$$

$$I_{KV} = \iiint_V \left| \overrightarrow{v(x,y,z)} \right| dV \tag{6}$$

According to our laboratory equipment possibilities the kinetic intensities of transformation in the plane are presented as transformations (deformations and flows) of the particular material along the moving blade at mixing process. The velocities of moving blades are step changed in its values during these mixing processes.

These results can be used e.g. for optimizing and for quality-process control of the selections of shapes, positions and velocities of movements of interactive

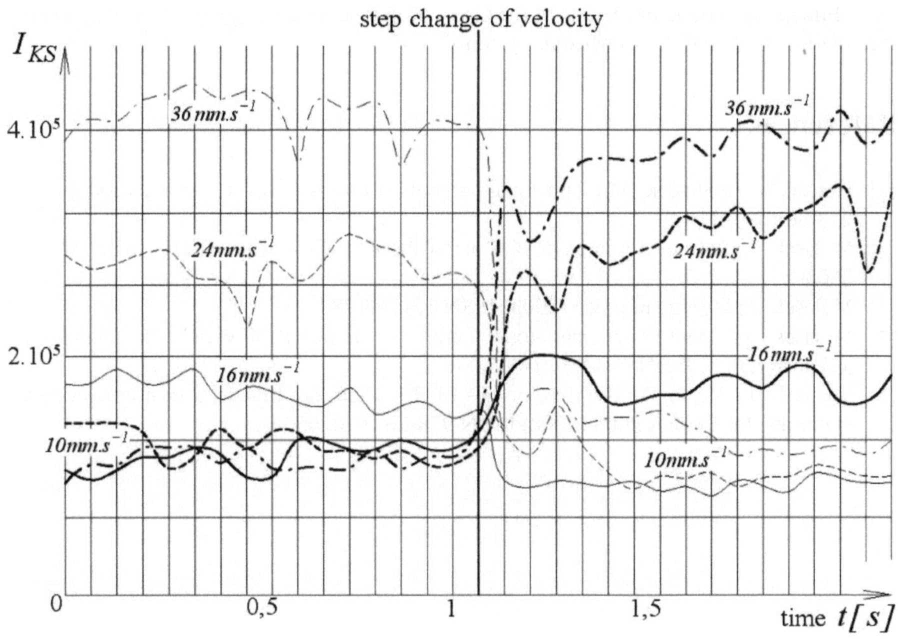

Fig. 6. Functional dependences of kinetic intensities of transformation I_{KS} [$mm^3.s^{-1}$] on the step changes of velocities $v_b = 10, 16, 24, 36\ mm.s^{-1}$ of moving blade

machine parts related to the processed non-homogeneous materials. These interactive machine parts are mechanical machine parts – blades and walls of mixers, jaws of compact-machines, parts of scrapers, spirals of conveyors, dump holes of silos.

4 Conclusion

Non-homogeneous materials cannot be regarded as continuum, they are partly continuous spaces, and the properties of these particles, including their size and shape, are unstable in time.

The transformed areas of non-homogeneous materials change their structural composition, shape and size according to the possibilities of their physical and constitutive characteristics to the form with minimal external forces effects. These changes of shapes and sizes cause simultaneous changes of physical and constitutive characteristics in time.

The mathematical image processing of pictures of transformed areas of non-homogeneous materials in forms of vector fields of velocities are chosen as the carrier of technological process-information.

Acknowledgements. Published results were acquired using the subsidization of the Ministry of Education, Youth and Sports of the Czech Republic, research plan MSM 0021630518 "Simulation modeling of mechatronic systems".

References

[1] Hlaváč, V., Sedláček, M.: Zpracování signálů a obrazů. ČVUT v Praze (2000) ISBN 978-80-01-03110-0
[2] Malášek, J.: Mísení a kompaktování partikulárních látek. Brno (2004) ISBN 80-214-2603-9
[3] Malášek, J.: Disertační práce (2003) ISSN 1213-4198
[4] Malášek, J.: Identifikace napjatosti, deformace a proudění v nehomogenních materiálech (2008) ISBN 978-80-214-3738-8
[5] Malášek, J.: Vector fields of velocities of flows and deformations in non-continuous materials. In: CHISA 2008 (2008) ISBN 978-80-02-02050-9

Acid Pickling Line Simulation

S. Simeonov, R. Hofman, and L. Krotký

Brno University of Technology, Faculty of Mechanical Engineering, Institute of Production Machines, Systems and Robotics, Technicka 2896/2, Brno, Czech Republic
simeonov@fme.vutbr.cz
Taurid Ostrava s.r.o., Hasičská 551/52, 700 30 Ostrava-Hrabůvka
robert.hofman@taurid.cz

Abstract. Presented paper deals with problems of discrete simulation within acid pickling lines. These types of lines are complicated mechanical systems required special algorithm for controlling. The paper discussed using of simulation approach both for projecting and controlling of acid pickling lines.

1 Introduction

The simulation modeling offers competitive advantages to design and production organizations. The system simulation can be performed within a relatively short time. Approaches related to manufacturing system design using simulation studies include a lot of applications. Simulation of production capacity of a shock absorber assembly line was studied to propose a modification of the current line [1]. Redesign of an injector assembly and calibration production was discussed [2]. Research in analyzing of concrete factors in a manufacturing system: the purpose of analyzing some concrete factor of manufacturing system was to improve the current system. Roser, Nakano, and Tanaka [3] studied a method for detecting the bottleneck in a manufacturing line. Duanmu and Taaffe [4] attempted to improve throughput of a manufacturing line using a combination of tact time and simulation analysis. Simulation of robotic welding system was investigated to show the impact of system failures and delays on the output and cycle time [5]. Man-machine ratios using simulation was studied to gain high resource utilization and output [6].

2 Description of Simulated System

Simulation objects are pickling lines. The topology of those lines can vary. According to the axis orientation of hook, which serves for manipulation with coils, can be distinguished the crosswise oriented system (the hook axis is perpendicular with regard to the production system axis), and the lengthwise oriented system.

Manipulation manners with the hook can also bring some limitations: The crane on a lower position is not structurally limited in its horizontal movement or the crane on a lower position is structurally limited in its horizontal movement. The empty hook return can be accomplished out of the line or by means of the line crane. In the second case - if the line includes more cranes - then the return is carried out through a re-hanging point.

The number of transport units (cranes) is a further parameter, which must be set so that the throughput (production rate) would be assured. The simplest variant has one crane only – but the line can be designed with two or more cranes. Because the crane is an expensive equipment then optimal crane number via the simulation brings considerable saving within investment costs.. In case when more cranes are used then two different control strategies are possible according to the set of positions which are served: Each crane has own set of positions only (the sector which crane serves), or cranes can serve all positions.

The sequence of pickling line position engagements can be changed according to the flow of material. There are possible two variants - when position engagements are performed just in the material flow direction, or when position engagements are performed also against the material flow direction (manifold engaged positions). This fact considerably influences the preparation of simulation model, which implicates product flow control algorithms (the flow of pickled coils). As to the technology two pickling technique can be applied – namely the cascade pickling at which all given positions must be engaged and simultaneously exposure durations cannot be summed, or summary pickling at which is sufficient to engaged one position and keep the summary exposure time. After the deviations from a given pickling plan the three alternatives can be distinguished: i. Without any deviation – that is the exposure cannot be exceeded; ii. The maximal deviation is defined – the exposure can be exceeded; iii. The maximal deviation is defined – the exposure can be exceeded only at some positions.

The pickling line configuration can be generally defined through the use of following parameters: a. the pickling line topology – the number of transport units TUn (the correct choice will be verified by a simulation), the number of transport hooks (the material is hung on the hook), the number and definition of technological posts (store, uploading, heating, pickling bath, rinsing bath, discharge, and likewise), b.the distance in horizontal direction (among positions), the distance in vertical direction (hook lift), c. the speed of travel, the speed of hoisting/lowering, d. the acceleration and deceleration of travel, the acceleration and deceleration of lift, e. the number of working positions, f. the pickling programme – that is entered by a technologist (the pickling programme is a sequence of positions to be engaged Sn for a chosen pickling programme according to the type of given technology – one or more pickling programmes), g. the sequence of position engaging is defined for instance by this string: S1-S2-S4-…, h. the technologist determines material exposure durations within positions, i. the technologist determines idle periods, for instance for draining away, uploading, discharging and the tolerances within exposure periods. During simulation studies these parameters are considered both as input parameters and subjects of simulating and optimising experiments.

3 Requirements for Simulation Model

The simulation model of the pickling line is to give answers mainly to following questions: i. with how many cranes can the production system be served?, ii. what is the production capacity of a given production system, that is the "Production Rate"?, iii. what are the requirements as to the number of hooks within the production system?, iiii, how will be cranes engaged?, iiiii what is the maximal deviation from required exposure periods?

The simulation serves for operational characteristic verifications of production systems with a certain probability measure. During the design of production system there is necessary to correlate those three following requirements: i. Technologic, ii. Technical, iii. Economic.

4 Simulation Model of Acid Pickling Line

From the previous elaboration large varieties of pickling line configurations can be derived. These varieties can be shown in quantities and arrangements of vats (baths), in means of transport numbers, in different algorithms for material flow control, and likewise. From such diversity naturally originates also different requirements with regard to the simulation model. To increase and effect works at the preparation of simulation models, the more common methodology is preparing, which guides to preparation of more universal simulation models of pickling lines. It deals with a parametric model, which enables with smaller changes to create the simulation model for a concrete line. The general structure of some universal model is shown in the example (Fig.1). The pickling line layout is derived from the proposal of line design that is elaborated in the Autocad environment. The line consists of individual vats according to the technological procedure. Vats are modelled as resources with capacity 1. That means that they can handle in one moment only one coil (Fault modelling problems will be discussed in another report). Analogously to vats also input/output positions and re-hanging stations are defined as resources.

The crane track creates a transport system; one or more cranes are modelled by means of the use of modelling entity AGV (Automated Guided Vehicle). The usage of sophisticated entity for transport system of AGV type is evincible if the pickling line will be retaining two or more cranes.

The AGV system consists of transport segments and control points. Control points define positions of individual vats, inputs/outputs to/from pickling line, transfer place, and likewise. In control points the decisions according to control algorithms of the transport system are performed. Hereinafter those following algorithms are defined: i. Sequence algorithms for load queue control – if the production load (coil) needs a resource (in that case a crane), which has been already allocated, then this load is placed into the queue according to the chosen algorithm, ii. Algorithms for load (coil) selection from the queue of waiting loads – this algorithm will be activated after the resource has been released and also can cause a new creation of the load queue, iii. Algorithms for crane selection – in case that

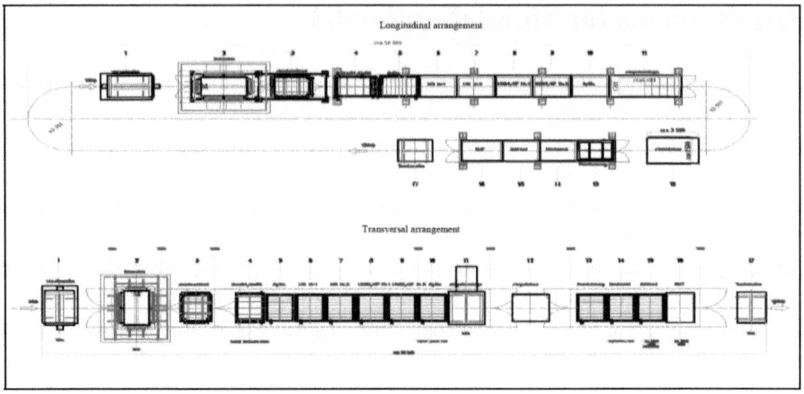

Fig. 1. Simulation model of pickling line

pickling line retains more cranes the track of which are shared, then those algorithms are applied, iiii. Algorithms to control the crane behaviour during blockage, fault, or idleness. From above we can see that control algorithms can influence the behaviour of pickling line and its characteristics. The help of simulating experiments can find out suitable algorithms. Other defined characteristics of transport system are the speed of crane with a load (it transports a coil), the speed of empty crane, and acceleration and likewise. Transport times are calculated by two methods: i. According to the length of segment, speed and acceleration; ii. Or according to the matrix of distances, speeds and accelerations. The choice of calculating method depends on what the simulation model will be used for.

The technological procedure is modelled under the use of process plan. The process plan consists of individual process steps. The technological procedure and the process plan are not identical, because the technological procedure is a subset of the process plan. In general the process plan describes the material flow (and also information and financial flow) through the production system. Due to that it contains besides the technology also control algorithms, bills of materials, and likewise. In the process plan two basic process steps (beside others) are used: "Manufacture" and "Move-Between". By means of the "Manufacture" process step the manufacturing operations can be modelled – in case of pickling lines the coil submersing to vats. The process step "Move-Between" is used to model transport operations. Algorithms for allocation and releasing of resources are applied in both process steps. The simulation model of technological process is processed within the development environment of licensed software SLVP by Infor.

5 Results of Simulation Experiments

Several experiments with the simulation model were performed to verify the production system design. In the following description will be mentioned only simulation experiments which balance two alternatives: alternative with one crane, and alternative with two cranes. The key factor to be followed was the throughput (number of produced coils per hour) of pickling line in a steady state. In case that one crane is used the throughput is 6,716pcs of coils per hour; when two cranes are used the throughput is already 9,061pcs of coil per hour (Fig. 2).

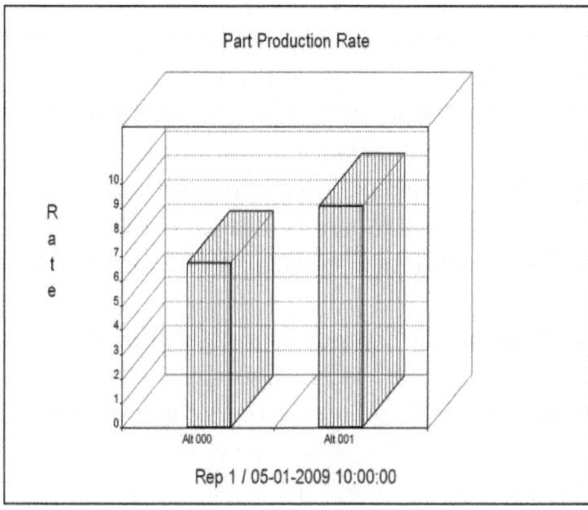

Fig. 2. Part production rate

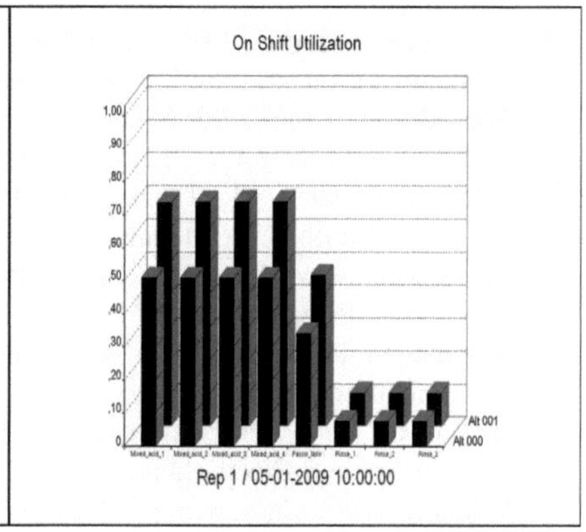

Fig. 3. Working station utilisations

The important factor the pickling line designs deals with is the source utilization, for instance cranes or working stations as shows fig.3. With the production throughput is also related the running time when coils remain within the pickling line in a steady state. To verify functions of the simulation model for analysis of pickling line operations, and to schedule the production, the Gantt's diagrams were generated.

6 Conclusion

The project benefits can be summoned to those following items: i. the demonstration of available technical parameters at the designed equipment, ii. savings at the equipment dimensioning and lower acquisition costs thereof (number of transport units, number of manipulation hooks, and others). The simulation implementation to the production (on-line version) will be proven at the customer within following spheres: the overall material transport acceleration and thus the production increase, to reduce delays necessary to return the transport units to normal state (recovery after faults), to lover the number of wasted products. From performed analyses can be expected the improvement within those economic parameters: the increase of work productivity within intervals 5-10%, the working time rationalisations by o 5-10%.

Acknowledgments. This work was supported by the CZECHINVEST, ICT and strategic services - Call Programme Name: Development and implementation of a multi-purpose simulation model, Registration Number: 2.2 ITS01/029.

References

[1] Gujarathi, N.S., Ogale, R.M., Gupta, T.: Production capacity analysis of a shock absorber assembly line using simulation. In: Ingalls, R.G., Rossetti, M.D., Smith, J.S., Peters, B.A. (eds.) Proceedings of the 2004 Winter Simulation Conference, pp. 1213–1217. Institute of Electrical and Electronics Engineers, Inc, Piscataway (2004)

[2] Grimard, C., Marvel, J.H.: Validation of the Re-design of a manufacturing work cell using simulation. In: Kuhl, M.E., Steiger, N.M., Armstrong, F.B., Joines, J.A. (eds.) Proceedings of the 2005 Winter Simulation Conference, pp. 1386–1391. Institute of Electrical and Electronics Engineers, Inc., Piscataway (2005)

[3] Roser, C., Nakano, M., Tanaka, M.: A practical bottleneck detection method. In: Peters, B.A., Smith, J.S., Medeiros, D.J., Rohrer, M.W. (eds.) Proceedings of the 2001 Winter Simulation Conference, pp. 949–953. Institute of Electricaland Electronics Engineers, Inc, Piscataway (2001)

[4] Duanmu, J., Taaffe, K.: Measuring manufacturing throughput using takt time analysis and simulation. In: Henderson, S.G., Biller, B., Hsieh, M.H., Shortle, J., Tew, J.D., Barton, R.R. (eds.) Proceedings of the 2007 Winter Simulation Conference, pp. 1633–1640. Institute of Electrical and Electronics Engineers, Inc, Piscataway (2007)

[5] Williams, C.R., Chompuming, P.: A simulationstudy of robotic welding system with parallel and seriál process in the metal fabrication industry. In: Yücesan, E., Chen, C.-H., Snowdon, J.L., Charnes, J.M. (eds.) Proceedings of the 2002 Winter Simulation Conference, pp. 1018–1024. Institute of Electrical and Electronics Engineers, Inc, Piscataway (2002)

[6] Ong, H.H.: Establishing man-machine ratio using simulation. In: Henderson, S.G., Biller, B., Hsieh, M.H., Shortle, J., Tew, J.D., Barton, R.R. (eds.) Proceedings of the 2007 Winter Simulation Conference, ed, pp. 1663–1666. Institute of Electrical and Electronics Engineers, Inc., Piscataway (2007)

Metrological Aspects of Laser Scanning System for Measurement of Cylindrical Objects

R. Jabłoński and J. Mąkowski

Warsaw University of Technology, Institute of Metrology and Biomedical Engineering, ul. Św. A. Boboli 8, Warsaw, 02-525, Poland
yabu@mchtr.pw.edu.pl

Abstract. In laser-scanning measurement of cylindrical objects the obtained fringe pattern is due to the superimposition of: reflected, unobstructed and scattered light. The proportions between these components vary in time and are affected strongly by instabilities of measuring system. The theoretical approach, based on Fresnel theory, leads to the conclusion that the existing solutions for diffraction of 3D bodies do not fit to engineering applications. Having the above in view, the close analysis of detector signal was carried out. The experiments concern: (1) warm-up time and stability of laser radiation, (2) the real structure of laser beam, (3) the influence of the active detector area on registered signal, (4) determination of the position of object edge in relation to the laser beam waist. The obtained results help to optimize the configuration of all interferometric laser measuring systems.

1 Introduction

In standard laser measuring scanners the measurement information is transformed several times and finally the detector output signal has the form of shadow of an object. When the accuracy of over 1 μm is sufficient this approach is satisfactory, but when more accurate measurements is required it is necessary to take into account the effects appearing on the edges of measuring object because the resulting interference pattern contains the information about the position of the edge of an object [3,4].

We proved that when using Fresnel theory for volumetric obstacles, the corrections depending on the shape of obstacle have to be taken into account. [1].

The measuring method was presented in [2]. Recently, in order to reduce the measurement error level, several modifications were introduced – resulting in obtained final accuracy 0.2μm. This up-dated configuration of measurement system is considered to be an optimum one.

2 Experimental Set-up

The main units of experimental set-up are presented in Fig.1. The laser *(1)* beam is transformed by optics *(2)* and directed to the object *(3)*. The unit *(2)* contains beam forming optics (beam expander with spatial filter) and scan lens. The intensity distribution pattern is measured by detector unit *(4)* and detector signal is processed by computer *(5)*. Computer controls the step motors *(6)* and *(7)* by means of interface in/out LC-012-1612. The motor *(6)* drives the laser unit and motor *(7)* connected to worm gear *(8)* drive the detector. Laser and detector are supplied by separate power sources *(9)* and *(10)*.

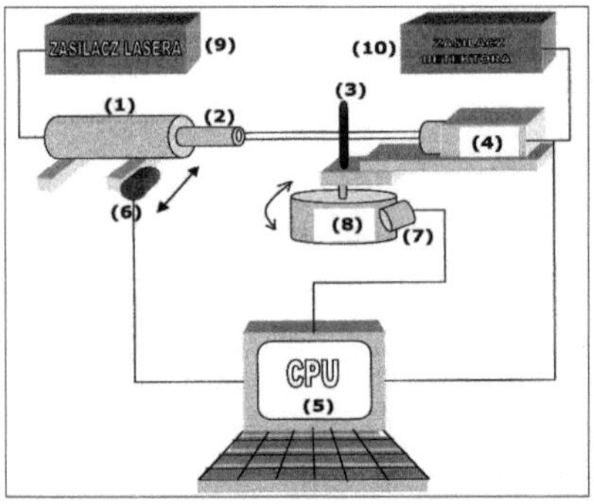

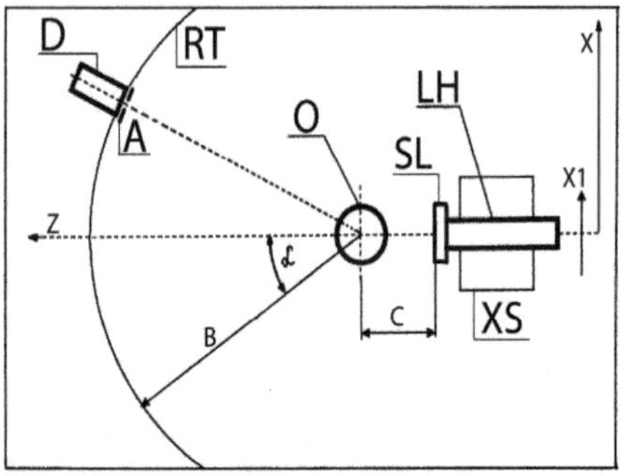

Fig. 1-2. Experimental set-up. a) Outlook (left), b) Scheme (right)

The experiment was limited to the edge effect – to the area located nearby the shadow border line (where the strong diffraction effect is expected).

The scheme of the set-up is presented in Fig.1b. An object ***O***, polished steel cylinder, is placed at the focal distance of scan lens ***SL***. He-Ne laser with beam expander fixed to the X-stage ***XS***, form the laser head ***LH***. Scan lens, of the focal length ***c*** = 65mm, forms the beam waist 60μm along the ***Z*** axis.

In order to avoid the inaccuracies caused by the instabilities of angular deflection, the entire laser head is scanned across the **Z**-axis (x1 movement, linear scanning).

The detector unit (fixed to the rotary table ***RT***, coaxially with object axis) is composed of photodiode ***D***, aperture 0.3mm ***A*** and electronic circuit. It was calibrated with the power meter (LaserMate-Q, Coherent) and obtained signal (in V) is proportional to the measured light intensity. It is driven by step motor and its smallest linear displacement is 0.026μm.

Fig.1 does not show many auxiliary components like: system assuring parallel laser travel, vibration protection set-up, dark chamber, detector electronics, etc.

3 Stability of Laser Power

The standard lasers have the output power stability on the level 0.1-10%. In designed measuring scanner, we decided to use the non-stabilized, low power and low cost He-Ne laser with catalogue stability 1%. The below presented figures

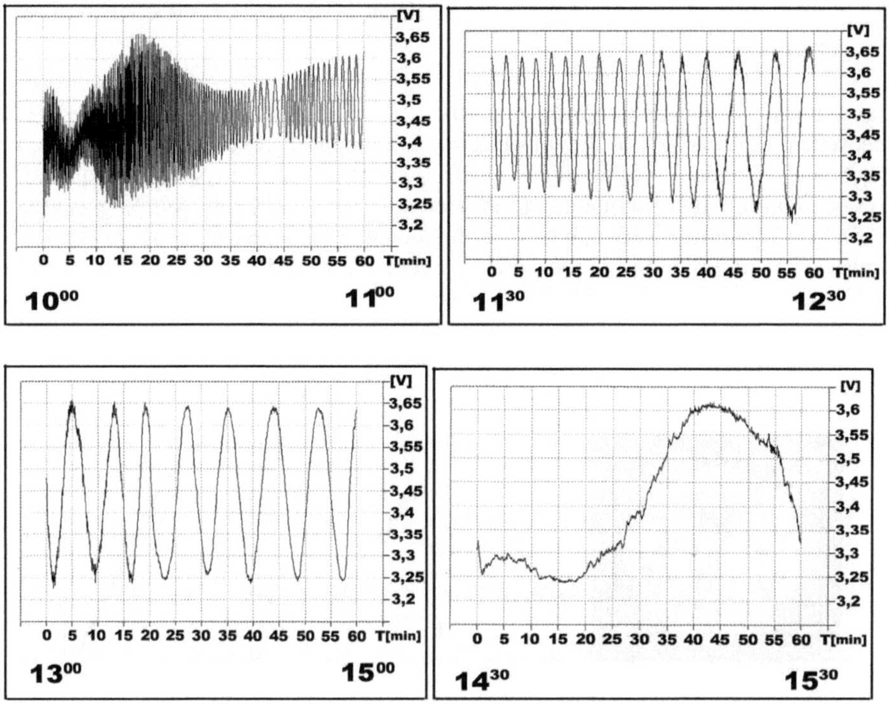

Fig.3-6. Four cycles of laser output power stability measurement

(Fig.3-6) show the output power fluctuations in time intervals 1 hour (4 cycles, in total nearly 6 hours). The experiments started after 10min warming-up time (measurement time is shown on time scale on each graph). The amplified signal was measured in volts with sampling rate 3s. The results are summarized in Table 1.

Table 1. Laser output power fluctuations

Measure. cycle	Measure. time[min]	Max [V]	Min [V]	Average [V]	Uncertainty [V]	Uncertainty %
1	**10-70**	**3,65**	**3,25**	**3,45**	**±0,2**	**±5,48**
2	**100-160**	**3,65**	**3,25**	**3,45**	**±0,2**	**±5,48**
3	**180-250**	**3,65**	**3,25**	**3,45**	**±0,2**	**±5,48**
4	**280-340**	**3,60**	**3,25**	**3,425**	**±0,175**	**±4,86**

The results (measurement the position of interference fringes, not their amplitudes) show how the frequency of power fluctuations changes in time and defines the power stability (with uncertainty 95%). It means that the designed system should be insensitive to the above fluctuations.

4 Structure of Laser Beam, Sampling Rate and Detector Active Area

Laser beam is considered to be Gaussian; however it departs from this simple theoretical function due to the imperfections in the laser. By placing the pinhole in the beam expander (spatial filtering) the fluctuations at high spatial frequencies can be significantly eliminated and theoretically the ideal beam can be generated. However it requires very small pinhole diameter and results in reducing laser power of about 90%. In the measuring system we solved that problem by matching the pinhole diameter with detector active area. Fig.7 shows the area of laser beam "seen" by detector of circular aperture 0.3mm. The observed pattern (speckles) changes, but averaged output signal is stable.

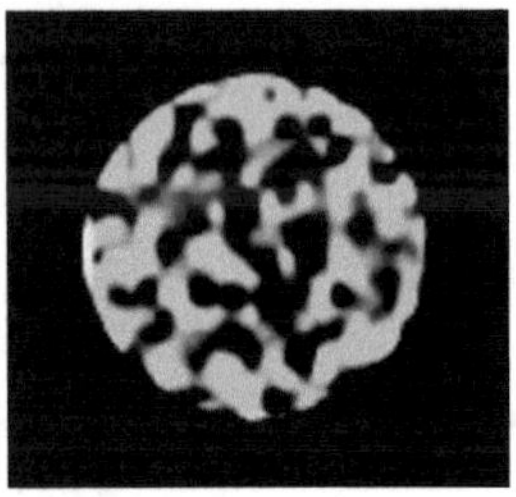

Fig. 7. The area „seen" by0.3mm detector

Fig. 8. Output signal for detector - 5 steps and 50 interval steps

In complement to the above discussion, the displacement (sampling rate) of detector has been optimized. The stability of 0.3mm detector signal was measured applying displacements in the range 1-200 steps. It corresponds to sampling intervals from 0.026μm to 5.2μm. The obtained output signal shows large fluctuations for small intervals (with maximum at about 5 steps, Fig.8, left) and with the increase of interval, the fluctuations are gradually reduced It was noticed that from 1μm interval (40 steps) the fluctuations level is stable.

For the further experiments the minimum interval 50 steps was chosen (Fig.8. right).

5 Position of Object Edge in Relation to the Laser Beam Waist

The determination of measuring area (location of measured object) is essential in all measuring instruments.

In the designed measuring system the beam waist is $2w_0$ = 60μm and its position in reference to the scanning lens is 65mm. In experiments ϕ1.3mm cylinder we used, placed perpendicularly to the laser beam axes in he distance 57-85mm from the scanning lens. The visibility of interference fringes was measured in 1mm step starting with 57mm distance.

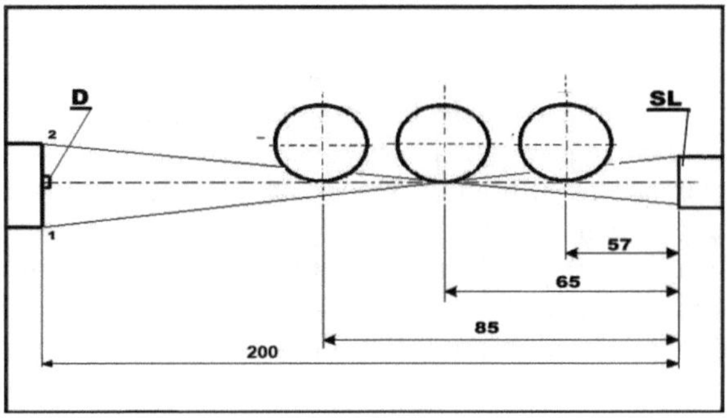

Fig. 9. Object position and general geometry for p.6 experiment

Fig.9 presents the geometry of experiment. In each position the laser beam is screened by the edge of an object and the fragment of resulting interference pattern is recorded. Detector moves between points 1 and 2.

Depending on the position of an object the interference pattern changes significantly (Fig.10). At the distances below 65mm the diffraction effects are visible on the left side of the graphs. The most uncertain signal was detected at the positions close to the beam waist (it did not show any interference). When the distance 65mm was exceeded the stronger and more distinct pattern appeared. The most of the information contains the graph for 75mm distance. There are clearly visible 6 interference fringes and also visibility of fringes is high.

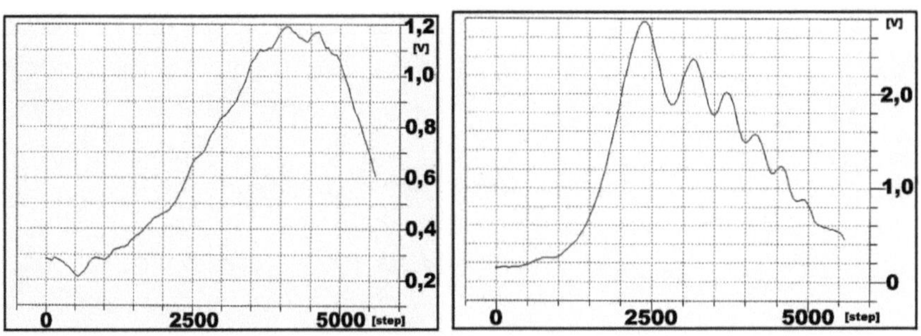

Fig. 10. Output signal for chosen object positions: 65mm and 75mm. Detector displacement from point 1 to point 2 is measured in steps.

6 Conclusions

Although the presented experiments were performed for the needs of laser scanning system, they have general aspect and can be applied to all laser instruments based on interferometry.

At first, it is extremely important to check the parameters of laser beam. In the paper the problem of power stability and speckles were discussed and it was solved by designing measuring system insensitive to these phenomena, (but in many cases it is necessary to use laser with stabilized power output). By optimizing the detector active area and detector sampling interval it is also possible to reduce significantly the detector signal noise.

Finally, in reference to diffraction theory on 3D objects, there is an absolute need to check experimentally the measuring area. The scientists know that the fringe pattern changes with the distance between object and beam waist, but these changes are so strongly dependent on laser beam parameters, object shape and optics used, that the optimization procedure does not exist. All must be just tested experimentally.

References

[1] Jablonski., R., Mąkowski, J.: Measurement of cylinder diameter by laser scanning. Recent Advances in Mechatronics, pp. 596–600. Springer, Heidelberg (2007)
[2] Jablonski, R., Makowski, J.: New approach to fringe pattern analysis obtained by scanning polished metal cylinders with Gaussian beam. In: ISMTII 2007, Sendai, Japan, p. 300 (2007)
[3] Meschede, D.: Optics, light and lasers. Wiley-VCH, Chichester (2006)
[4] Rubinowicz, A.: Die Beugungswelle in der Kirchhoffschen Theorie der Beugung. Springer, Heidelberg (1966)

Continuous Quality Evaluation: Subjective Tests vs. Quality Analyzers

A. Ostaszewska, S. Żebrowska-Łucyk, and R. Kłoda

Warsaw University of Technology, Faculty of Mechatronics, Institute of Metrology and Biomedical Engineering, sw. Andrzeja Boboli 8, 02-525 Warsaw, Poland
{a.ostaszewska,s.zebrowska,r.kloda}@mchtr.pw.edu.pl

Abstract. The paper presents two ways of continuous quality assessment of compressed video: subjective and objective. The attempt to comparison of the Single Stimulus Continuous Quality Evaluation (SSCQE) results with SSCQE emulator is performed.

1 Introduction

In case of video, the use of lossy compression is almost a rule, as it enables for significant decrease of the file size. The side effect of this process is the loss of quality, which should be measured to meet the requirements of a consumer.

The compressed video quality measurement is conducted either on the way of experiment which involves a group of observers, or with the use of quality analyzers – the algorithms which emulate the human observers' behavior.

The growing number of quality analyzers available on the market raises various questions among the potential consumers. The most important is: how perfect is the imitation of human audience?

The paper presents the first attempts to comparison of the signal generated by Rohde&Schwartz® DVQ Digital Video Quality Analyzer (which is claimed to emulate a subjective method known as Single Stimulus Continuous Quality Evaluation (SSCQE) [1, 2] scores) with the result of the SSCQE itself. Statistical parameters are used for this purpose.

2 SSCQE Method, Test Equipment and Experimentation

In SSCQE method a series of video sequences is presented once to a viewer. Video sequences are coded with different parameters, so the impairments may be more or less visible. A panel of viewers follows the temporal variations of quality in the video using a slider device with a continuous scale from "excellent" (100) to "bad" (0) attached. Ratings are sampled at regular time intervals, so the time variations in perceived quality are precisely recorded. Viewers are not explicitly shown the reference sequences, thus their ratings are absolute. This corresponds well to an actual home viewing situation, where the reference is not available to the viewer either.

Test equipment used by the authors consisted of a 20" professional-grade monitor (SONY PVM-20M4E) and a professional DVD player (Pioneer DVD-V7300D) (Fig. 1).

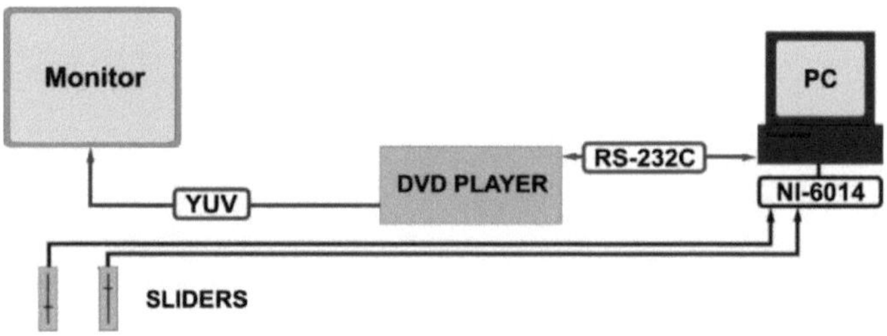

Fig. 1. A diagram of the experimental station

SSCQE ratings were sampled at a rate of 2 samples per second and entering the PC directly through the NI 6013 card. All data was synchronized with the timecode. Due to the dedicated application there was a possibility of displaying all incoming signals, so all data was being monitored by a supervisor.

Test material was built on the basis of four source sequences: bbc3, mobl, cact and susi (Fig. 2) that were obtained from Tektronix, each of 15 seconds duration, 40 Mbps and 25 frames per second. They contained elements difficult to decode: flat areas, complex patterns, masking effect, object and/or camera motion (zoom, pan) at different speeds, objects appearing, crossing the scene, moving in different directions, faces and landscapes. Soundtrack was not included.

Source sequences were processed to the most popular standard MPEG-2. There was a possibility of changing three coding parameters: bitrate, length and a structure of a group of pictures (GOP). Because the duration of the test was limited, only 13-frame GOP was taken into account. The bit stream was rated as follows: 2, 2,2, 2,6, 3, 3,2, 3,6, 4, 4,2 4,6 and 5 Mbps. For those 10 levels of bitrate, each sequence was coded with all three possible structures of GOP: with two B-frames, one B-frame and without them. This way of coding resulted in 30 variants of each of source scenes. Each observer was shown the same test material.

Fig. 2. The screenshots of test sequences

The experiment was performed in accordance with ITU-T P.911 Recommendation [1]. 45 non-expert viewers – mostly university students (aged from 20 to 25) – participated in test sessions. Each of them was screened for normal visual acuity or corrective glasses and normal color vision and all spoke polish so could comprehend instructions given in that language. A training process was used to instruct subjects about the task they were to perform during the subjective test and to become acquainted with the quality range to be seen in the test.

The test was limited to two 15-minute sessions (separated by a 15-minute break), in order to keep the limit of a viewing period up to 30 minutes [1].

The same test material was applied to R&S®DVQ Digital Video Quality Analyzer.

3 R&S®DVQ Digital Video Quality Analyzer

Digital Video Quality Analyzer R&S DVQ enables for objective realtime measurement based on the analysis of DCT-coded video data applied to DVQ in a MPEG2 transport stream [3, 4]. It needs no reference signal and quality values are assessed according to the subjective masking effects produced by high temporal and/or spatial activities of the picture. The result of analysis is a reproducible quality level (DVQL-W) on a SSCQE scale.

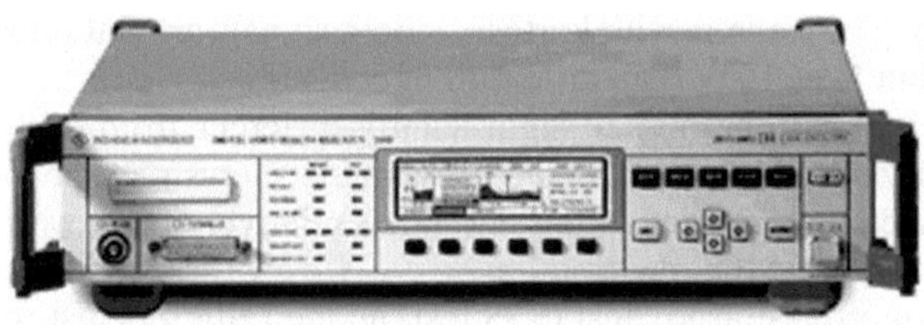

Fig. 3. R&S®DVQ Digital Video Quality Analyzer [4]

4 Results

In case of SSCQE the mean of all scores given in time by all audience was computed to obtain Mean Opinion Score (MOS) signal as a function of time. Fig. 4 shows the SSCQE MOS signal and the DVQL-W.

SSCQE MOS signal uses a wider range (64,7 points of the total scale) and is positioned almost in the middle of scale (Tab. 1). DVQL-W signal amplitude is two times smaller (measures 32 points) and the mean is 90% of the maximum of the scale. Standard deviation of the SSCQE MOS signal is 2,7 times bigger than standard deviation of DVQL-W.

Table 1. Statistical analysis of SSCQE MOS and DVQL-W signals

	mean	min	max	Standard deviation	Skeweness	Kurtosis
SSCQE MOS	56,2	17,6	82,3	12,8	-0,52	0,21
DVQL-W	90,2	67,0	99,0	4,7	-1,25	2,21

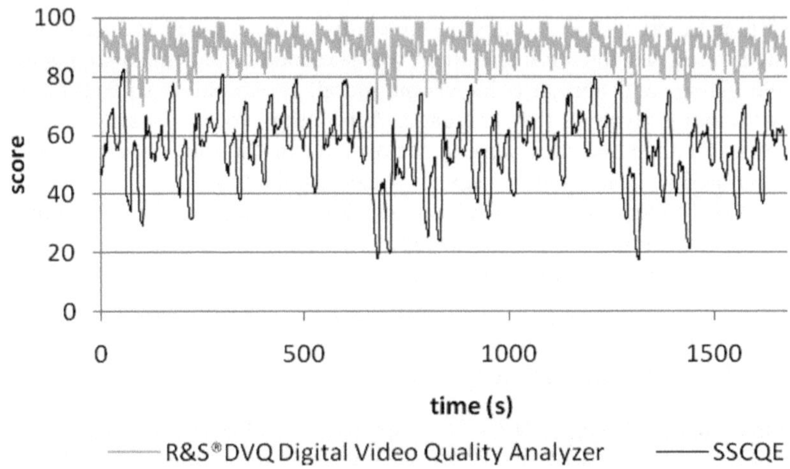

Fig. 4. SSCQE MOS vs. DVQL-W

Both signals are characterized by negative skeweness, which indicates that data are skewed left, but in case of DVQL-W the effect is more evident than in case of SSCQE MOS. The same is with kurtosis – both signals are flat relative to a normal distribution, but DVQL-W data differ stronger again.

The shapes of histograms are significantly different from each other (Fig. 5). The probability distribution of DVQL-W variable shows that more of the variance is due to isolated extreme deviations and it has relatively few lower values in comparison with SSCQE MOS distribution.

The Kolmogorov-Smirnov test (K-S test) an the Lilliefors test were used to verify normality of probability distributions of both signals. K-S test as well as Lilliefors entitled to reject the null hyphothesis that samples came from normal populations (with significance level $\alpha = 0{,}01$).

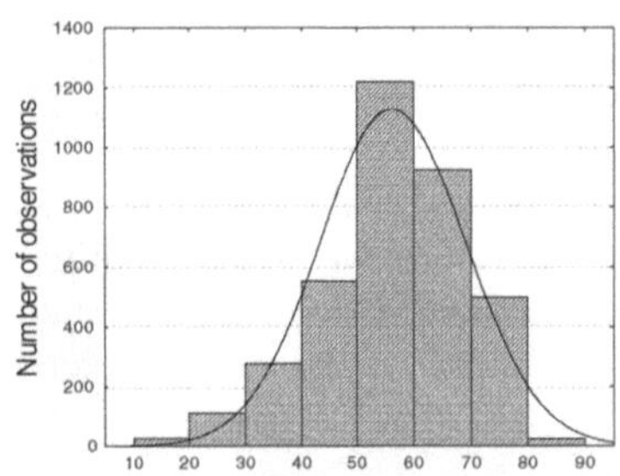

a)

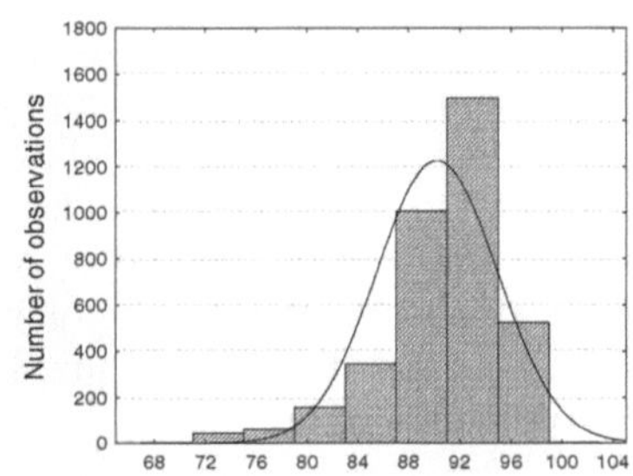

b)

Fig. 5. Probability distributions of variables: a) SSCQE MOS (K-S test: $d = 0{,}05808$, $p < 0{,}01$; Lilliefors test: $p < 0{,}01$), b) DVQL-W (K-S test: $d = 0{,}13778$, $p< 0{,}01$; Lilliefors test: $p < 0{,}01$).

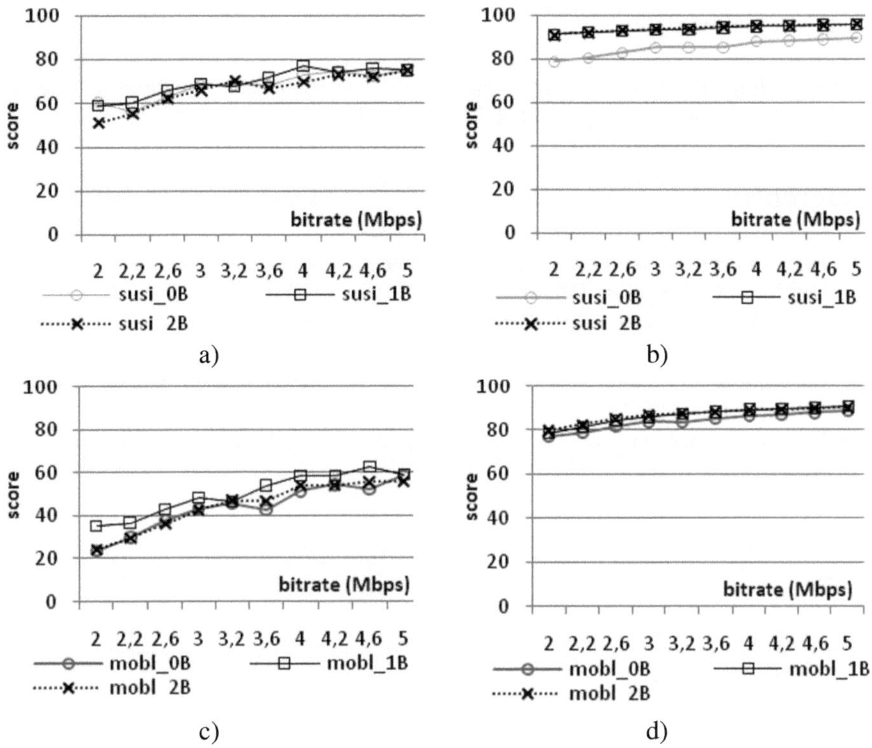

Fig. 6. Mean Opinion Score (MOS) computed on the basis of SSCQE ratings (a and c) and on the scores generated by DVQL-W (b and d)

Diagrams of Mean Opinion Scores shows that both SSCQE MOS and DVQL-W are positively correlated with bitrate, but SSCQE MOS is more prone to changes of video data stream (Fig. 6). Additionally, the SSCQE MOS susceptibility to bitrate strongly depends on the scene content, whilst in case of DVQL-W the scene subject seems to exert no influence. Then again DVQL-W scores show difference between the quality of video coded with B-frames and coded without them, whilst for tested range of bitrate the audience seem not to perceive any evident improvement.

5 Conclusion

The analyzers are especially useful for continuous quality measurement (for example for monitoring digital television signals, where it would be impossible to engage a group of observers), consume less time and costs, but the results obtained automatically differ from the results obtained on a way of subjective tests. Therefore there is an incontestable need to enhance the analyzers. This also means that the subjective methods should be under constant improvement, as they provide both the knowledge on Human Visual System (HVS) and the basis for analyzers' calibration.

Acknowledgments. This scientific research work was sponsored by the funds for science in years 2007-2009 as research program N N505 4282 33.

References

[1] ITU-T Recommendation P.911, Subjective audiovisual quality assessment methods for multimedia applications (1998)
[2] ITU-T Recommendation P.910, Subjective video quality assessment methods for multimedia applications (2008)
[3] Lauterjung, J.: Picture Quality Measurement. In: International Broadcast Conference 1998, pp. 413–417 (1998)
[4] `http://www2.rohde-schwarz.com/product/DVQ.html` (30.05.2009)

Measurement of the Temperature Influence on NiMH Accumulator Characteristic

M. Synek, V. Hubík, and V. Singule

Brno University of Technology, Faculty of Mechanical Engineering, Institute of Production Machines, Systems and Robotics, Technicka 2896/2, Brno, Czech Republic
{synek,singule}@fme.vutbr.cz, vhubik@nbox.cz

Abstract. The work deals with influence of the temperature on cylindrical type nickel-metal hydride (further NiMH) accumulators. The basic aim of the research is to study charging and discharging characteristics of accumulator depending on the working temperature from 20 °C to 45 °C. According to the reached results it is effort to predict the capacity, lifetime and other parameters of battery during duty cycle especially in mobile devices such as robots, RC models, etc. The first measurement on the intelligent battery charger Robitronic Overloader2 has been performed and the special measurement card for data collection has been created.

1 Introduction

The sealed NiMH accumulators are one of the most used power supply and these days have become a mainstay in rechargeable accumulators. They have higher capacity (energy density) than nickel-cadmium ones and are more friendly to the environment that the other types because contain no cadmium, mercury or lead.

The main advantages of NiMH accumulators are:

- Up to 40 % higher capacity / energy density than the NiCd or Lead-acid ones
- High drain current, that means low internal resistance
- Compact size (flexible to user demand)
- Acceptable cost
- Contain no heavy metals

Due to this positive qualities NiMH are also commonly used, except mobile devices and robots, in all kinds of consumer electronics, multimedia devices, mobile phones, notebooks, etc.

Basic characteristics of NiMH in numbers are:

- Nominal voltage per one cell 1,2 V
- Voltage of fully charged cell 1,35 – 1,4 V (without load)
- Voltage of flat cell 1,1 – 1,0 V
- Energy density 70 Wh / kg (250 kJ / kg)
- Volumetric energy density 300 Wh / l (360 MJ / m^3)

Positive electrode: $$Ni(OH)_2 + OH^- \underset{\text{Discharge}}{\overset{\text{Charge}}{\rightleftarrows}} NiOOH + H_2O + e^-$$

Negative electrode: $$M + H_2O + e^- \underset{\text{Discharge}}{\overset{\text{Charge}}{\rightleftarrows}} MH_{ab} + OH^-$$

Overall reaction: $$Ni(OH)_2 + M \underset{\text{Discharge}}{\overset{\text{Charge}}{\rightleftarrows}} NiOOH + MH_{ab}$$

(M: hydrogen-absorbing alloy; H_{ab}: absorbed hydrogen)

Fig. 1. Electrochemistry of NiMH [4]

General NiMH cells consist of a positive electrode, made of nickel hydroxide as an active material, a negative electrode composed of hydrogen absorbing alloys, a separator between electrodes and an alkaline electrolyte. All this parts are composed to the metal case with sealing plate and nowadays equipped with safety vent. The charge and discharge reactions in NiMH are on the Fig. 1, construction of the cylindrical cell on the Fig. 2.

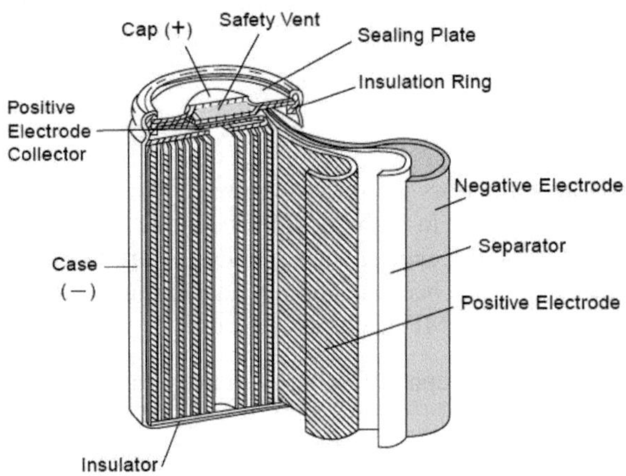

Fig. 2. Schematic of the cylindrical cell construction of NiMH [4]

2 Charging and Discharging Characteristics

This text works with influence of temperature on charging and discharging characteristic of NiMH. Used sample of NiMH accumulator was LPR VTEC NiMH-battery. This battery is made of 6 cells and output voltages is 7,2 V, capacity 1400 mAh.

All characteristics are measured by Robitronic Overloader2, connected through RS-232 or USB interface to notebook with appropriate measuring software LogView ver.2.4.5.203. The Overloader2 is a performance computerized system, which is capable of charging, discharging, cycling and conditioning of battery types – NiCD, NiMH, Li-Pol, Lead-acid and Li-XX and other features (especially for RC models). It can be programmed for many different characteristics depending on type, manufacturer or status of connected accumulator.

Setting of Overloader2 for mentioned LPR battery was:

- Type – NiMH, 6 cells, capacity 1400 mAh
- Charging characteristic was linear
- Charging current 1,4 A and 0,7 A (1C and 0,5 C)
- End of charging was set up on delta peak 15 mV
- Maximal temperature of the battery cell 42 °C
- Maximal discharging current 5 A

Several experiments have been made with mentioned configuration of the battery charger. Temperature of the cell was measured by thermal sensor, connected to the Overloader2. Cooling was provided by small electric PC fan and pad made of aluminum.Measured characteristic for 1 C (1,4A) charging without external cooling depicts the trigger-stop after reaching about 70 % of the battery capacity due to temperature cross the limit of 42 °C. On the other side, if the external cooling fan is used, charging stops at 75 % of the battery capacity due to delta peak voltage (15 mV) drop out.Another important parameter of accumulator is discharging characteristic. Discharging current was constant with value of 5 A. In the both cases, with or without external cooling, the interruption was occurred by reaching the limit of 5,4 V for battery cell. The difference between cooled discharge was the end temperature about 6 °C lower than uncooled.The next experiment will be with discharging of the battery with variable load by means of real application in any mobile device, such as RC model of off-road car. With the view of fully independency of the mobile device, the special data logging card has been created, which is described in the following chapter.

3 Data Logging Card

Data logging card is used for analysis of discharging characteristic of the NiMH battery cell for RC models, described above. The aim is to put the card into the mobile device and study the battery qualities during variable load, which simulates the real duty cycle of the accumulator.

The equipment, depicted in the Fig. 3, is consisted of the several parts – modified battery cell, CPU board and SD card data logger. This all (several PCB modules) are located on the special main board with appropriate analog interface circuits and DC power supply. Every part will be more described separately.

Modified battery pack

Modified battery pack is obtained by 6 temperature sensors KTY81-1 by Philips, that are able to measure temperature from -55°C to +150°C. For the battery temperature measurement purpose it is not necessary to obtain such a wide range. The input instrumentation amplifiers are tuned for range from 10°C to 50°C with the resolution of 0,1°C. Temperatures above this scale are fatal for used NiMH accumulators. Amplified and translated signals from sensors are fetched in the analog-to-digital converters located on the CPU board.

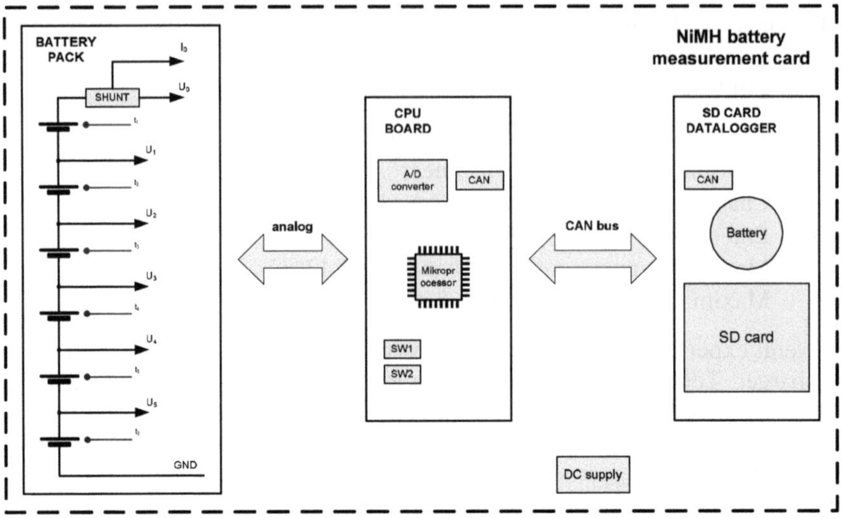

Fig. 3. Schematic diagram of the battery measurement card

The whole battery pack is additionally obtained by voltage measurement of every cell. It is possible to measure with resolution of 0,1 V on all six channels. The basic analog conversion circuits are located on the main board as well and then fetched to the internal ADC of the main processor.

It is also necessary to measure the actual current form the battery. Smart Hall effect sensor, instead of some cheap shunt resistor, is used and it is possible to determine current with the resolution of 0,1 A and maximal current 20 A.

All sensors described above, except the voltage ones, are calibrated with original battery pack and cannot be used with another one. Calibration of the temperature sensors were made in the special climate chamber. Voltage and current channels were calibrated by accurate measurement instruments.

The main CPU board

The main CPU board consists of the powerful 16 bit DSP processor by Microchip, which is clocked by the external crystal oscillator up to 116 MHz. Relatively high frequency is used due to the high number of measured analog signals (14 channels), that are subsequently sampled and averaged several times per second. Every channel has to be re-counted by the appropriate calibrating constants (gain and offset), that are stored in the internal EEPROM memory.

The board is connected with the other peripherals, such as SD data logger or personal computer, by the CAN serial data bus. Data bus has about 100 Kbits/s transfer rate and CANaerospace high layer protocol is used. This type of protocol is convenient to use for the application, where is not necessary to transfer huge amount of data and is suited for control with short response time delays.

SD card data logger and CPU board are internally connected through CAN bus on the main board. CPU sends several times per second measured and re-counted data to the SD card for saving. It is possible to set the period of sending on the multiple of 10 ms. Whole message consists of 14 frames with actual temperature, voltage and current on every cell.

Due to maintenance it is possible to connect the measurement card through CAN bus to the personal computer with appropriate software. Several calibrating constants, repeating period, etc. are set with this software also.

The board is powered by 5 V DC from the power supply located on the main board.

SD card data logger

This card was created for general purpose of data logging on the CAN bus, which is relatively often used in industry, automotive and aerospace environment. With advantage of small diameters, low power consumption, big storing capacity and low prize of memory media it is possible to use this solution in our battery analyzing project. The SD card data logger listens on the CAN bus and stores all incoming packets to the internal SD card with capacity up to 2 GB. Commercial and cheap SD card, well known from many multimedia applications (such as cameras, phones, MP3 players, etc.), is used. For worse working environment, that means low or extremely high temperature, it is usually better to use special type of industrial SD card, which is much more reliable.Received packets are added by exact time mark from implemented RTC (real-time clock) circuit. The resolution of time determination is 100 ms.

Used SD card must be formatted in FAT16 file system and must contain any data file. Prepared data packets are then stored in this data file. This configuration allows very comfortable data post processing by means of easy upload the data file to the computer by any commercial data reader. Analysis can be performed by appropriate software, such as Microsoft Excel, MATLAB, etc.Consistence of the data file in the SD card is check by internal firmware automatically for higher reliability.In case of 2GB card and average 42 packets per second (that means all message is sent 3 times per second), the data could be collected for more than 2 month continuously.

4 Conclusion

Temperature influences most of the operating characteristic of NiMH accumulators as it is described above in this paper. External cooling provides more efficient charging. Due to trigger-stop by delta-peak voltage, we can use almost full capacity of accumulator. For discharging by invariable current is external cooling much less efficient. Keeping accumulator temperature near to the value of 30 °C extends their lifetime. The next step is providing measurement on the real RC model or other mobile device with the various loads (various current) and simulates the real duty cycle. Stand-alone measurement and data logging card has been created for this purpose. Collected data from several working cycles will be analyzed and according to the reached results will be predicted optimal working cycle, lifetime and temperature influences. The analyzing software, which enables the appropriate data post processing and graphical outputs, has been created in the MATLAB environment. Component part of the analyzing software is graphical user interface – GUI as well. Obtained results and behaviors of the NiMH batteries could be used in the several part of industry environment, mobile devices, medicine and other part of the technical branches, where these types of accumulators are plentifully applied.

Acknowledgments. Published results were acquired using the subsidization of the Ministry of Education, Youth and Sports of the Czech Republic, research plan MSM 0021630518 "Simulation modeling of mechatronic systems".

References

[1] Cenek, M.: Akumulátory od principu k praxi. FCC Public, 248 pages (2003)
[2] LogView software webpage (2009), http://www.logview.com
[3] LRP webpage (2009), http://www.lrp.cc
[4] Panasonic electronic (2009), http://www.panasonic.com
[5] Cobasys energy storage (2009), http://www.cobasys.com
[6] Duracell webpages (2009), http://www.duracell.com

Synthetic Method of Complex Characteristics Evaluation Exemplified by Linear Stepper Actuator Characteristic Comparison

K. Szykiedans

Warsaw University of Technology, Faculty of Mechatronics,
Institute of Micromechanics and Photonics, Sw.A.Boboli 8, Warsaw, Poland
k.szykiedans@mchtr.pw.edu.pl

Abstract. Characteristics of electric stepping micro-machines could be changed or shaped by using many methods other than simple re-design of their electromagnetic circuit. Many possibilities of characteristics forming give us idea of using special methods of steering. For planning, simulation and evaluation of new modified characteristics objective apparatus that allows comparing complex characteristic curves is a must. During development of new stepper actuators steering method such requirement occurred. Because of complexity of stepper characteristic curve, its incapability being analytically described and its irregularity – synthetic method was proposed and developed. This method is based at analyses of geometric parameters of an area under a characteristic curve. Because it does not use particular features of stepper motors characteristics, it can be used to comparisons of all characteristic types. Especially these ones, in which an area under a curve is important, what is happening i.e. when estimating absorption.

1 Introduction

Starting his studies [1] on influence of various parameters on stepper actuator characteristics author has faced the problem how to measure or estimate this influence. Rules of setting point of work for stepping machines are widely known and described [2,3,4], but there are no clue to judge which characteristic is better. It is more important because this problem if more often to choose pair of a control unit and a stepping motor not just simple comparison. Such feature analyses have to be made on the features that cannot be described by a single value. It has to be based on aggregated data or characteristics that describe the device. In many cases selection is made by computer systems with use of simulation data. That's why the comparison result should be calculable numerical value.

Comparing two items that can be described by a single value it is easy to judge which is better. If we call this value as n than its gain Δn can be used for comparison. It can be stated as absolute or relative measure.

$$\Delta n = n_2 - n_1; \Delta n = \frac{n_2 - n_1}{n_1} \cdot 100\% \tag{1}$$

It is possible to make such comparison also for two selected parameters of stepper actuators i.e. maximum force, maximum frequency, or to compare maximal frequency of pull-out work under set load for a different values of driving voltage. When analyzed device is described by a nonlinear characteristic case became to be more complex, especially when characteristic is an irregular curve or this curve is just a border of area that describes device behaviour. This is a case of all stepper motor and actuators which can work in any point of area bounded by characteristic curve and both axes of coordinate system. For a pull-in characteristic it is true in all conditions, for a pull-out characteristic only when following proper acceleration and braking conditions.

Typical pull-in or pull-out characteristic is irregular curve. Simple comparison of maximal force or maximal frequency is not including any information about characteristic shape. Let's consider case of two actuators. One - call it A- has maximum force of 100 units, second - B – of 90 units. Which one is better - A? But when comparing characteristics of these two devices it is plausible that the case will be like presented below.

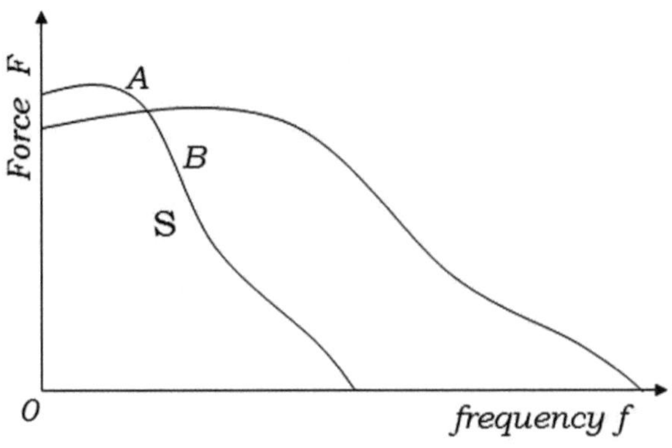

Fig. 1. Hypothetical characteristics of two stepper actuators A and B

As it can be seen simple comparison in this case could lead to misinterpretation. Actuator A can give more force but machine B characteristic is much wider, more useful when flexibility of drive if needed.

2 Proposed Method of Evaluation

As it was stated above characteristic of stepper actuator is a figure bounded by a characteristic curve and both axis of coordinating system. This obvious phrase was an inspiration for author to dissolve problem of characteristics evaluation from the point of view of geometry not the electrical machines.

Let's consider few simple equations describing some features of planar figure. As a first what's comes to mind - area of a figure. In considering coordinate system of frequency f and force F area S of figure bounded by both axes and a curve can be calculated as follows [5]

$$S = \int F(f)df \tag{3}$$

And relative change of area can be our first comparison value.

$$\Delta S = \frac{S_B - S_A}{S_A} \cdot 100\,\% \tag{4}$$

Index A will be used for referential characteristic, index B is for characteristic being compared. Use of area as the only comparator can be misleading. Figure 2 shows the case when the same shape figures are compared, only one of them was mirrored. Areas of this figures are equal but as they placement differs they also describes different devices.

To avoid such misleading as showed in this simple example, use of complementary comparisons is a must. There are also some other geometrical parameters:

The second moment of area (area moment of inertia) about the axis f

$$J_f = \int_S F^2 dS \tag{5}$$

and its relative change

$$\Delta J_f = \frac{J_{fB} - J_{fA}}{J_{fA}} \cdot 100\% . \tag{6}$$

Coordinates (f,F) of center of mass C

$$f_C = \frac{\int_S f dS}{S} ; \quad F_C = \frac{\int_S F dS}{S} \tag{7}$$

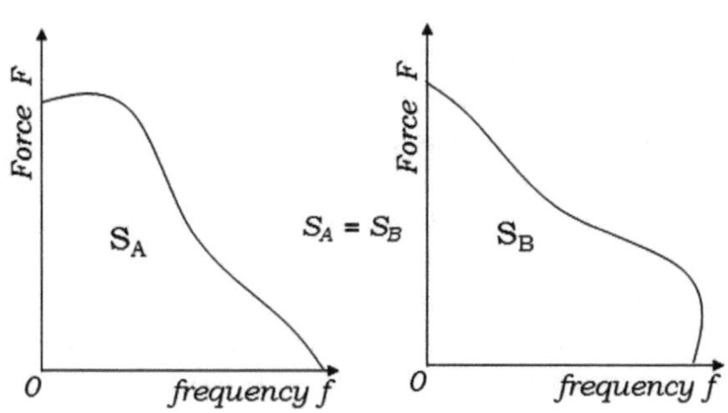

Fig. 2. Hypothetical characteristics having the same area

and their relative change

$$\Delta f_C = \frac{f_{CB} - f_{CA}}{f_{CA}} \cdot 100\%\,; \Delta F_C = \frac{F_{CB} - F_{CA}}{F_{CA}} \cdot 100\%\,. \tag{8}$$

Having all three presented measures it possible to correctly evaluate changes of shape of pull-in or pull-out characteristic of a stepper actuator. This evaluation can be made on these three measures separately or having them totalized as a one weighted measure ΔQ.

$$\Delta Q = w_S \cdot \Delta S + w_J \cdot \Delta J_f + w_f \cdot \Delta f_C + w_F \cdot \Delta F_C \tag{9}$$

As proposed method is not strait based on stepper features explanation of possible changes and it influences on actuator will be presented.

Table 1. Listing and short explanation of presented measures changes

Measure change	Explanation
$\Delta S>0$	General increment of area of characteristic. Desirable.
$\Delta S<0$	General reduction of area of characteristic. Undesirable unless special requirements.
$\Delta J_f>0$	Increment of area or change of shape of characteristic. Must be considered with others measures. Large increment with $\Delta S=<0$ means increasing of range on ordinate axis and reduction of range on abscissa axis
$\Delta J_f<0$	Reduction or change of shape of characteristic. Must be considered with others measures. Followed by $\Delta S>=0$ indicates increasing of range on abscissa axis and reduction of range on ordinate axis
$\Delta f_C>0$ $\Delta F_C>0$	General increment of area of characteristic. Follows with $\Delta S>0$. Desirable.
$\Delta f_C>0$ $\Delta F_C<0$	Characteristic expanded in direction of abscissa axis. If followed with $\Delta J_f>0$ characteristic figure means that characteristic was not lowered on higher values of ordinates. Other cases - characteristic was lowered.
$\Delta f_C<0$ $\Delta F_C>0$	Characteristic expanded in direction of ordinate axis. If followed with $\Delta J_f>0$ characteristic figure means that characteristic was not lowered on higher values of abscissas. Other cases - characteristic was lowered.
$\Delta f_C<0$ $\Delta F_C<0$	Generally caused by reduction of characteristic area. When occurs with $\Delta S>0$ and/or $\Delta J_f>=0$ then characteristic curvature was changed from convex to concave

3 Examples

Method described in proceeding chapter was tested by author on synthetic data. It was also used to estimate influence of new method of acceleration [1], simulation and tests result of nonlinear acceleration (b) were compared to the data received during standard acceleration (a).

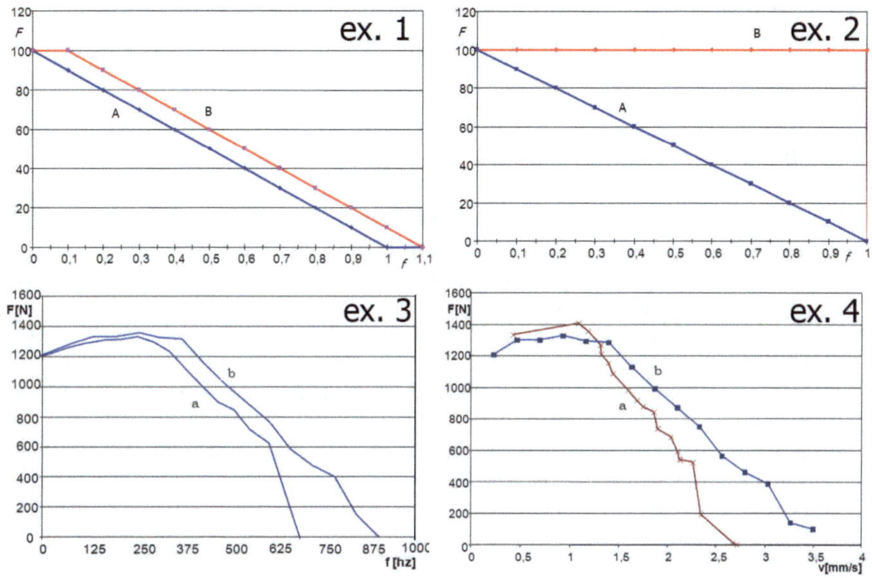

Fig. 3. Examples of characteristics comparisons ex.1, ex.2 synthetic data, ex.3 simulation data, ex.4 test results.

For the presented examples evaluation method described in chapter 2 gave following results.

Table 2. Evaluation results.

Example	ΔS	ΔJ_f	Δf_C	ΔF_C
1	18%	32%	9%	24%
2	100%	258%	25%	42%
3	25%	29%	8%	0,5%
4	18%	11%	14%	9%

For a unknown application or for automated comparing during optimization or during training of neural networks or genetic algorithms synthetic target as shown in examples 1 and 2 should be considered.

4 Conclusions

There were presented method of quantitative evaluation of a complex characteristic. Method based on geometrical parameters of a figure bounded by axes of coordinate system and a characteristic curve can be easily used in numerical calculation during i.e. simulation. Because there is no use of any specific feature of figure presented method is universal and can be used in many technical areas. Depending of used coordinates system each used measure can be interpreted differently (i.e. as power, flux or work). Use of this method, especially interpretation of it results can be debatable when strait connection to physical phenomena is not obvious. This method is a comparative tool that can be used by an engineer with other ones.

References

[1] Szykiedans, K.: Method of increasing performance of stepper actuators. Recent Advances in Mechatronics, pp. 253–258. Springer, Heidelberg (2007)

[2] Acarnley, P.P.: Stepping Motors: a guide to modern theory and practice. IEE and Peter Peregrinus Ltd., London (1982)

[3] Kenjo, T.: Stepping motors and their microprocessor controls. Clarendon Press, Oxford (1984)

[4] Yeadon, W.H., Yeadon, A.W.: Handbook of small electric motors. McGraw-Hill, New York (2001)

[5] Bronshtein, I.N., Semendyayev, K.A., Musiol, G., Muehlig, H., Mühlig, H.: Handbook of Mathematics. Springer, Heidelberg (1997)

Aircraft Sensors Signal Processing

J. Bajer, R. Bystřický, R. Jalovecký, and P. Janů

Brno, University of Defence, Faculty of Military Technology,
Department of Aerospace Electrical Systems, Kounicova, 66210 Brno, Czech Republic
{josef.bajer,radek.bystricky,premysl.janu}@unob.cz
rudolf.jalovecky@unob.cz

Abstract. The paper deals with possibilities and methods of processing of signals from aircraft sensors. Because of a large amount of proceeded data the subject of the paper is limited only to data processing from electrical subsystem of aircraft. Designed data acquisition and processing avionic system is based on CANaerospace communication network. Primary function of the system is to collect all available information from all avionic blocks. Proposed system will be employed in being designed system NEC (Network Enabled Capability) in the Czech Armed Forces in future.

1 Introduction

Information Technologies (IT) play a dominant role in development of Network Enabled Capability (NEC) and in professional Czech Armed Forces in general. NEC is network communication and information system, which combines strategy, recent tactics, ways, procedures and organization methods, which military powers can use to achieve superiority over the enemy. Each commander is able to make right decision and issue orders only on the basis of enough necessary information. Network communication and information system NEC is being developed for this reason. This system collects data from ground troops and air forces too. Afterwards each commander could effectively use these data.

Any comprehensive study on required structure of information from aircraft, their collecting, processing and providing to the NEC hasn't been published yet. How-ever, well and logically ordered information from aircraft on-board systems and sensors is essential prerequisite for effective NEC realization at aviation area.

In avionic systems of modern aircraft the trend in collecting of information from all on-board systems steadily grows. Development of new avionic systems pre-supposes implementation of new systems into existing, sometimes obsolete, avi-onic network like ARINC 429, RS-422, RS-485, MIL-STD-1553B. Some of men-tioned networks have been even classified.

2 Data Acquisition from Flying Vehicles

A basic hierarchical structure of proposed Aircraft on-board Electronic System (AES) is described in full details in [2]. AES is a modular system composed from several subsystems. Communication among particular modules is based on CA-Naerospace protocol, which is an enhancement of the Controller Area Network (CAN) [4]. CANaerospace is communication protocol specially developed for use in avionic systems. CANaerospace offers a lot of additional functions and also ensures better data transfer security [5]. An effort to certificate CAN for use in avionic systems [3] was made in the Czech Republic in the past. CAN hasn't been certified as a standalone system, but only as a part of system SAM [6] from UNIS company. However, the capability for use in avionic systems has been confirmed by Federal Aviation Administration (FAA) and Civil Aviation Authority Czech Republic (CAA CZ). European Aviation Safety Agency (EASA) certificate was issued in December 2005.

Electrical subsystem blocks (Fig. 1) process information concerning inertial meas-urement of the aircraft space position (Earth's magnetic field, geographic posi-tion), airframe and engine operation conditions (operational pressures at systems of oil, fuel, air and oxygen), temperature at important points of the aircraft, sur-rounding temperature, air pressures, voltage of on-board power supply system and positions of selected actuators.

Diagnostic equipment could be also added into electrical subsystem.

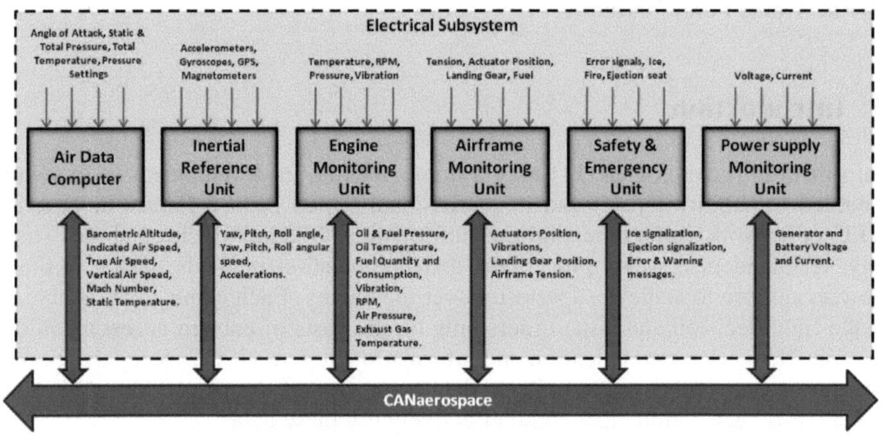

Fig. 1. Proposal of electrical subsystem

3 Signal Source Characteristic on an Aircraft

3.1 Signal Parameters

Characteristics and parameters of signals occurring on an aeronautical vehicle are clearly shown with aid of several signals mentioned at the table (Tab. 1). First of them with CAN message identifier (CAN-ID) 300 is acceleration in aircraft longitudinal axis. Range and scale of measured value generally depends on applied sensor. In this presentive case ±10 g with 14bits resolution is used. Sensor transmits measured data through SPI bus. A period between two measurements is controlled by microcontroller.

Quantity with CAN-ID 326 is static pressure performing for altitude calculation. While using commercially widespread available sensor MPX4115 voltage at theo-retical range from 0.2 to 4.8 V with altitude corresponding 15÷115 kPa is meas-ured. When internal A/D converter with 0-2.5 V voltage reference is used nor-malization of measured voltage to this range is necessary. Afterwards resolution will be 10bits. A period between two measurements is again controlled by micro-controller.

Quantity with CAN-ID 324 represents static air temperature. Sensor with tempera-ture dependent resistance value is used. Temperature dependent resistor is a part of flip-flop circuit. Measurement of resistance is converted to measurement of pulse width. AVR microcontroller's counters/timers allow measuring with 16bits resolution from approximately 1 μs.

3.2 Signal Processing

There are lots of signal sources on aeronautical vehicle as mentioned at previous text. For signal processing is necessary to divide them to several groups according to way of physical principle of sensing quantity (employed sensor) and according to way of its analyzing at measuring chain.

First group represents sensors whose input quantities are non-electrical signals (e. g. pressures, RPM, temperatures). Typical processing chain of non-electrical analogue quantity is depicted in Fig. 2. Signals from sensors are impedance matched and eventually also separated. Anti-aliasing filter for suppression of unnecessary spectrum part follows. All signals are consequently normalized, thus DC offset is cut off. By amplification (attenuation) their magnitude is adapted to useful range of measuring converter. Afterwards precision is given by converter bit resolution.

In the case of electrical quantities (voltage, current) is possible to leave out sensor itself and use the same measurement chain. Second group represents sensors working with time quantities. Here is also necessary to implement impedance match eventually also separation and sequentially to adapt microcontroller voltage possibilities. In these cases parameters of monostable circuit are varied by measur-ing signal. Then pulse width or frequency is consequently analyzed in microcontroller

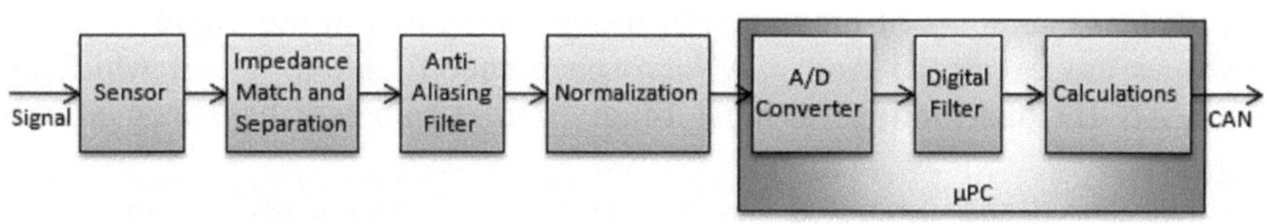

Fig. 2. Non-electrical quantities processing chain

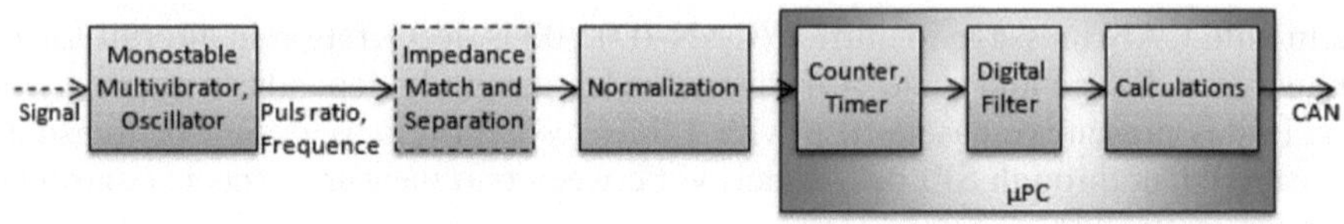

Fig. 3. Time quantities processing chain

with help of timer/counter s. Fig 3. Resolution is subsequently defined by timing rate of microcontroller core and its hardware architecture. Maximal range of measured quantity is determinate by bit size of timers and counters.

Third group represents signals from digital sensors using buses and communica-tion protocols. Afterwards in this factual case microcontroller is not a part of measuring chain but its function is as a converter among different buses and protocols. Two-state signals usually representing with help of two voltage levels logical "0" and "1" could also be included in this group. In the both cases to ensure compatibility of microcontroller with used buses is necessary.

Then digital filtering and implementation of necessary computation operation in order to get particular physical quantity virtually always follows at microcontrol-ler. Consequently microcontroller provides gained data by CAN with CANaero-space protocol for other usage into the NEC environment.

3.3 Data Fusion from Sensors

Following procedure is data sharing among all participants in the system and inte-grated environment NEC. The CAN with CANaerospace protocol, mentioned in the Chapter 2, is used for this purpose. There are two basic ways of communication controlling on the CAN. It is event-triggered approach and time-triggered ap-proach. The CAN is natively event-triggered system. Data transmission is driven by internal events in the particular module, e.g. data are transmitted when specific parameter value get changed or when module received the request for data from another bus participant. In case of time-triggered systems, there each data have exactly predefined time interval, when data have to be send and the receiving participant exactly knows, when receive data addressed to it. The time-triggered approach has been chosen for proposed system, because it is more deterministic and more dependable way. Important parameter is CAN message identifier (CAN-ID), which denotes content of each message and indicates its priority. Low CAN-ID value means high message priority. A few examples of message definition are shown in the Table 1. CAN-ID values are excerpted from CANaerospace specifi-cation [5], in which default CAN-ID distribution of most often used parameters has been already suggested.

All measured and computed signals and data are periodically (or on request) pro-vided to NEC via the CAN.

Table 1. CAN-ID distribution

CAN-ID distribution				
CAN-ID	Flight state parameter name	Suggested data types	Units	Sensor
300	Body longitudinal acceleration	Float Short2	[g]	ADIS 16350
315	Indicated airspeed	Float short2	[m/s]	MEMS barometer
321	Heading angle	Float short2	[°]	Magnetometer, GPS
324	Static air temperature	Float short2	[K]	Resistance element
326	Static pressure	Float short2	[hPa]	MPX4115
520	Engine exhaust gas temperature (EGT)	Float short2	[K]	Thermocouple

4 Conclusions

Conception and features of proposed Aircraft Electronic System (AES) have been shown on the example of electrical subsystem. AES can be used in any transport or combat aircraft or e.g. unmanned aerial vehicle (UAV). AES is a modular and universal system, that's why only few modifications in certain modules are necessary during implementation to concrete aircraft. Only selections of measured quantities (needed by other modules in the network) and optimization of their scale are required. The benefit of this system lies in utilization of proven, flexible and relatively high speed CAN enhanced by CANaerospace protocol for data fusion from all available aircraft sensors and their real-time providing to the integrated environment NEC.

Acknowledgments. The work presented in this paper has been supported by the Ministry of Defence of the Czech Republic (Project No. OVUOFVT200802 defence research).

References

[1] Janu, P., Bystricky, R., Bajer, J.: Proposal Of A Time-Triggered Avionic Electrical Subsystem Using Canaerospace. In: Jalovecky, R., And Stefek, A. (eds.) Proceed-Ings Of The International Conference On Military Technologies 2009, pp. 387–393. University Of Defence, Brno (2009)

[2] Jalovecky, R., Bajer, J.: Development Of The Aircraft Electronic System Using Can With Canaerospace Protocol. In: Jalovecky, R., Stefek, A. (eds.) Proceedings Of The International Conference On Military Technologies 2009, pp. 360–365. University Of Defence, Brno (2009)

[3] Koukol, O.: Analysis And Certification Method Of Cots (Commercial/Cost-Off-The-Shelf) Networks For Aviation Usage. Prague, 64 P. University Of Defence Brno. Ph.D. Thesis Supervisor: Assoc. Prof. Rudolf Jalovecky (2007) (in Czech)
[4] Bosch, Can Specification Version 2.0, `http://www.Semiconductors.Bosch.De/Pdf/Can2spec.Pdf` (Cit. 2008-12-04)
[5] Stock, M. Canaerospace, `http://Www.A2tech.Eu/Pdf/Canas_17.Pdf` (Cit. 2009-05-04)
[6] Unis, Sam – System Of Aviation Modules, `http://www.Unis.Cz/Level2.Aspx?Page=Sam.Aspx` (Cit. 2009-06-01)

Demonstration Model of the Passive Optoelectronic Rangefinder

V. Čech[1], J. Jevický[2], and M. Pancík[3]

[1] Oprox a.s., Vnitrni 10, Brno, Czech Republic
cech-vladimir@volny.cz
[2] University of Defence, Faculty of Military Technology,
Department of Mathematics and Physics, Kounicova 65, Brno, Czech Republic
Jiri.Jevicky@unob.cz
[3] Student of Masaryk University, Faculty of Informatics, Botanicka 68a, Brno
mpancik@gmail.com

Abstract. A demonstration model of the passive optoelectronic rangefinder was presented to the opponent committee of the Ministry of Industry and Trade of the Czech Republic within the final opponent proceeding in March 2009. The committee stated that POERF is fully functional and recommended continuing in its next research and development. This contribution gives basic information about the demonstration model of POERF.

1 Introduction

The passive optoelectronic rangefinder (POERF, see Fig. 1) is a measurement device as well as a mechatronic system that measures geographic coordinates of objects (targets) selected by an operator in real time. In the case of a moving object, it also automatically evaluates its velocity vector and simultaneously extrapolates its trajectory (Fig. 2).

Active rangefinders for measurement of longer distances of objects (targets), e.g. pulsed laser rangefinders, emit radiant energy, which conflict with hygienic restrictions in many applications. In security and military applications there is a serious defect that the target can detect its irradiation. The use of POERF eliminates mentioned defects in full.

All required information is sent to external users (clients) by Internet in near-real-time whereas the communications protocol is preconcerted [5, 6]. On the present, the communication with the system ECC®s (Emergency Control Centre System) of the firm Z.L.D., s.r.o., Praha is ensured.

Presumed users of the future system POERF are the police, security agencies and armed forces.

2 POERF Principles

The measurement principle [1, 4, 6] is based on the evaluation of information from stereo-pair images obtained by the sighting (master) camera and the metering (slave) one.

Their angles of view are relatively small and so a spotting camera with zoom is placed alongside of the sighting camera. This spotting camera is exploited by operator for targets spotting. After operator's steering the cameras towards a target, the shots from sighting camera serve to evaluate angle measured errors and to track the target automatically – Fig. 1, 3.

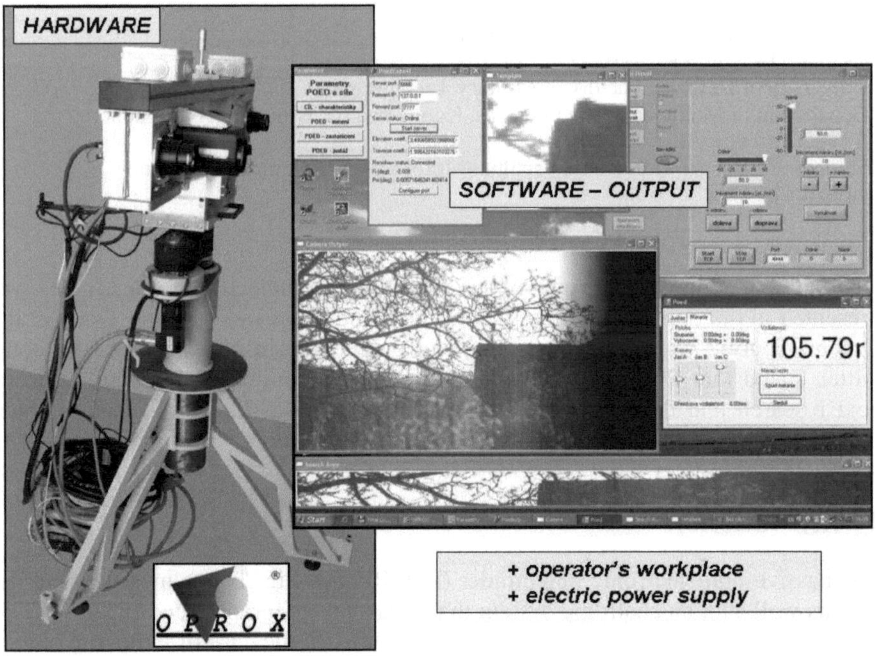

Fig. 1. Passive optoelectronic rangefinder: demonstration model 2009 [6]

POERF consists of hardware and software. The author of POERF conception is Vladimir Cech. The hardware incorporates the main construction of POERF, the operator workplace with computer and the electric power supply (electrical power network or car battery 24 V). Authors of the complete software package (except for servomechanism software) are authors of this contribution.

From the system view, the POERF as a mechatronic system is composed from three main subsystems:

- the range channel,
- the direction channel and
- the system for evaluation of the target coordinates and for their extrapolation.

The task of the range channel [4, 6] is on the one hand automatic recognition and tracking of the target which has been selected by the operator in semiautomatic regime and continuous measuring of its slant range D_T (c. 10 measurements per second at the present) and on the other hand the evaluation of angle measured errors that are transferred on input of the direction channel.

The direction channel [6] (its core consists of two servomechanisms) ensures continuous tracking of the target in the automatic and semiautomatic [3] regime and measuring of angle coordinates of the target (the elevation φ and the traverse ψ) – Fig. 3.

The spherical coordinates of the target (D_T, φ, ψ) are transformed into the UTM coordinates by the system for evaluation of the target coordinates and their extrapolation [5, 6]. Withal, the knowledge of the POERF geographic coordinates $(E, N, H)_{RF}$ and the POERF individual main direction α_{HS} (Fig. 3) is utilized. In the case of moving target, required extrapolative parameters are consecutively evaluated (coordinates of the measurement midpoint, corresponding time moment and the velocity vector of the target). The extrapolative parameters (UTM coordinates of the target are transformed into geographic coordinates) are sent periodically to a user in near-real-time (at the present with the period 1 second, i.e. the data "obsolescence" is c. 0.5 seconds) [5, 6] – Fig. 2.

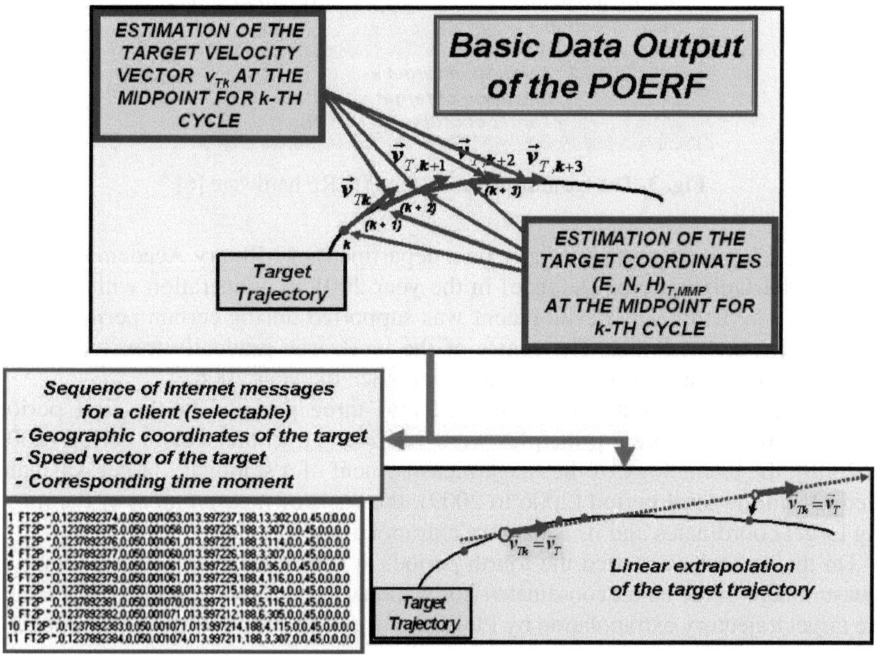

Fig. 2. Principle of measurement of the target trajectory and of data export to users [5, 6]

The principle of passive optoelectronic rangefinder is known minimally since the 80's of the 20th century. The development was conditioned primarily by progress in the areas of DSP (digital signal processor) and by progress in miniature computers with ability to work in field conditions (target temperature limit from – 40 to +50 °C, dusty environment, etc.) and to realize the image processing in the real-time (frame rate minimally 5 to 10 frames per second, ideally 25 to 50 fps).

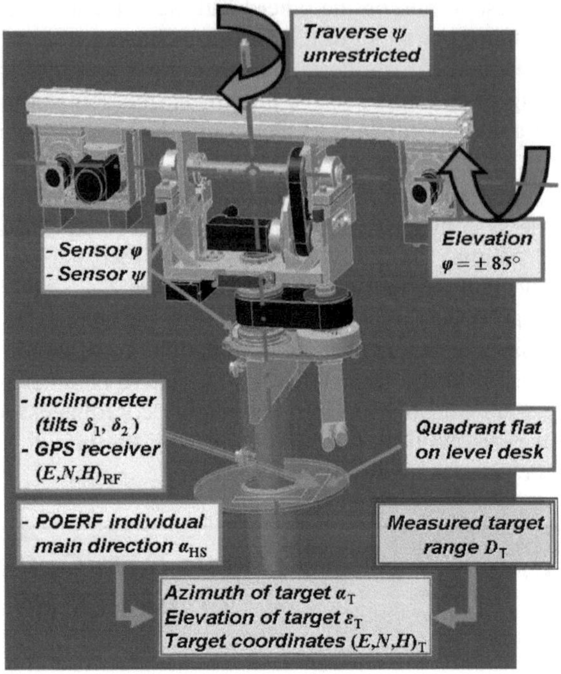

Fig. 3. The basic structure of the POERF hardware [6]

Our development started initially on a department of Military Academy in Brno (since 2004 University of Defence) in the year 2001 in cooperation with the firm OPROX, a.s., Brno. The development was supported during certain period by the firm Z.L.D., s.r.o., Praha. The centre of the work was gradually transferred into OPROX that is practically the pivotal solver since the year 2006.

Existing development can be divided into three periods. In the first period (2001 to 2003), the basic principles were verified [1]. In the second period (2003 to 2006), the technology of the range measurement of a stationary target was handled [2]. In the third period (2006 to 2009), the basis of measurement of the moving target coordinates and its trajectory extrapolation [6].

On the present we started the fourth period, in which we intend full handling of measurement of the target coordinates (for stationary and moving target) inclusive of the target trajectory extrapolation by POERF that can be set on a moving platform.

3 POERF Range Channel

As mentioned, the main task of the range channel is on the one hand automatic tracking of the target which has been selected by the operator in semiautomatic regime and continuous measuring of its slant range D_T and on the other hand the evaluation of angle measured errors that are transferred to the input of the direction channel.

The core of hardware consists of three digital cameras fixed through adjustable suspensions to the cameras girder – Fig. 1, 3.

The algorithm for computation of estimate of a slant range D_T is based on solution of the telemetric triangle that lies in the triangulation plane. The input data is ordinal numbers c_{T1}, c_{T2} of columns of matrix sensors in which are projected images T'_1, T'_2 of the target point T. In the concrete, it is sufficient to determine their difference Δc_T that is proportional to the appropriate parallactic angle γ. Therefore algorithms for computation of estimate of difference Δc_T are crucial. We work with algorithms for an estimate of Δc_T, which involve the definition of *2D model of the target image* (shortly "target model"). We use a rectangular target model for the present. This size (in pixels: rows × columns = $(2m_M + 1) \times (2n_M + 1)$) is adjustable.

The rangefinder power (constant) $D_{RF1} = (b \cdot f_a) / \rho$ is the basic characteristics of potential POERF accuracy (b – geometric base of the telemetric triangle, ρ – characteristic pixel size in camera matrix sensor, f_a – absolute value of image focal length of the objective of digital camera). With increasing value of the power, the accuracy of measurement increases too. The power of POERF demonstration model is $D_{RF1} = 9627$ m.

A choice of size of the image focal length results from the requirement that the target must be identifiable in the requisite maximal working range D_{Tmax} of the rangefinder. In accordance with Johnson criterion (50 % successfulness of the target identification), the target has to be displayed minimally on 16 times 16 pixels [1]. In practice, the resolution of the target image should be minimally 30 times 30 pixels. See more closely in [1, 4, 6].

The basic indicator of accuracy of a range measurement by means of the *whole* range measuring system is *the relative value s_{DR} of the sample standard deviation s_D of a measured range D_T*. In security and military applications of rangefinders there is often required that s_{DR} should be less than 3 to 5 % within a specified interval (D_{Tmin}, D_{Tmax}) for measured target ranges. This condition is usually satisfied for the target ranges $D_{Tmax} < (D_{RF1} / d)$, $d \in \langle 3, 5 \rangle$, if Johnson criterion is satisfied together. So, it depends significantly on the real sizes of the target. The maximum effective range D_{Tmax} of the POERF demonstration model is respectively 500 to 800 m for persons and 1800 to 2200 m for trucks and buildings.

The real maximum measurable range of the target depends simultaneously on the horizontal meteorological visibility S_M. The previous specifications are valid for circa $S_M \geq 10$ km.

4 Conclusion

The extent of this paper does not allow dealing with the direction channel. Let us only remark that the system is able to track a moving object continuously in the range c. 1000 m, if its traverse speed is greater than c. 1 m/s. The maximum attainable speed was not systematically probed yet due to limited time and space possibilities, but it is greater than c. 10 m/s.

The complete information about the POERF demonstration model can be found in the final research report [6], which contains also a representative overview of utilized references (143 items).

Acknowledgements. This work has originated under the support of financial means from the industrial research project of the Ministry of Industry and Trade of the Czech Republic – project code FT–TA3/103: "Research of high-end technologies and methods for recognition of moving objects, for determining of objects movement parameters and for automatic tracking systems of moving objects".

References

[1] Balaz, T.: Utilization possibilities of the passive optoelectronic rangefinder in current tank fire control systems (in Czech) Habilitation dissertation. Military Academy in Brno, 139 pages (2003)

[2] Skvarek, J., et al.: Research and development of technology and technical devices for both passive optoelectronic tracking and objects measuring (in Czech). Final research report of industrial research project of MIT CR FD – K3/99. Oprox, a.s., Brno, Epitome 50 pages (2004)

[3] Cech, V., Jevicky, J.: Simulators of the Direction Channel of the Passive Optoelectronic Rangefinder. In: Proceedings UKSim Tenth International Conference on Computer Modelling and Simulation EUROSIM / UKSim2008, pp. 132–137. IEEE Computer Society, Cambridge (2008)

[4] Cech, V., Jevicky, J.: Program Test POED for Simulation of Properties of Range Channel of the Passive Optoelectronic Rangefinder II. In: Proceedings of 30th Autumn International Colloquium Advanced Simulation of Systems – ASIS 2008, MARQ, Roznov pod Radhostem, pp. 90–95 (2008) ISBN 978-80-86840-42-0

[5] Cech, V., Jevicky, J.: Algorithms of the Target Trajectory Extrapolation. In: Proceedings of 43rd Spring International Conference Modeling and Simulation of Systems – MOSIS 2009, MARQ, Roznov pod Radhostem, pp. 144–153 (2009) ISBN 978-80-86840-45-1

[6] Composite authors: Research of high-end technologies and methods for recognition of moving objects, for determining of objects movement parameters and for automatic tracking systems of moving objects. Final research report of industrial research project of MIT CR FT – TA3/103. Oprox, a.s., Brno, 175 pages, appendices 404 pages (2009)

An Ultrasonic Air Temperature Meter

A. Jedrusyna

Wroclaw University of Technology, Faculty of Power Engineering,
Institute of Aviation, Processing and Power Machines Engineering,
Wybrzeze Wyspianskiego 27, 50-370 Wroclaw, Poland
arturj@pwr.wroc.pl

Abstract. In the paper a design of ultrasonic air temperature meter is presented. The instrument is suitable for non-contact measurement of air temperature inside a confined space (room, boiler, tank). The temperature of air is derived from changes in sound velocity in the space between transmitter and receiver. Changes in the speed of sound are detected by phase shift variations of a 40 kHz pulsed ultrasonic wave (binary frequency shift keyed signals). The instrument uses modern RISC microcontroller, no expensive FPGA circuits are necessary. Such an approach results in a very compact construction of the meter and its fast operation (response time below 500 ms).

1 Introduction

One of the most important physical parameters of a gas or liquid is its temperature. The classical methods of temperature measurement require a physical contact between the sensor and the medium. When gas temperature is measured, another problem is often generated by a long time constant of the sensor.

One possible solution of the problem is based on fact that a velocity of sound wave propagation in the air is sensitive to changes of temperature.

2 Velocity and Temperature Measurement

The speed of sound c according to Boyle's law in an ideal gas at a constant pressure is given by the following equation:

$$c = \sqrt{\frac{\gamma RT}{M}} \tag{1}$$

Where γ, R, T, and M are the specific heat ratio, universal gas constant, the absolute temperature in Kelvin and molar mass, respectively.

In practical applications a simplified formula [1] is often used:

$$c = 331.4\sqrt{\frac{T}{273}} \tag{2}$$

Where c is in m/s and T is in Kelvin. To measure T to a practical accuracy of 1°C, c must be measured with relative accuracy about 1 x 10 $^{-3}$ [5]. The most common measurement technique is to measure the time of flight of the acoustic pulse between transmitter and receiver located at the known distance from each other (classical TOF method).

The precision of TOF measurement is very limited because it is difficult to detect the exact beginning of the received pulse (Fig.1). Various methods that may improve precision of the measurement are used [1]. However, more accurate information can be obtained by using information from phase shift comparison of the continuously transmitted/received waves. Most often the frequencies in the range 20..100 kHz are used.

The older implementations of instruments using such approach required relatively complex and expensive hardware. For instance, phase shift measurement for single test frequency of 40 kHz and distance between transducer and receiver about 0.1 meter required a FPGA-based digital phase detector [2]. The modern microcontrollers with fast built-in counters and timers make possible to construct an ultrasonic thermometer without the FPGA circuits.

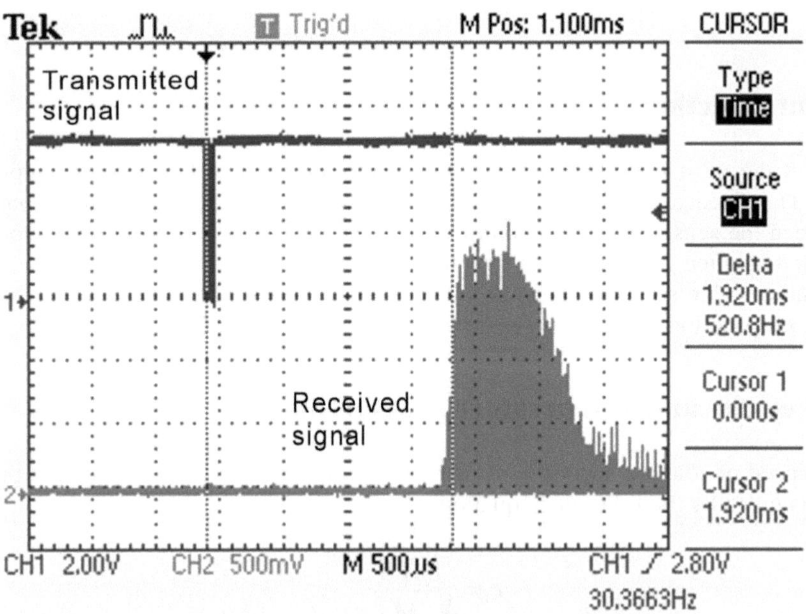

Fig. 1. TOF measurement – transmitted and received signals. Please note slow turn-on of the receiver (an envelope effect)

3 BFSK Method

Another possible measurement method is to use a multi-frequency test signal, where phase shift is computed for two or more discrete frequencies. In case of two frequencies, the method is known as Binary-Frequency Shift Keying (BFSK). During the transmission the phase shift between the transmitted and received signals is measured for different frequencies. When phase shifts for different frequencies (and for constant distance between transmitter and receiver) are compared, then the information about the sound velocity can be derived [3],[5].

4 Test Setup

To check the feasibility of microcontroller-based BFSK instrument idea, a simple meter was built. The block diagram of the instrument was shown in Fig. 2. The device was based on 8-bit RISC microcontroller ATmega32 by Atmel. The frequency measurement was performed using internal timers of the microcontroller. The precision of the measurement was extended using the external HCMOS counters. The transmitted signal was 40/41 kHz sine wave. The cheap ultrasonic 40kHz transducers were used as both the transmitter and receiver. The additional 64 MHz clock signal was used for precision evaluation of the phase shift. The digital phase meter was built with EX-OR logic gates and internal timers of the microcontroller.

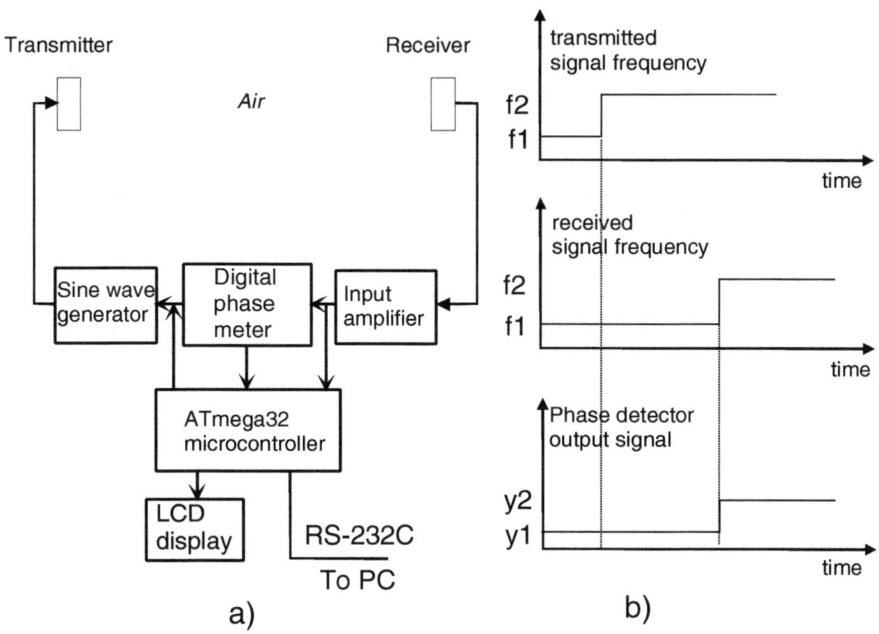

Fig. 2. The BFSK method-based ultrasonic thermometer a) the block diagram of the instrument, b) time dependencies during the measurement

The firmware of the meter was written in C language. The device was tested inside a controlled temperature chamber where temperature was controlled in the range 0..80°C. The reference temperature meter was using Pt100 sensor. The space between the ultrasonic transmitter and receiver was changed between 200 and 400 mm for different series of measurements.

5 Test Results

For the whole measurement range 0..80°C, the measurement error was below 1°C. An average response time of the meter was about 400 ms. The main problem concerned the echo signal that could introduce additional errors. A simple damping system containing a plastic foam wave deflectors located near the receiver reduced the level of the echo and other spurious signals.

6 Summary

It was proved that it is possible to measure a small change of temperature inside a confined space using the ultrasonic measurement. The total cost of components for the prototype presented was below 30 Euro. Such a result was possible due to the utilization of a modern microcontroller. The instrument described here is particularly well suited for temperature checking inside confined spaces, ranging from refrigerators to tanks and boilers. After the installation, the instrument requires a single calibration, from that moment it can work as an independent device. The following changes and improvements can be introduced in future:

- The number of test frequencies could be increased,
- A correction function for humidity of the environment should be introduced [4]
- A more complex filtering algorithm could be implemented in order to reduce the noise level for the received signal.

It is interesting to note that after minor changes in software the instrument described might be used as a precise ultrasonic distance meter.

References

[1] Olmos, P.: Extending the accuracy of ultrasonic level meters. Meas. Sci. Technol. 13, 598–602 (2002)
[2] Huang, K.N., Huang, C.F., Li, Y.C., Young, M.S.: High precision, fast ultrasonic thermometer based on measurement of the speed of sound in air. Review of Scientific Instruments 73(11), 4022–4027 (2002)

[3] Huang, C.F., Young, M.S., Li, Y.C.: Multiple-frequency continuous wave ultrasonic system for accurate distance measurement. Review of Scientific Instruments 70(2), 1452–1458 (1999)
[4] Tsai, W.-Y., Chen, H.-C., Liao, T.-L.: An ultrasonic air temperature measurement system with self-correction function for humidity. Meas. Sci. Technol. 16, 548–555 (2005)
[5] Liao, T.-L., Tsai, W.-Y., Huang, C.-F.: A new ultrasonic temperature measurement system for air conditioners in automobiles. Meas. Sci. Technol. 15, 413–419 (2004)

Optical Torque Sensor Development

P. Horváth and A. Nagy

Széchenyi István University, Department of Mechatronics and Machine Design,
1 Egyetemtér, H-9026 Győr, Hungary
horvathp@sze.hu

Abstract. The purpose of this study is to develop a contactless torque sensor in the mNm range. The applied optical method is based on the birefringe effect of photoelastic materials. The novelty of the modified reflective photoelastic method is the application of a photoelastic tube as a coupling and measuring element between shafts. Change in intensity of polarized light is proportional to the torque to be measured. Basics of operational principle and practical considerations are also discussed.

1 Introduction

Torque sensors are often used at various fields such as testing, condition monitoring and research. The commonly used torque sensors apply an elastic element whose deformation is transformed to electrical signal for example with strain gauges. The highest difficulty during operation is taking off the signal from a rotating shaft which can happen with slip rings or a telemetric device [4]. The friction moment of slip rings can decrease the accuracy of measurement considerably while the dimensions of a telemetric device are too high to mount on a slim shaft. There are also contactless torque sensors with optical operational principle [1]. In this case two disks with slits are mounted on a shaft a distance L apart. The twist angle of shaft controls the overlapping between the slits and thus, pulse width modulates the transmission for a LED light source to a photodetector.

2 Basics of Photoelasticity

When a linearly polarized light beam of intensity I_0 and velocity c_0 reaches a transparent, optically active material under load, the light intensity vector can be

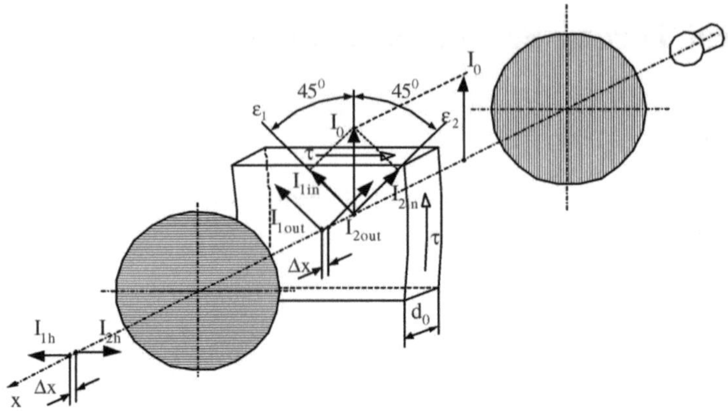

Fig. 1. Principle of transmissive photoelasticity

divided into two components in the direction of the principal strains. In Fig.1 one can see a portion of a thin wall tube made of photoelastic material under torsion. In this case the directions of principal strains are inclined 45 degrees to the geometrical axis of the tube.

Light beam components travel in the x-direction with different velocities c_1 and c_2 depending on the magnitudes of principal strains ε_1 and ε_2 and after leaving the sheet of thickness d_0 an optical path difference (relative retardation) Δx will occur between them:

$$\Delta x = k(\varepsilon_1 - \varepsilon_2)\frac{d_0}{c_0}. \tag{1}$$

In plane stress-field the principal strains can be expressed by the principal stresses and after simplification we get the following equation:

$$\Delta x = m\lambda = Cd_0(\sigma_1 - \sigma_2) \tag{2}$$

where C denotes the photoelastic constant as well as σ_1 and σ_2 the principal stresses arising in the photoelastic material respectively. The retardation can be expressed as a product of wavelength-number *m* and the wavelength of light λ. After leaving the photoelastic layer the light beam components pass through an analyzer with perpendicular direction of polarization to the first one. The resulting intensity of light beams in the horizontal plane is then

$$I_h = \frac{I_0}{2}[\cos 2\pi(ft - m) - \cos 2\pi ft] = I_0 |\sin m\pi| \cdot \cos 2\pi(ft - \gamma) \tag{3}$$

The magnitude of light intensity changes according to a sinusoid as it can be seen in Figure 2.

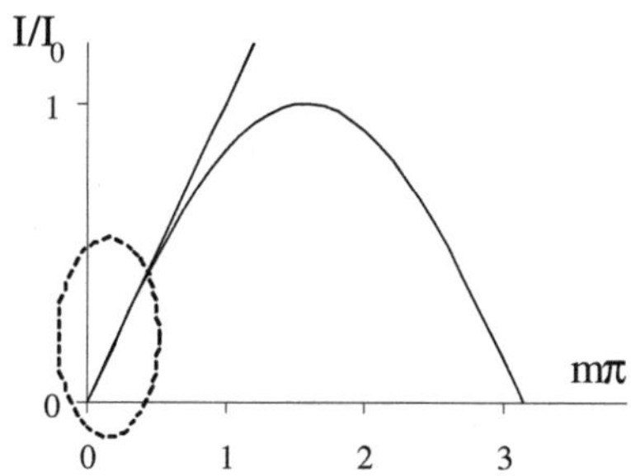

Fig. 2. Change of intensity vs. retardation

Unfortunately the transmissive photoelastic method is not suitable for torque measurement because the resulting phase shift of a polarized light beam passing through the opposite sides of a twisted, photoelastic shaft of circular cross section is always zero, independently on the torque applied.

3 Applying Reflective Photoelasticity

Instead of the transmissive method, the so called reflective photoelastic method can be applied for torque measurement. (Fig. 3).

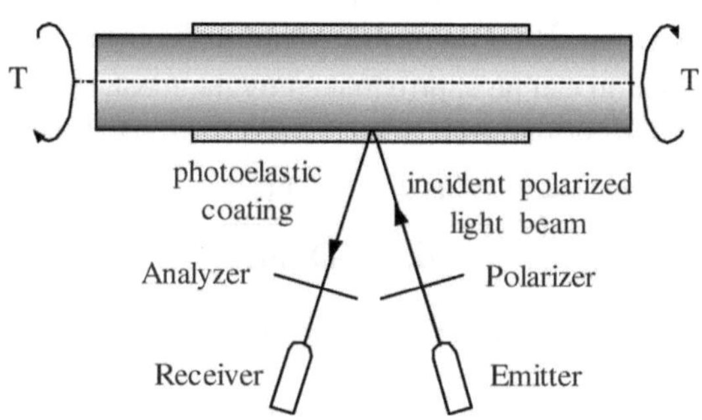

Fig. 3. Principle of reflective photoelasticity

The surface of the body to be measured is coated by a thin photoelastic layer, which deforms together with the body. The light beam travels the distance d_0 twice, first passing through the layer from outside to a mirrored surface and second, from the reflector to outside. In practice the reflective element is the shaft's surface itself. As the twist angle of the steel shaft and the photoelastic coating is the same, the ratio of torque in the shaft (T_s) and coating (T_c) is the following

$$\frac{T_s}{T_c} = \frac{I_{ps} G_s}{I_{pc} G_c} \tag{4}$$

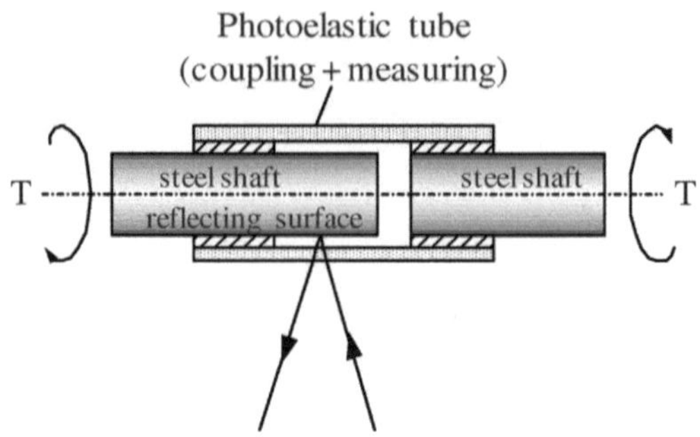

Fig. 4. Proposed solution to measure small torque

Since the polar inertia of shaft's cross section $I_{ps} > I_{pc}$ and the shear modulus of a steel shaft $G_s > G_c$, this method is suitable to measure only high torque.

For measuring small torque it is proposed to omit the rigid steel shaft as a coupling and use only a thin wall tube made of a more elastic photoelastic material as a coupling and measuring element all in one (Figure 4). Since the whole torque is transmitted by the photoelastic tube, due to the higher principal strains the sensitivity of measurement can be increased. A similar but more robust solution can be found in [3].

In case of reflective photoelasticity the expressions derived above for transmissive method can be applied with the substitution $d=2d_0$. Assuming pure torsion of the coupling tube, the difference of principal stresses (the diameter of Mohr's circle) equals to $\sigma_1 - \sigma_2 = 2\tau$. For further simplification the maximum value of shear stress during torsion can be approximated by the mean shear stress calculated by Bredt's formula valid for thin wall tubular cross sections of mean diameter D

$$\bar{\tau} = \frac{T}{2Ad_0} \approx \frac{2T}{D^2\pi d_0}. \tag{5}$$

After substituting (5) to (2) one can find the wavelength-number as follows:

$$m = \frac{4C}{\pi\lambda D^2} \cdot T \tag{6}$$

It is to be noted, that the wavelength-number is proportional to the torque to be measured but does not depend on the wall thickness of cross-section. To maintain linear condition of operation with error of linearity less than 1.6%, the domain of wavelength-number $0<m<0.1$ is desired, see Figure 2. One has to make sure that shear stress in the photoelastic tube must not exceed the allowable shear stress τ_{all}. From these conditions follow the next design criterions:

$$d_0 \geq \frac{0{,}05\lambda}{C\tau_{all}} \tag{7}$$

$$D \geq \sqrt{\frac{40C}{\pi\lambda} \cdot T} \tag{8}$$

For a given photoelastic material one has an opportunity to influence the range of measurement by choosing the diameter D and wall thickness d_0 of tube as well as the wavelength of light λ. In order to evaluate change of light intensity easily, monochromatic light is preferred.

4 Test results

For testing the principle presented above a test rig has been implemented consisting of a photoelastic tube of outer diameter D_o=6 mm and wall thickness 1 mm. The tube was statically loaded by weights of 50, 100, 150 and 200 gram respectively k=50 mm apart from the center of shaft. As light source a red color LED was applied. Polarizer and analyzer sheets were purchased from an old calculator display. Ten series of measurement were performed and their average are shown in Figure 5. As it can be seen, nearly linear operation occurred. The error of linearity is about 2%.

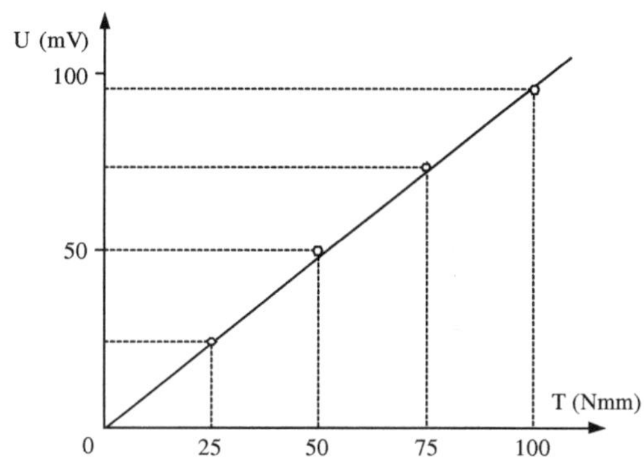

Fig. 5. Diagram of measured output voltage vs. applied torque

5 Summary

This paper dealt with the development of a contactless torque sensor applying the principle of reflective photoelasticity. The implemented test equipment and measured data verified that this simple solution is suitable for measuring torque of small and medium range. Some problem areas however need further investigation, such as:

- optoelectronic components need fine mounting because of the small dimensions of the measured area of the photoelastic tube
- ambient light, rotational speed and additional bending effect of the photoelastic tube decrease the accuracy of measurement.

References

[1] Nwagboso, C.O.: Automotive sensory systems. Springer, Heidelberg (1993)
[2] Chung, D., Merat, F.L., Discenzo, F.M., Harris, J.S.: Neural net based torque sensors using birefringent materials. Sensors and Actuators A 70(3), 243–249 (1998)
[3] Discenzo, M.F.: Photoelastic neural torque sensor. US patent No. 5723794
[4] Brenner, W., Suemecz, F., Vujanic, A.: Principles of micro torque measurement-an overwiev. In: Proceedings of XVII IMEKO World Congress, Dubrovnik, pp. 277–280 (2003)

The Temperature Effect of Photovoltaic Systems with dc-dc Converters

J. Leuchter[1], V. Řeřucha[1], and P. Bauer[2]

[1] University of Defence, Faculty of Military Technology,
Kounicova 65, Beno, Czech Republic
{vladimir.rerucha,jan.leuchter}@unob.cz
[2] Delft University of Technology, Electrical Power Processing,
Mekelweg 4, 2628CD Delft, The Netherlands
p.bauer@ewi.tudelft.nl

Abstract. This paper describes the temperature effect of photovoltaic modules and the mathematical modelling due to Matlab-Simulink. The temperature has effect to reduction of total efficiency which cannot be overlooked. Paper shows the application of maximum power tracker techniques (MPT) to achieve of decreasing of temperature effect to get the maximum power of photovoltaic cells. The maximum power tracker one of the major trends in photovoltaic technology towards system modular design based on the optimum power control with low cost and optimal energy yield. The algorithms of MPT are shown in the paper and it is discussed. The results of modelling of the temperature analyses are verified by experimental measurements on the photovoltaic set up.

1 Introduction

The photovoltaic power sources is very attractive source of electricity. The solar power can be utilized in the form of heat or electrical energy. Considering the case of harnessing the energy of the sun in the form of solar electric energy, the high cost and low efficiency of the photovoltaic cells, that convert the sun light to direct current electricity, stands as a major constraint for its full utilization. However, as the solar energy is believed as the real source of sustainable and clean energy; a lot is being done to improve the efficiency of the solar cells to use it in large scale. The configuration of photovoltaic system with dc-dc and dc-ac converters is shown in Fig. 1. Most good converters are capable of operating at conversion efficiencies greater than 90% over most of their output power range. A set

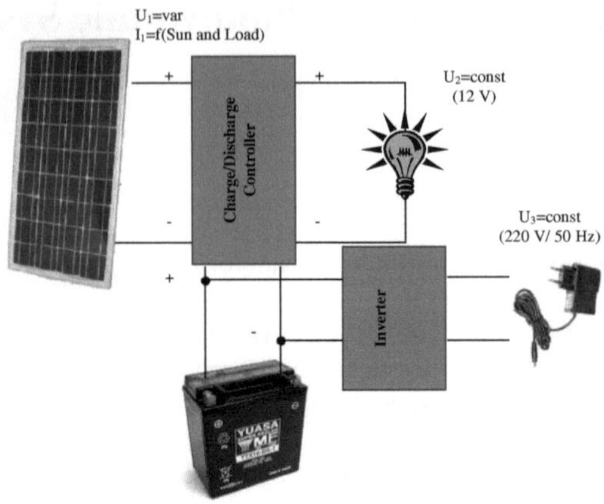

Fig. 1. The photovoltaic system

of a small PV power photovoltaic power system can be easily connected in parallel to yield higher output power. When photovoltaic modules (PV) are set up with a number of the small power systems, the total system cost while increase and will be a matter of concern.

2 Mathematical Modeling

Photovoltaic module (FVM) can be express as system to transformation of sun energy *G* to electrical *P*. Input of photovoltaic model (FVM) are sun energy *G*, load current *I*, temperature *T*, output of model is voltage *U* (Fig. 2). Common FVM can be expressed by equals 1.

$$F(I,U,G,T)=0 \qquad (1),$$

where $I \in (0, I_k)$, $U \in (0, U_0)$ and $G \in (0, \infty)$. I_k is short current, U_0 is no-load voltage, and both are outside points. G is parameter of sun intensity.

The results of I-V charateristics of FVM modeling is shown in Fig 3. The temperature effect have a influence on shape I-V characteristics. The results of

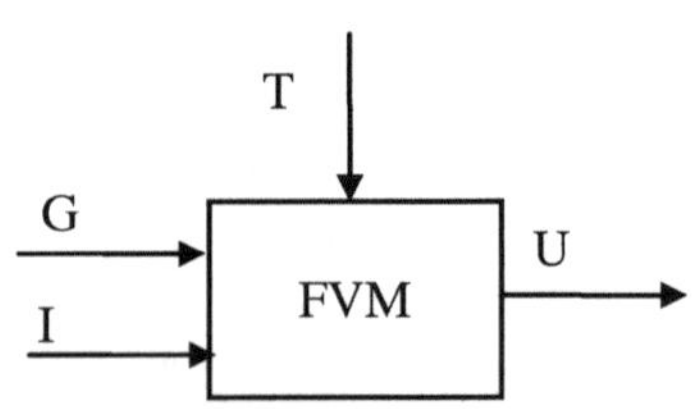

Fig. 2. Model of FVM

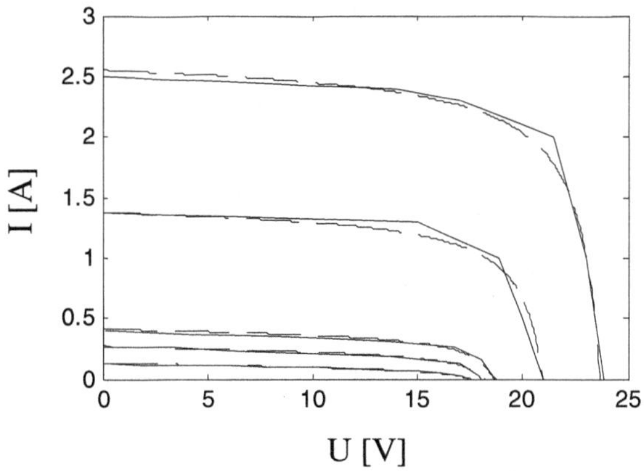

Fig. 3. The results of I-V characteristics of FVM modeling

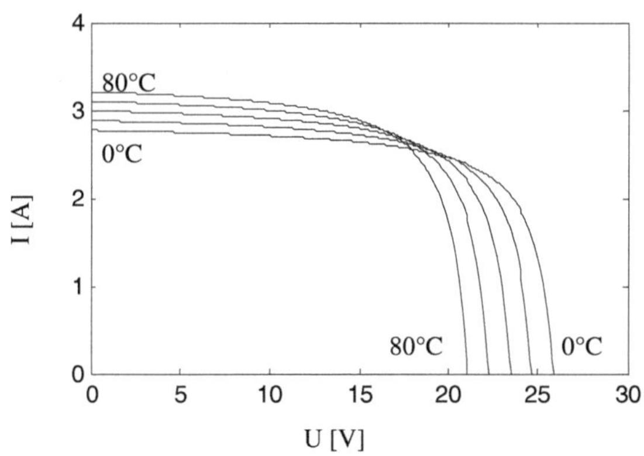

Fig. 4. I-V characteristics as a temperature function modeling

temperature effect modeling is shown in Fig. 4, the value of no-load voltage decreasing with value temperature (0.24 % per °C).

3 Experimental Verification

The results of experimental verification of a temperature effect of FVM are shown in Fig. 5. Figure 5a show the experimental workplace of photovoltaic cells (TU Delft) and results of I-V characteristics as a function of temperature of photovoltaic cells is shown in Fig. 5b. The results of modeling from Fig. 4, where a shape of characteristics is changing, was confirmed. If the temperature is increasing the output voltage go down and current up.

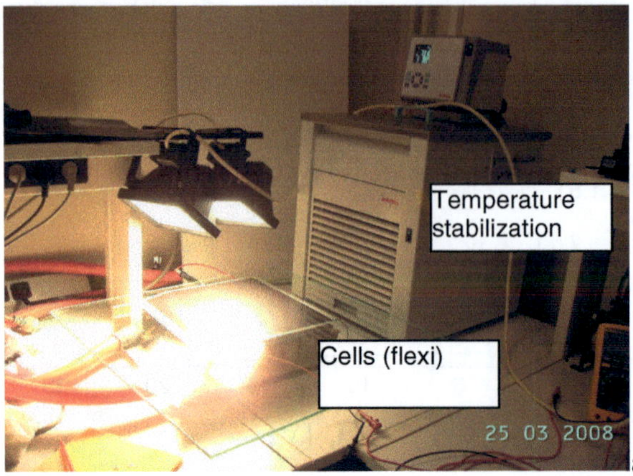

a)

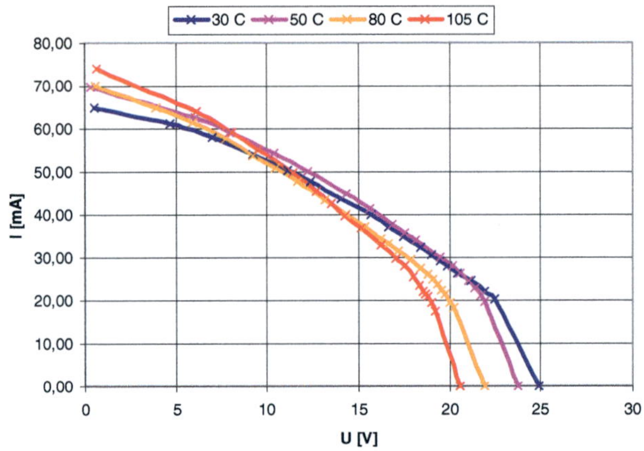

b)

Fig. 5. a) experimental workplace of photovoltaic cells b) I-V characteristics

4 Photovoltaic System with Maximum Power Tracker (MPT)

The main idea of photovoltaic system with maximum power tracker is to achieve the maximum available power every time. I-V characteristic of a modeling of photovoltaic modul (50W) as a function of sun intensity G is shown in Fig. 6. According to the results which were mentioned above (Fig. 4 and 5b) it is necessary to apply MPT techniques to achieve the maximum efficiency of photovoltaic systems. A variety of techniques has been developed in recent years to determine an operating point of MPT.

The analyses indicate that the optimal operating voltage of a photovoltaic module is very close to a fixed percentage of the open-circuit voltage. This implies that MPT could simply use the open-circuit voltage to predict the optimal operating condition. This is called voltage-based MPT [3]. Similarly to the previous method can be the short-circuit current [3]. Both methods are encumbered of the flaw during temperature effects.

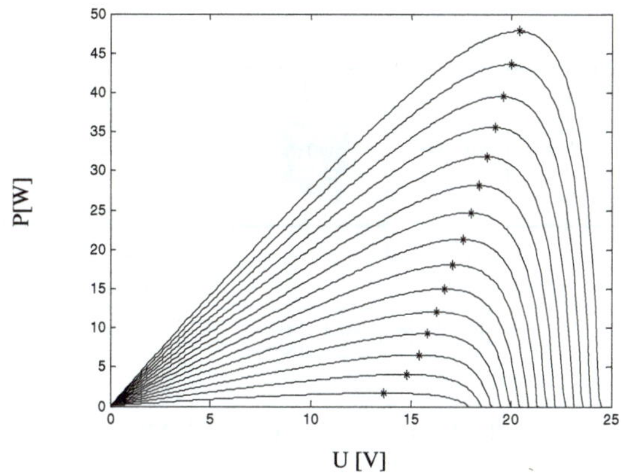

Fig. 6. Power output as a function of sun intensity

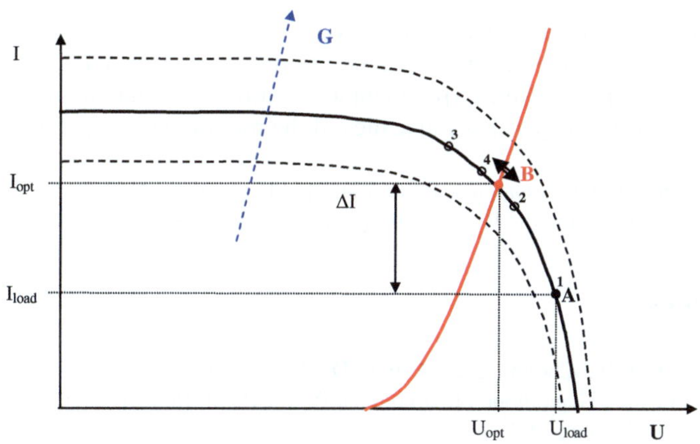

Fig. 7. Principe of dynamic setting of MPT

Another technique was developed as dynamic setting according power output. Output current and voltage are measured and next step are apply to feed-back to find the maximum power. The principle of setting MPT is shown in Fig. 7. Dynamic setting of point *B* from point *A*, which is give by the load, must be change. In figure is shown a method of interval halve, which is very common. Power which we catch and it is not required by the load (over-abundance) must be charge to battery or supercapacitor.

Figure 8 shows general scheme of photovoltaic system using of MPT and battery concept, which was suggest, build and verified. The maximum power can be archived by feed back control, where inputs are current and voltage of FVM according figure. There are used two semiconductor power converters, where dc-dc^{*1} converter set it the maximum power tracker point by means of duty control to achieved maximal efficiency of photovoltaic panel and dc-dc^{*2} converter set up a constant output voltage. Both converters systems must be coordinated and control.

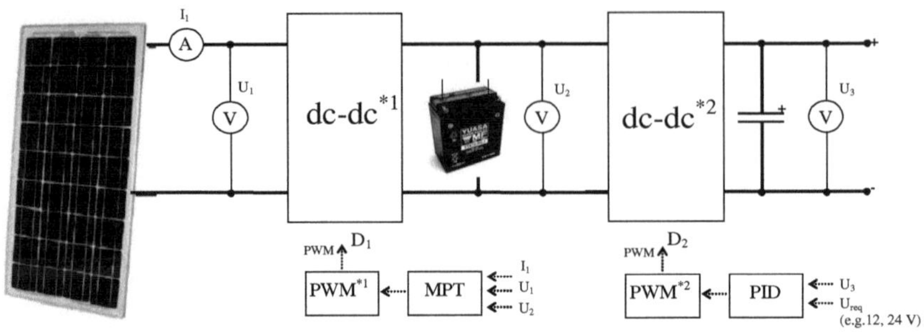

Fig. 8. Photovoltaic system with MPT

5 Conclusions

The paper presented experimental photovoltaic system and results of temperature effect on the decreasing of total efficiency of photovoltaic modules. Mathematical modeling was verified using experimental measuring and maximum power tracker have been shown in paper that can achieved the maximum power for every load.

Acknowledgment. The research work is supported by the GA of the Czech Republic (project no. 102/09/0013) and the Research program no VZ-0000403.

References

[1] Benjamin, V.P., Chong, L.Z., Abbas, D.: Modeling and Control of a Bidirectional Converter for a Standalone Photovoltaick Power Plant. In: EPE Conference 2007, p. 8. EPE press, Aalborg (2007)

[2] Diver, R., Andraka, C., Rawlinson, K., Moss, T., Goldberg, V., Thomas, G.: Status of the Advanced Dish Development System Project, paper no. ISEC2003-44237. In: Diver, R., Andraka, C., Rawlinson, K., Moss, T., Goldberg, V., Thomas, G. (eds.) Proceedings of the ASME International Solar Energy Conference, Kohala Coast, Hawaii, March 15–18 (2003)

[3] Chimento, F., Musumeci, S., Raciti, A., Sapuppo, C.: A Control Algorithm for Power Converter in the Field of Photovoltaic Application. In: EPE Conference 2007, p. 8. EPE press, Aalborg (2007)

Design of Capsule Pressure Sensors Thermal Compensation

R. Vlach[1] and J. Kadlec[2]

[1] Brno University of Technology, Faculty of Mechanical Engineering,
Institute of Solid Mechanics, Mechatronics and Biomechanics,
Technicka 2896/2, Brno, Czech Republic
vlach.r@fme.vutbr.cz
[2] Brno University of Technology, The Faculty of Electrical Engineering and Communication, Department of Microelectronics, Udolni 53, Brno, Czech Republic
kadlecja@feec.vutbr.cz

Abstract. This paper describes development and design of new type of pressure sensing. One of the main parameters of this sensor is pressure measuring independence on operating temperatures. Significance of this is design of thermal compensation capsule warranting accuracy of pressure measuring. Software ANSYS was used for verify of capsule thermal compensation. The main goal of the project is the research of complex smart pressure sensor based on the new sensing system with galvanic fiber optic isolation including research and experimental validation of this new pressure sensing principle.

1 Introduction

This project deals with development of a new unique method for scanning deflection of deformation membrane and applying it in practical applications. The new method is based on the optical measuring system with optical fiber. Main advantages of this principle consists in excellent pressure sensitivity, galvanic isolating of whole sensor via optical fiber, very good accuracy of static and dynamic measuring, maximum immunity against electrical and magnetic interference and miniature size of resulting pressure sensor.

Thermal compensation of pressure sensor capsule is necessary for warranty of measuring accuracy. The pressure sensor must be operating for temperature range from -20°C to 150°C.

2 Principle of Sensor

The pressure sensor idea is measuring membrane deflection, which is linearly adequate of applied pressure. The mechanical deflection must be converted to electrical signal. This pressure sensor is based on optical sensing membrane deflection using optical fiber. We decided to use amplitude-optical principle [1]. Signal from optical fiber is converted to digital electric signal, which is easily applicable to PC. Conception of pressure sensing capsule prototype is on the figure 1.

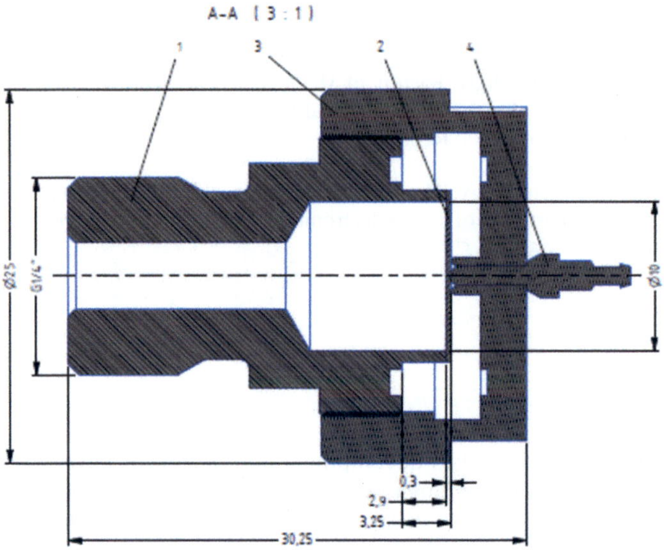

No.	Qty.	Description	Drawing No.	SAP-No.
1	1	Pressure connection - thermal compensation	OP.198.111	
2	1	Membrane prum.10	OP.198.105	
3	1	Screw female - thermal compensation	OP.198.112	
4	1	Ferule		

Fig. 1. Capsule of optical pressure sensor

The disadvantage of the capsule is dependency on the temperature. Size of this capsule, thermal expansion coefficient of used material and sensing range in nanometres result in expansion of this capsule and in the deviation in the pressure sensing. This parasitic influence of temperature was eliminated through the special thermo-compensated package.

3 Thermal Compensation

The first idea of thermal compensation method is based on assumption of one dimensional thermal expansion. Illustration of temperature expansion elimination is on the figure 2.

Temperature compensation arise from the following equation

$$L_{mat} = L_{dr} + t_{mem} + \delta , \tag{1}$$

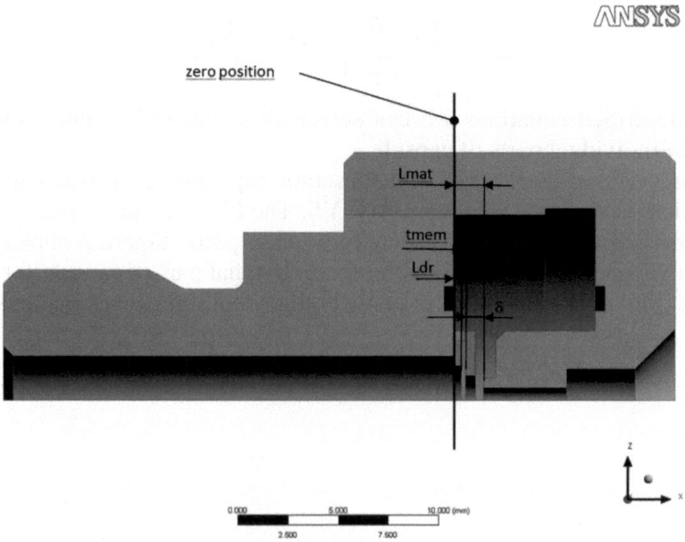

Fig. 2. Temperature compensation of the pressure sensor capsule

where L_{mat} is length of screw female from zero level, L_{dr} is fiber holder from zero level, t_{mem} is thickness of membrane and δ is measured distance between optical fiber end and membrane. This distance must be independent on the operating temperature. Equation (1) is changed with respect operating temperature on the following equation

$$L_{mat} \cdot (1 + \alpha_{mat} \cdot \vartheta) = L_{dr} \cdot (1 + \alpha_{dr} \cdot \vartheta) + t_{mem} \cdot (1 + \alpha_{mem} \cdot \vartheta) + \delta, \quad (2)$$

where ϑ is operating temperature and α_i is thermal expansion coefficient of used materials. Therefore is possible derived following equation

$$L_{mat} \cdot \alpha_{mat} \cdot \vartheta = L_{dr} \cdot \alpha_{dr} \cdot \vartheta + t_{mem} \cdot \alpha_{mem} \cdot \vartheta. \quad (3)$$

From thence temperature separation and using equation (1) result in

$$(L_{dr} + t_{mem} + \delta) \cdot \alpha_{mat} = L_{dr} \cdot \alpha_{dr} + t_{mem} \cdot \alpha_{mem}. \quad (4)$$

Temperature compensation can be done by two ways:

- Choose material and most of required dimensions. Then we can calculate size of the fiber holder according to equation

$$L_{dr} = \frac{(t_{mem} + \delta) \cdot \alpha_{mat} - t_{mem} \cdot \alpha_{mem}}{\alpha_{dr} - \alpha_{mat}}. \quad (5)$$

- Choose dimensions and most of the used materials and compute thermal expansion coefficient of the screw female according to the following equation

$$\alpha_{mat} = \frac{L_{dr} \cdot \alpha_{dr} + t_{mem} \cdot \alpha_{mem}}{L_{dr} + t_{mem} + \delta}. \tag{6}$$

From the described equations we can select ideal material combination and dimensions of individual parts of capsule.

The final version of thermal compensation capsule was verified using FEM model, which was created in system ANSYS. The FEM model respects also effect of radial thermal expansion of capsule individual parts. Figure 3 shows distribution of equivalent stress (vonMisses) in the individual part of capsule for temperature 150°C. Maximal stress is sufficiently below yield limit of all used materials.

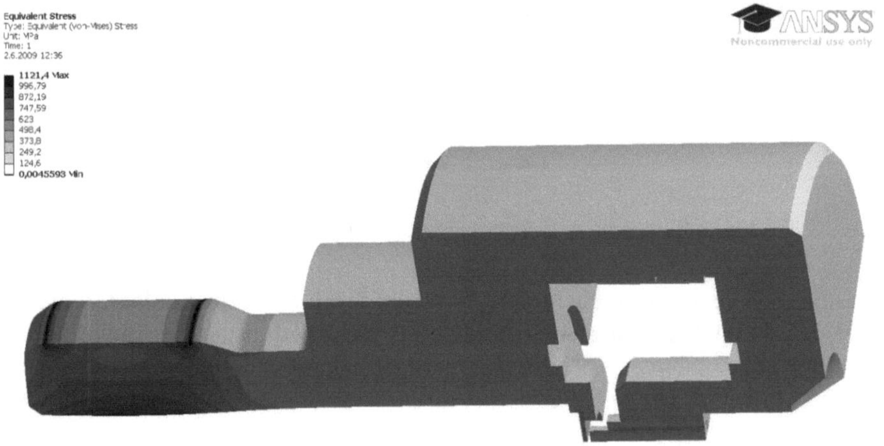

Fig. 3. Stress distribution in the pressure sensor capsule

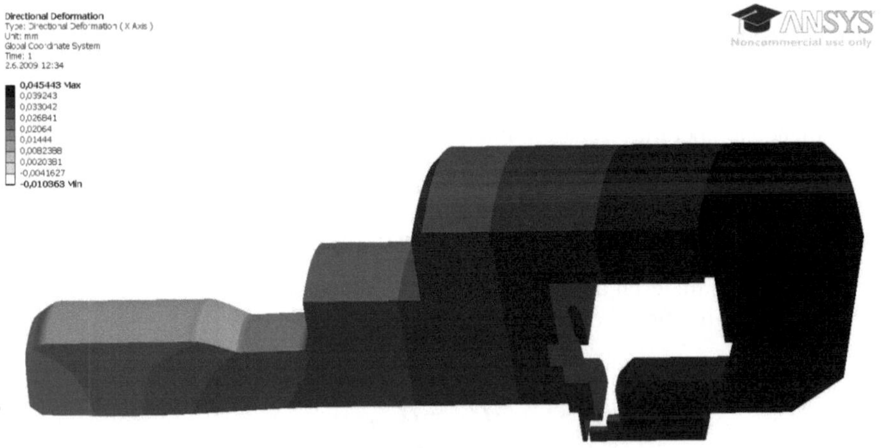

Fig. 4. Displacement distribution in the pressure sensor capsule

Figure 4 shows distribution of displacement in the capsule axis direction for temperature 150°C. The results show constant distance between membrane and optical fiber end for different operating temperature, which was the aim.

4 Conclusion

The new pressure sensor based on the optical principle was developed. Main disadvantage of this sensor was eliminated by the thermal compensation capsule. Optimal material combination can be selected and used for final capsule realization. We tested all of described sensor parts and results show us final pressure measuring precision under 1% over pressure range from -20°C to +70°.

The greatest benefit of this pressure sensor consists in new optical sensing principle. Output characteristic and used materials guarantee excellent parameters for using this type of pressure sensor in industrial environment, explosive environment and environment with high electrical and magnetic interference.

Main advantages of this principle consists in excellent pressure sensitivity, galvanic isolating of whole sensor via optical fiber, very good accuracy of static and dynamic measuring, maximum immunity against electrical and magnetic interference and miniature size of resulting pressure sensor. Those attributes of pressure sensor based on optical sensing principle are optimal for using in explosive gas environment and in environment with high electromagnetic interference.

Acknowledgments. Published results were acquired using the subsidization of the Ministry of Education, Youth and Sports of the Czech Republic, research plan MSM 0021630518 "Simulation modelling of mechatronic systems" and by the Czech Ministry of Industry and Trade in the projects FT-TA2/087.

References

[1] Kadlec, J., Helán, R., Uban, F., Vlach, R., Vrba, R.: Temperature calibrated design of the optical pressure sensor capsule. In: Proceedings of the EDS 2008 conference. Brno, Ing. Zdeněk Novotný CSc 2008, pp. 1–6 (2008); ISBN 978-80-214-3717-3

[2] Kadirvel, K., Taylor, R., Horowitz, S., Hunt, L., Sheplak, M., Nishida, T.: Design and Characterization of MEMS Optical Microphone for Aeroacoustic Measurement. In: 42nd Aerospace Sciences Meeting & Exhibit, January 5-8, 2004, pp. 5–8 (2004)

[3] Švéda, M., Vrba, R., Beneš, P.: A System Architecture of Networked Pressure Sensors. In: Proceedings of the 5th WSES/IEEE World Conference on Systems (CSCC2001), WSES/IEEE, Rethymno, Greece, pp. 3541–3546 (2001) ISBN 960-8052-33-5

The Cavitation Effect on the Electromagnetic Field

F. Pochylý and S. Fialová

Brno University of Technology, Faculty of Mechanical Engineering,
Institute of Energetics, Technicka 2896/2, Brno, Czech Republic
{pochyly,fialova}@fme.vutbr.cz

Abstract. It is generally known, that liquids, including water, are electrically conducting. The electric conductivity magnitude among others depends on the minerals concentration. Using the additives it is possible to increase the electric conductivity.

1 Introduction

The flowmeters principle is also based on the electromagnetic conductivity. However the electromagnetic voltage magnitude caused by the movement of current carriers that are drifted in the flow is very small. It is possible to enhance the voltage by cavitation, which increases velocity of the moving charges by several orders of magnitude compared to non-cavitating flow. Following figures show the moving cavitation bubbles.

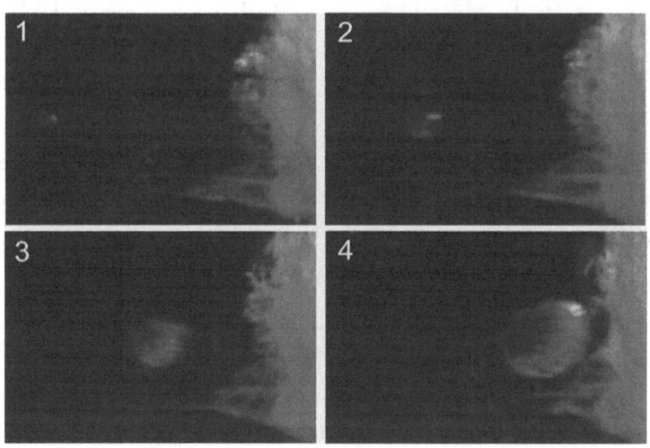

Fig. 1. Cavitation bubbles movement

The collapse and velocity of the bubbles is a dimension higher than the moving liquid.

The cavitation bubble (of spherical shape – see Fig. 3) collapse is described by the Rayleigh- Plesset equation:

$$\frac{3}{2}R^{\bullet 2}+RR^{\bullet\bullet}+2\frac{\sigma}{\rho R}-\frac{1}{\rho}\left[p(0)+\frac{2\sigma}{R_0}\right]\left[\frac{R_0^3}{R^3}\right]^{*}+4\eta\frac{R^{\bullet}}{R}=+\frac{1}{\rho}\left[p(0)-p_N(t)\right] \quad (1)$$

Input parameters	
p_0	100000 Pa
p_a	120000 Pa
f_{ext}	26500 Hz
p_N	$p_a \sin(2\pi f_{ext} t)$
R_0	0,0000045 m
σ	0,0725 N.m^{-1}
κ	1,33
ρ	998 kg.m^{-3}
η	0,01 Pa.s

Fig. 2. The input parameters

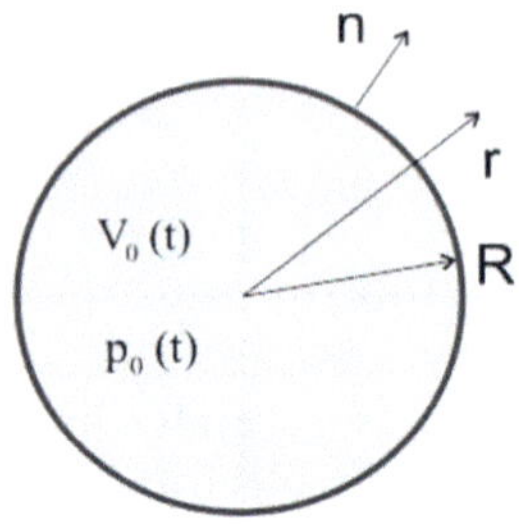

Fig. 3. The spherical shape cavitation bubble

Fig. 4 shows the cavitation bubble diameter change as a function of time calculated by the Rayleigh- Plesset equation (1).

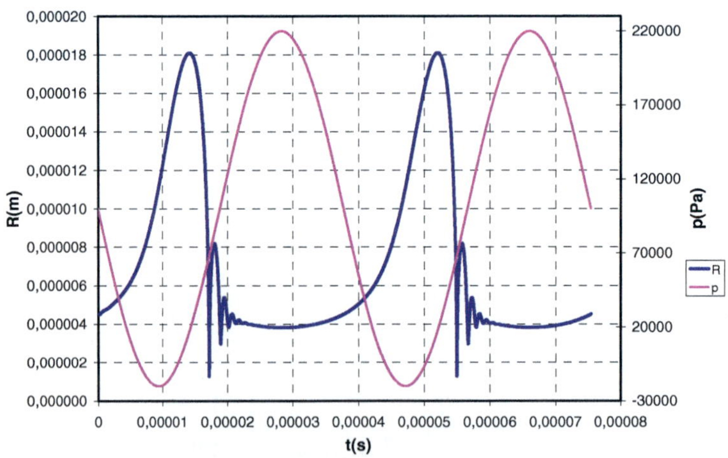

Fig. 4. Cavitation bubble diameter change as a function of time

Next figure (Fig. 5) shows the experimental device with the cavitation pipe with the permanent magnets and the point electrodes for the electromotive tension induced by the cavitation. Fig. 6 shows the cavitation bubbles creation.

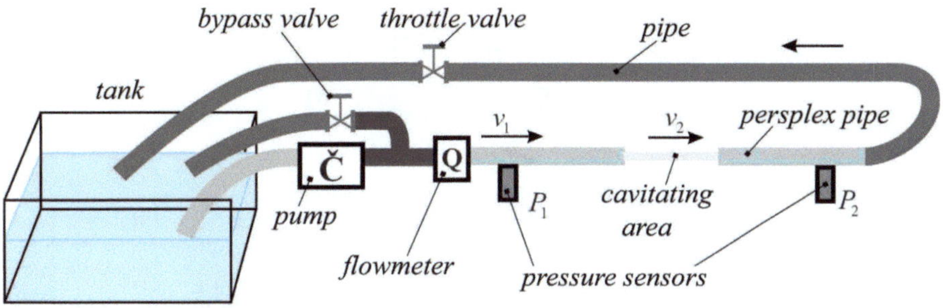

Fig. 5. The experimental device with the cavitation pipe

One the following figures already visualized the bubbles creation and cavitation cloud that wholly fulfilled the cavitation pipe.

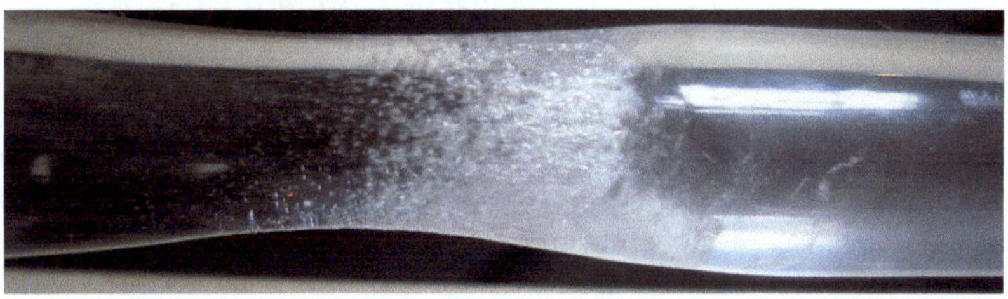

Fig. 6. The cavitation bubbles creation (Pochylý, Fialová, Kantor [4])

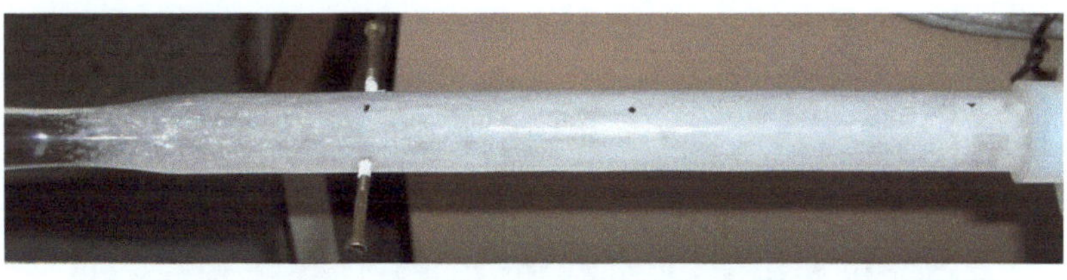

Fig. 7. The cavitation cloud (Pochylý, Fialová, Kantor [4])

2 Electromotive Tension

The equation for the electromotive tension induced by the current carrier of the liquid in the magnetic field is defined by the following form:

$$\varepsilon = \int_{S_1} \frac{d}{dt}(\mathbf{H}\cdot\mathbf{n})dS_1 + \int_{S_2} \mathbf{H}(\mathbf{c}\cdot d\mathbf{l})dS_2 \tag{2}$$

The creation principle is shown on the Fig. 8.

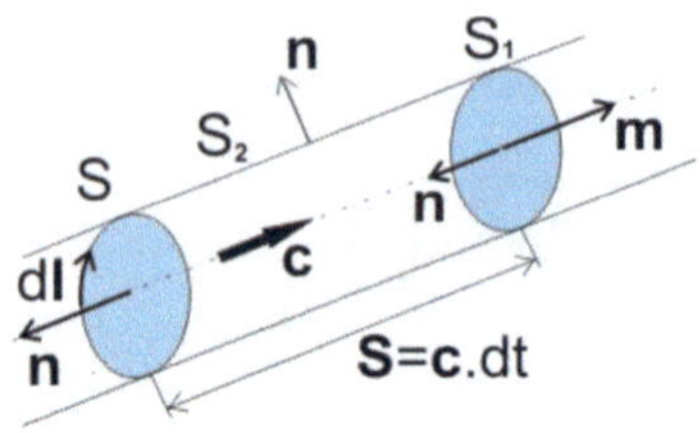

Fig. 8. Cavitation creation principle (Kantor [4]), *where **n** is the unit normal vector to the surface*

The magnitude of the magnetic field intensity is possible to set by the Maxwell equations:

$$\frac{\partial^2\mathbf{H}}{\partial t^2} + \frac{\delta}{\varepsilon}\frac{\partial\mathbf{H}}{\partial t} - \frac{\delta}{\varepsilon} rot(\mathbf{c}x\mathbf{H}) + \frac{1}{\varepsilon\mu} rotrot\mathbf{H} = 0 \tag{3}$$

Hence it is clear that **H** depends on the specific conductivity of the liquid and the liquid velocity vector **c**. It is appropriate that the **H** and **c** are orthogonal. In the cavitation in a tube of circular cross-section, however, particles can oscillate in all directions, which may result in impaired electromotive tension.

It is therefore necessary to optimize the experimental device in this respect. One of the possible solutions is flat cavitation tube.

Next figure (Fig. 9) shows the arrangement of the point electrodes in the case of circular cross-section tubes.

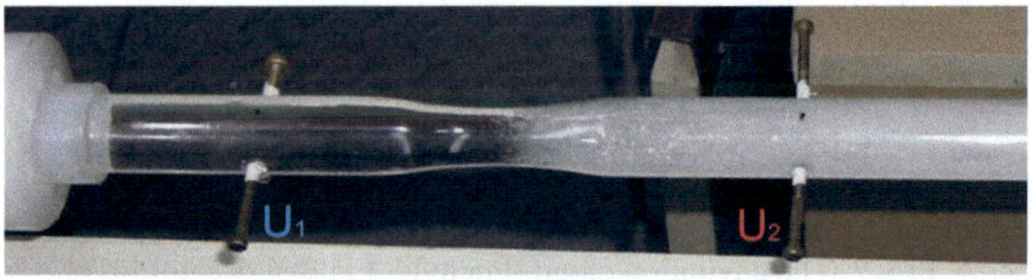

Fig. 9. Point electrodes arrangement

Fig. 9 shows two pairs of electrodes. Before the cavitation area are the electrodes that measure the voltage U_1, after the cavitation area the voltage U_2.

It is obvious that velocity of cavitation bubble collapse can significantly affect the values of the magnetic induction, magnetic field, permanent magnet close to the pipe.

Electromagnetic voltage magnitude is result of cavitation effect on the magnetic field.

The cavitation effect on the magnetic field results the electromotive tension magnitude. See Fig. 10. This magnitude has step changed in the cavitation cloud (Fig. 7). It depends on the electrical conductivity of the liquid. The conductivity magnitude was changed by the salt addition. The results are shown on Fig. 11.

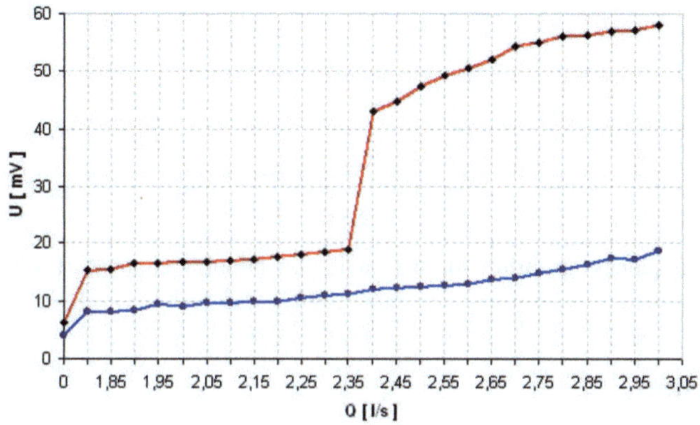

Fig. 10. The flow - electromotive tension dependence (Pochylý, Fialová, Kantor [4])

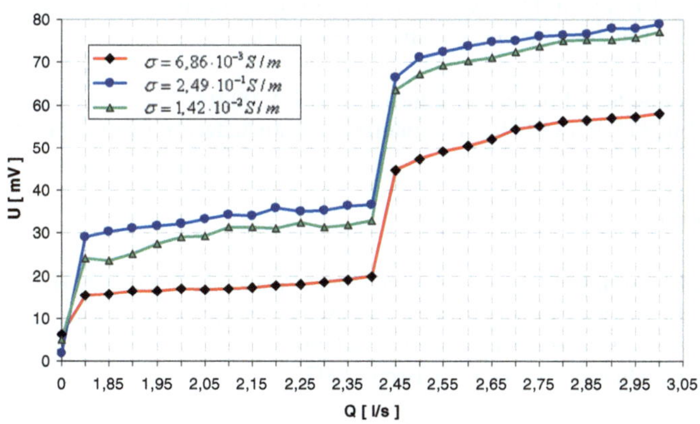

Fig. 11. The electrical conductivity effect on the electromotive tension magnitude (Pochylý, Fialová, Kantor [4])

The cavitation bubbles collapse is connected with the high frequencies see Fig. 12.

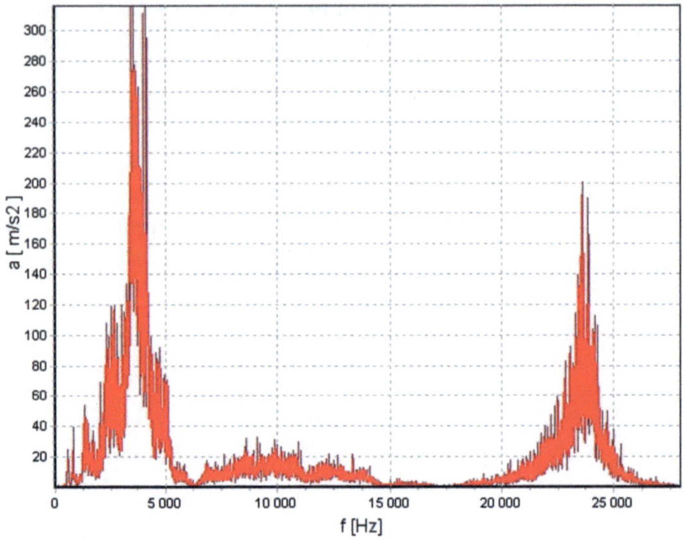

Fig. 12. The frequency spectrum for the cavitation bubbles collapse (Pochylý, Fialová, Kantor [4])

3 Conclusion

By the performed experiment it was acknowledged that the cavitation bubbles bear as the electric charge with the high speed of movement. Its movement in the magnetic field is possible to use both for identification of the cavitation creation beginning and the electric energy micro resources.

Acknowledgements. Ministry of Education MSM 0021630518 and Grant Agency of Czech Republic GA 101/09/1715 are gratefully acknowledged for support of this work.

References

[1] Sedlák, B., Stoll, I.: Elektrina a magnetismus. Academia Praha (1993)
[2] Alesko, P.: Mechanika zidkosti i gaza.Charkov (1977)
[3] Noskievic, J.: Kavitace, Akademia Praha (1969)
[4] Kantor, D.: Diploma thesis. BUT Brno (2007)

Identification of MR Fluids Properties in Mechatronic Damping Elements

J. Roupec, I. Mazůrek, M. Klapka, and P. Číž

Brno University of Technology, Faculty of Mechanical Engineering,
Institute of Machine and Industrial Design,
Technicka 2896/2, Brno, Czech Republic
jroupec@hotmail.com

Abstract. Mechatronic systems often use a magnetorheological fluids for the damping of movement. A significant dependence of their rheological properties on a magnetic field is the main reason. Therefore these fluids are relatively well controllable in mechatronic systems. For correct control of these elements it is important to assign to MR fluids a suitable rheological model of behavior. This model describes flow and viscosity dependences by various magnetizing currents and various shear rates. This paper is concerned with measurement problems of flow and viscosity curves on standard and original rheometers. Based on measured data, a behavior of MR fluid can be mathematically described, a relevant model can be compiled, a behavior of different MR fluids can be compared or changes of mechanical properties during the measurement can be observed.

1 Introduction

The first MR fluid was made by Rabinow as early as 1946 [1] and the following by Winslow [2]. Since then up to the nineties a research in this discipline stagnated. In the nineties unique properties of MR fluids were rediscovered for mechatronics. Particularly their ability to change from a fluid state to a semi-solid or plastic state upon the application of a magnetic field. For rheological model creation an experimental identification of its basic parameters is necessary. In semi-solid state the MR fluid proves a viscoplastic behavior [3]. It can be described by a field-dependent yield stress. The yield stress, a fast response time and the equally fast reversibility enable the MR fluids an extensive application in semi-active control area. Rheological properties as the yield stress and the response time were investigated only in a narrow range of shear rates. This range does not correspond to real conditions of an operation. If we want to use MR fluids for damping of impacts or shock loading it is necessary to test the fluids at shear rates from 10^3 to 10^6 s^{-1} [4]. The only commercial device which enables a measurement of flow and viscosity curves in a magnetic field is provided by Anton Paar company. The maximum measurable shear rate in a magnetic cell has got a value of 10^2 s^{-1}.

This device is totally unsuitable particularly for intended durability measurement. For that reason our team has designed and constructed a slit-flow rheometer that enables to evaluate the flow curves at high shear rates on the basis of Force-Velocity characteristic. The rheometer allows measurement at different intensities of a magnetic field and an accurate observation of MR fluid temperature. This paper describes the most serious problems encountered.

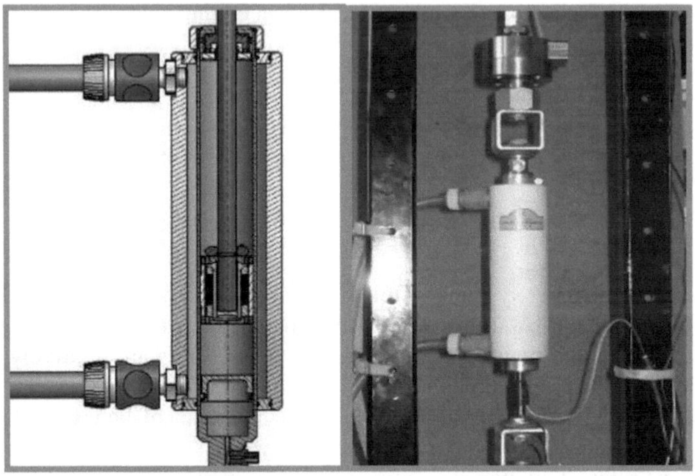

Fig. 1. Design of a new slit-flow rheometer

2 Problems of Rheometer Construction

A rheometer construction comes from a commercial MR damper made by Delphi company. A reconstruction of a common damper to a rheometer requires overcoming several constructional and metrological problems:

- Reconstruction for easy assembling, disassembling and change of oil filling [5]
- Pressurization of hydraulic system by gas for prevention of cavitation [5]
- Water cooling of rheometer for perfect stabilization of temperature regime [5]
- Reading of tested oil temperature in a slit
- Determination of magnetic induction in dependence on coil current
- Correction of measured curve from influence of elastic gas filling
- Identification and removal of influence of seal shear friction

2.1 Temperature Measurement in a Slit

Due to the piston construction it would be constructionally difficult to place a temperature sensor directly to the choke gap. Therefore a linear temperature dependence of electrical resistance of copper coil winding was utilized. The measurement was performed for current range from 0 to 2A in increments of 0,25A.

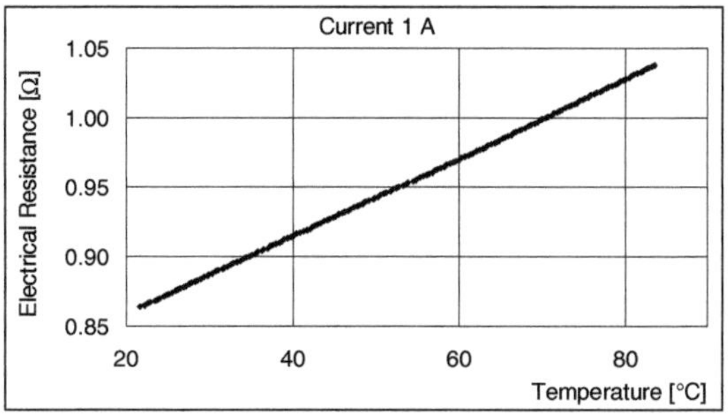

Fig. 2. Temperature dependence on coil resistance and magnetizing current

From these measurements the coil resistance dependence on temperature and magnetizing current was found.

2.2 Magnetic Field Intensity in a Slit

Some preparatory measurements were performed on a dismounted piston with the aid of magnetometer F.W.BELL 5070 with a probe for measurement in a gap. A course of magnetic flux density in an air gap was being measured for different magnetizing currents along all piston length (Fig. 4). A magnetic flux density is the highest in the place where the lines of force perpendicularly intersect the choke gap. For real conditions in rheometer it is necessary to adjust a magnetic flux density by different permeability of environment (oil, MR fluids).

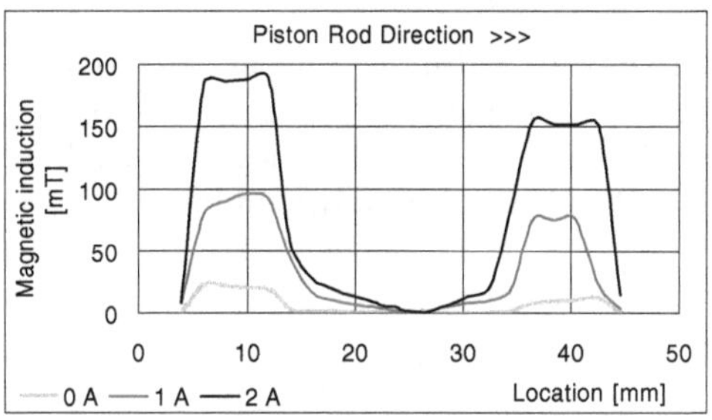

Fig. 3. Course of magnetic flux density for all length of gap on the air

2.3 Selective Criterion for Measured Data

A stand is solved like a crank mechanical pulsator with almost sinusoidal course of velocity. Our intention is to observe a fluid behavior preferably during steady-state flowing. Data acquisition system had to be adjusted so that the recorded data are considered as *valid* only when the piston acceleration does not exceed the set limits (Fig.4). Problems with the elimination of a dynamic force influence faced with Carlson and Goncalves designed pulsator were our motivation for finding this solution [6].

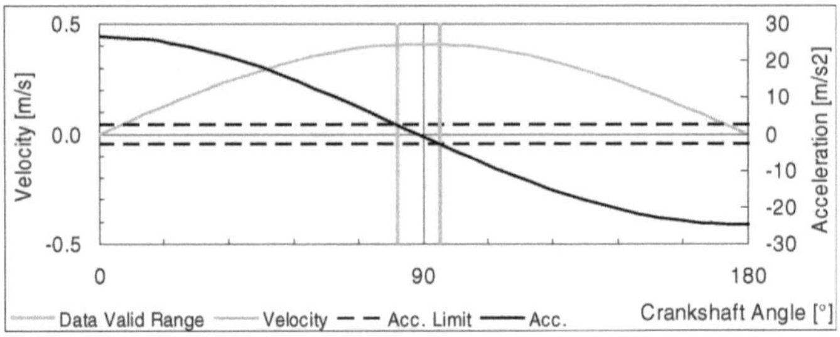

Fig. 4. Course of crank mechanism motion and selective criterion

2.4 Calibration on Newtonian Fluids

A rheometer calibration was performed on Newtonian fluids, concretely on gearbox oils SAE 90 and SAE 85W-140. A formula for shear stress τ computation was derived:

$$\tau = \frac{3k\eta Q}{\pi rh^2} = \frac{3kS_p v_p}{\pi rh^2}\eta, \tag{1}$$

where η is dynamic viscosity, Q flow rate, r mean radius of piston gap, h width of choke gap, S_p piston area, v_p piston velocity and k is correction factor. Shear rate value can be determined from the formula :

$$\dot{\gamma} = \frac{du}{dh} = \frac{3S_p}{\pi rh^2}v_p = \frac{6u_m}{h}, \tag{2}$$

where u is local fluid velocity and u_m is a mean velocity of volume rate of fluid flow through a choke gap. The purpose of the calibration was to find correction factor k. A correspondence of measured values after calibration with curves acquired on a viscometer RotoVisco 1 is shown on Fig.5.

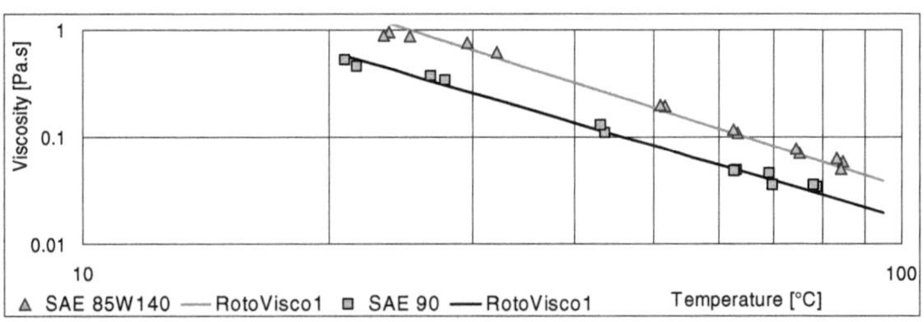

Fig. 5. Verification of rheometer sensitivity on gearbox oils

3 Measured Results

An analyzer Dewetron Dewe-2010 reads a force (dynamometer HBM U2AD1), a piston motion (inductance sensor SD2), a piston temperature (sensor LM35), a voltage and a current in a piston coil. A control software created in LabView for commercial testing of automotive dampers ensures a pulsator control and data recording with a sampling frequency of 500Hz. Immediate piston velocity is acquired by a numerical derivation of position. A selected test procedure is a frequency sweep of crank revolutions from n_{min} = 1,5RPM to n_{max} = 150 RPM, which corresponds to piston velocities from 0,001 to 0,1m.s^{-1} at the stroke of 13mm. A special software created in MS Excel only converges output data into flow and viscosity curves and provides all above mentioned computations and corrections.

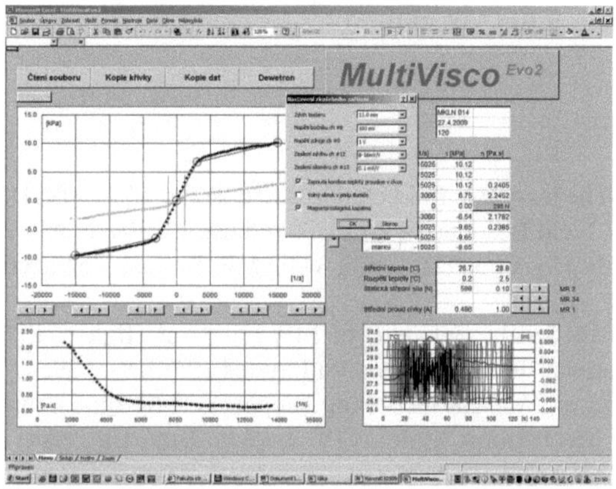

Fig. 6. Program mask of MultiVisco Evo^2

A viscosity dependence of MR fluid Lord 132DG on shear rate is plotted on the graph in Fig. 7. The left side of the graph is measured to 10^2 s^{-1} by Anton Paar company and the right side is a measurement from our new slit-flow rheometer. It is visible there that the curves on the right side are a logical continuation of the curves from the left side. The bottom curves are curves measured in zero-magnetic field and the upper curves correspond to a magnetic induction at the value of 0,8T.

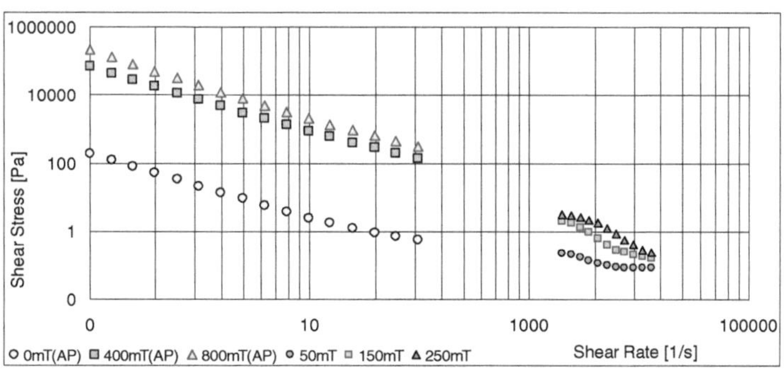

Fig. 7. Comparison of measured curves from Anton Paar and from the new slit-flow rheometer

4 Conclusion

It was succeeded to construct the slit-flow rheometer that enables an investigation of MR fluid behavior in real working conditions of linear MR devices. A creation of this rheometer is from economical point of view very advantageous and rather easy. Measured values are in accord with expected results. Existing results of measurement indicate a progressive shear thinning of MR fluid also in high shear rates. This rheometer can also be conveniently used for durability testing.

Acknowledgments. This work was developed with the support of the research plan MSM 0021 630 518 and grant BD 138 3001.

References

[1] Rabinow, J.: The magnetic fluid clutch. AIEE Trans. 67, 1308 (1948)
[2] Winslow, W.: Field Responsive Fluid Couplings, US Patent No. 2 886 151 (1959)
[3] Lord Materials Division, Designing with MR Fluids, Engineering Note (November 1999)
[4] Goncalves, F.D., et al.: Behavior of MR Fluids at High Velocities and High Shear Rates. International Journal of Modern Physics 19(7,8,9), 1395–1401 (2005)
[5] Mazůrek, I., et al.: Virtual modelling support of MR fluid properties. In: Simulation Modelling of Mechatronic Systems IV, 1st edn. Brno University of Technology, Brno (2009)
[6] Koo, J.H., et al.: A comprehensive analysis of the response time of MR dampers. Smart materials and structures 15(2), 351–358 (2006)

Influence of External Magnetic Field on Measuring Characteristics of the Magnetoelastic Sensors

A. Bieńkowski, R. Szewczyk, and J. Salach

Institute of Metrology and Biomedical Engineering, Warsaw University of Technology, sw. A Boboli 8, 02-525 Warszawa, Poland

Abstract. Paper presents results of investigation on influence of external magnetic field on measuring characteristics of magnetoelastic sensor with ring-shaped core. The core of sensor was made of a high permeability amorphous alloy. External magnetic field up to 4 kA/m was generated by Helmholtz coils. Presented results confirm significant role of the demagnetization in analyze of influence of the external magnetic field on characteristics of the magnetoelastic sensors. Due to demagnetization even strong magnetic field has relatively small influence on the functional properties of the sensor.

1 Introduction

In magnetoelastic sensors external force generate mechanical stresses in ferromagnetic core. These stresses change magnetic characteristics of sensor's core, what can be measured electrically. In the case of magnetoelastic sensors the use a transitional elastic element is not necessary, as it is required in the strain gauge sensors [1, 2].

Magnetoelastic sensors exhibit significant change of the magnetic core parameters. As a result obtained a measuring signal is convenient for further processing [3]. Moreover convenient choice of sensor's operation range by the change a diameter of the core is next advantage. It should be also indicated that application of novel magnetic materials in production of the cores of magnetoelastic sensors enable its application in high temperatures range. Some of recently developed nanocrystalline magnetic materials can operate in temperature up to 600 °C [4].

For development of magnetoelastic sensors based on the ring-shaped sensing elements, the most important is methodology enabling achievement of a uniform stress distribution in the core [5]. Mechanical construction device which enable uniform stress generation due to application of external compressive force F is presented in figure 1.

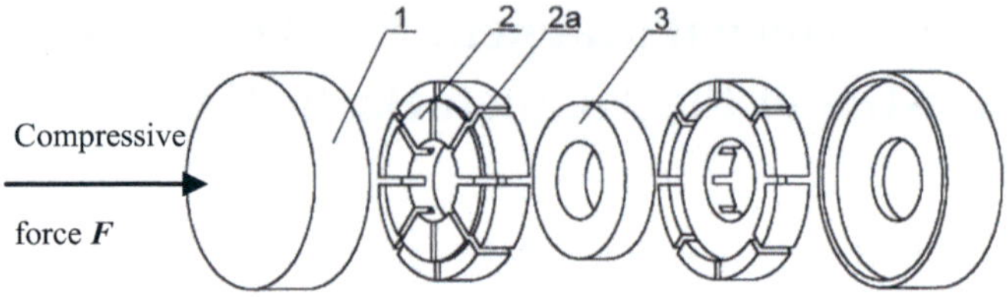

Fig. 1. Schematic diagram of the device for applying the uniform compressive stress to the ring core of the magnetoelastic sensor [6]. 1 – magnetoelastic ring-shaped core, 2 – non-magnetic cylindrical backing, 2a – grooves for magnetizing and sensing windings, 3 – base backings.

Magnetoelastic sensors utilizing method of application of the force *F* presented in figure 1 can be applied to monitoring burning process in engines of diesel locomotives. As it is presented in figure 2 the magnetoelastic sensors is located under a nut of binding-screw. In such a case magnetoelastic sensor gives possibility of real-time detection of ignition in cylinder of the engine and monitoring of the burning process. Is should be stressed that such method of monitoring of diesel engine is not invasion. Is not necessary to mount the sensor inside of the cylinder, as well as this method of measurement is very chip in comparison with traditional method utilizing pressure sensors in engine cylinder.

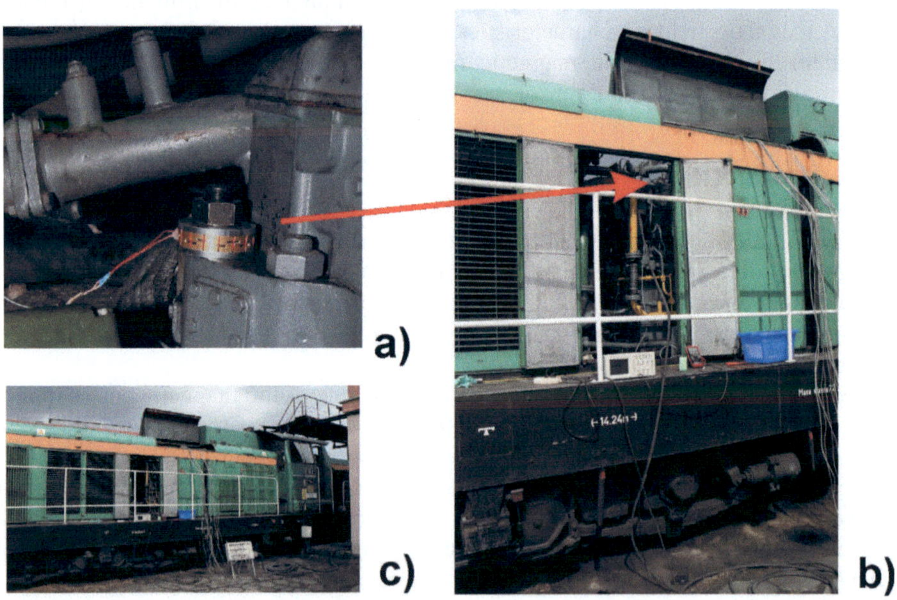

Fig. 2. Application of the magnetoelastic sensor for real-time monitoring of burning process in engine of the diesel locomotive: a) sensor mounted under binding-screw, b) location the sensor in engine chamber c) diesel locomotive on test place

Application of magnetoelastic sensors on locomotives may be connected with expository of the sensor on external magnetic field. Such external field may disrupt normal operation of the sensor and be a reason of measuring error. Due to this fact it is necessary to investigate the influence of external magnetic field on the sensors and taking into account possible disturbance.

In the case of external magnetization of the ring core by magnetic field H_m, the demagnetisation H_{dem} occur, as it is shown in figure 3.

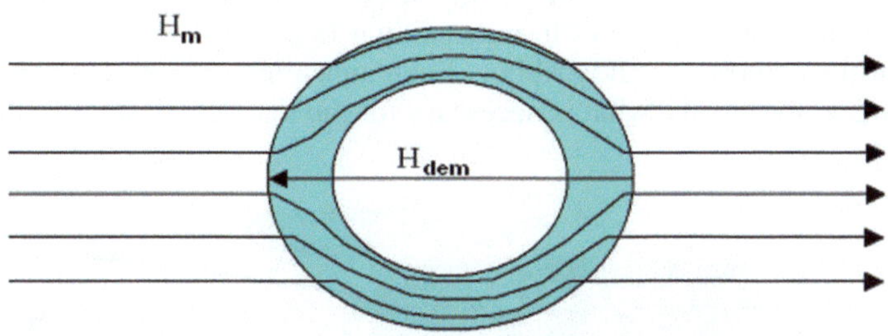

Fig. 3. Demagnetise H_{dem} of ring shape sample in magnetic filed H_m in the direction of diameter

Effects of demagnetization may be described by the demagnetization coefficient N of the core. In such a case effective magnetization H_{eff} is given by the following equation:

$$H_{eff} = H_m - H_{dem} = H_m - N \cdot M \tag{1}$$

where M is magnetization of the core.

In the case of a ring-shaped core, the demagnetization factor N may be determined for two directions of magnetic field: parallel and perpendicular to the core diameter. In the case of magnetization which is perpendicular to the diameter of the core, the demagnetisation coefficient N depends on the sensor core dimensions and can be determined from special diagrams [7]. In the case of sensor for monitoring of the diesel engines, demagnetization factor in this case exceed 0,7. In the case when core magnetized by the external magnetic field applied parallel to the diameter (as it is presented in figure 3), demagnetisation coefficient is equal 0,33 [8].

From practical point of view it can be expected, that ring-shaped core of magnetoelastic sensors will be magnetized by external magnetic field in the direction of diameter. In such a case demagnetization coefficient is smaller, but still high. As a result the demagnetization will play a very significant role in the reduction of the influence of the external magnetic field on operation of the magnetoelastic sensors. Accordingly to equation (1) this reduction is especially important in the case of high permeability materials, such as amorphous alloys.

2 Method of Investigation

Experimental setup for investigation of the influence of the external magnetic field H_m on the magnetoelastic sensor with ring-shaped core is presented in figure 4. External magnetic field H_m is generated by Helmholtz coils (2). Strain gauge force sensor (3) is used as an reference for measurements of the force applied to magnetoelastic sensor by the head of hydraulic press (4). Magnetoelastic force sensor (1) is used for tests, utilizes ring-shaped core made of $Fe_{81}Si_4B_{15}$ amorphous alloy after thermomagnetic treatment. Core of the sensor was annealed in 350 °C for one hour. Annealing was carried out in magnetizing field 150 kA/m perpendicular to direction of the amorphous alloy ribbon. Such annealing reduce value of the magnetomechanical hysteresis, what is necessary for the magnetoelastic sensors.

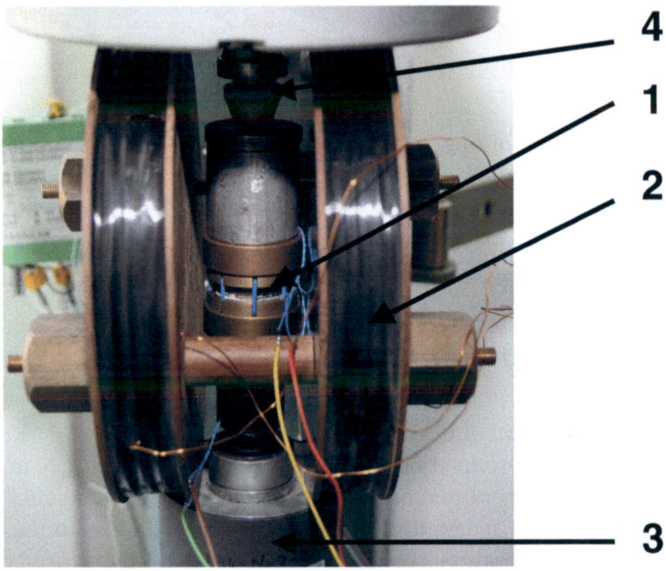

Fig. 4. Measuring setup for testing of the influence of external magnetic field on magnetoelastic force sensors: 1 – magnetoelastic force sensor with ring-shaped core, 2- Helmholtz coils, 3 - reference strain-gauge sensor, 4 – head of press for application of stresses to sensors

3 Results

Figure 5 presents the influence of the external magnetic field H_z on the shape of quasistatic hysteresis loop of the core made of the $Fe_{81}Si_4B_{15}$ amorphous alloy after thermomagnetic treatment. The results of the measurements show both the influence of external stresses σ up to 10 Mpa, as well as influence of external magnetic field H_z on the shape of the hysteresis loop $B(H)$.

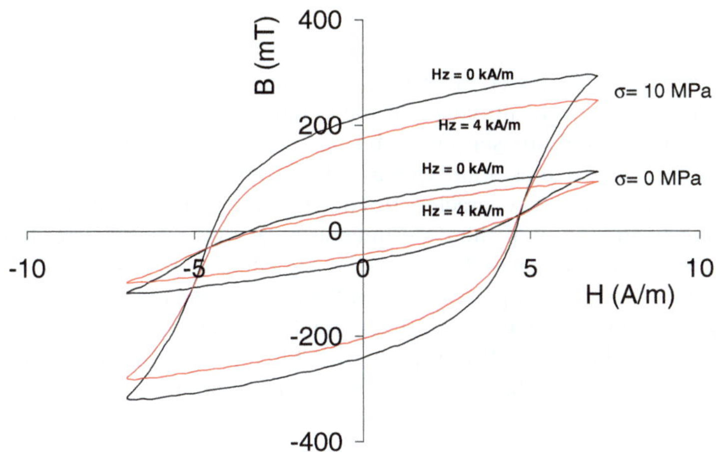

Fig. 5. Influence of the compressive stresses σ and external magnetic field H_z on the shape of histeresis loop B(H) for the sensor made of the $Fe_{81}Si_4B_{15}$ amorphous alloy

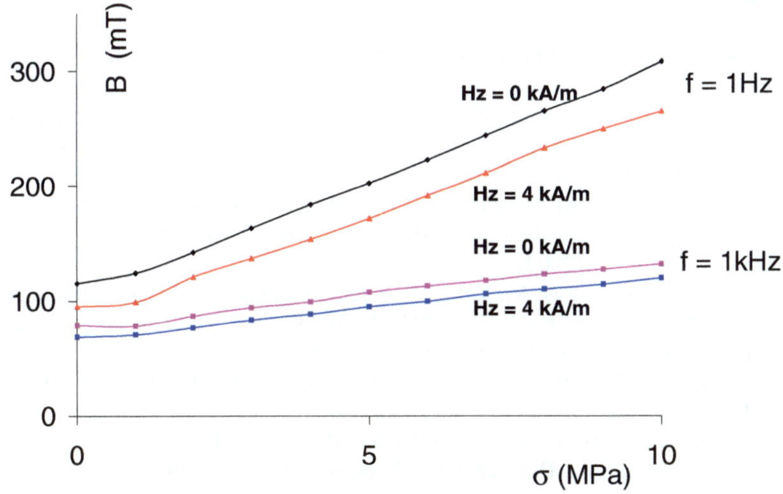

Fig. 6. Magnetoelastic *B(σ)* dynamic characteristics of the amorphous core measured for different steering frequencies *f* and different external magnetizing field H_z

Figure 6 presents the results of the measurements of the magnetoelastic characteristics *B(σ)*. These characteristics determine characteristics of the magnetoelastic sensors. For this reason measurements were carried out for steering frequencies *f* up to 1 kHz and external magnetizing field H_z up to 4 kA/m.

4 Conclusion

Analyses presented in the paper indicate, that demagnetization coefficient plays significant role in the case of influence of the external magnetic field on the performance of the magnetoelastic sensors with ring-shaped cores. Especially for high permeability materials, such as a amorphous alloys, demagnetization significantly reduces changes of the sensors characteristics due to the external magnetic field.

On the other hand results of the experimental measurements indicate, that the influence of the external magnetic field can't be neglected, both in the case of quasistatic magnetic hysteresis loop as well as in the case of dynamic measurements. External magnetic field up to 4 kA/m causes about 10 % changes of the flux density B in amorphous alloy core. These results should be taken into consideration in the metrological analyses of the sensors for real-time monitoring of the diesel engines of locomotives.

Acknowledgements. This work has been supported by the European Union in the framework of European Social Fund through the Warsaw University of Technology Development Programme.

References

[1] Jiles, D.: Introduction to Magnetism and Magnetic Materials. Chapman&Hall, London (1998)
[2] Bieńkowski, A., Szewczyk, R.: The possibility of utilizing the high permeability magnetic materials in construction of magnetoelastic stress and force sensors. Sensors and Actuators A113, 270 (2004)
[3] Bydzovsky, J., Kollar, M., Jancarik, V., Svec, P., Kraus, L.: Strain sensor for civil engineering application based on CoFeCrSiB amorphous ribbons". Czechoslovak Journal of Physics 52A, A117 (2002)
[4] Bieńkowski, A., Szewczyk, R., Kulik, T., Ferenc, J., Salach, J.: Magnetoelastic properties of HITPERM-type Fe41.5Co41.5Cu1Nb3B13 nanocrystalline alloy. Journal of Magnetism and Magnetic Materials 304, e624 (2006)
[5] Bieńkowski, A.: Magnetoelastic Villari effect in Mn–Zn ferrites. J. Magn. Magn. Mater. 215-216, 231 (2000)
[6] Bieńkowski, A., Szewczyk, R.: Patent Pending P-345758 (2001)
[7] Du-Xing, C., Brug, J.A.: Demagnetizing Factors for Cylinders. IEEE Trans. Magn. 27, 3601 (1991)
[8] Macintyre, S.A.: Magnetic Field Measurement. CRC Press, Boca Raton (2000)

Mechatronic Lighting Pole Testing Device

P. Steinbauer and M. Valášek

Czech Technical University in Prague, Faculty of Mechanical Engineering,
Department of Mechanics, Biomechanics and Mechatronics,
Karlovo nám. 13, 12135, Praha 2, Czech Republic

Abstract. Monitoring of street lighting pole health is of increasing importance. Extensive erosive damage of lighting poles may lead to unexpected falls of poles which may cause dangerous injuries and accidents. There are already existing many methods how to determine actual state of the construction. Their application is however costly, time consuming and often provide local information only (related to the position of measurement). Requirement of regular inspection of many lighting poles needs approach, which enables rapid and unsophisticated measurement device installation and usage, with automatic result evaluation. Experimental investigation of many lighting poles with several measurement methods was carried on. Especially method based on low frequency harmonic force excitation and compliance measurement gave very promising results corresponding with actual state of the pole. Based on it, automatic measurement device was proposed. The paper deals with the description of the measuring procedure and the automatic measuring device for the health monitoring of lighting poles.

1 Introduction

As lighting poles are usually installed on places with heavy personal traffic load and they are used for many years in severe climate condition, including aggressive chemical and mechanical ones, the knowledge about their actual health status is of high importance.

Currently the pole inspection is done, if at all, by visual inspection and endoscopy of the pole internals. The pole failure is anticipated by regular replacement each n-th year. This is however not sufficient as poles at less exposed places are replaced in quite good status and pole at very aggressive places should be replaced sooner.

Many nondestructive methods of construction health evaluation were taken into account, e.g. ultrasound, eddy currents, roentgen, relative comparison of electromagnetic properties, acoustic emissions. Very promising group of methods is based on measurement of mechanical properties change. They can be stiffness, modal properties, modal damping, acoustic emissions. Their common big advantage is they are nondestructive and they provide global information about evaluated construction / pole. Initial experiments really did show, that damaged poles have different modal properties from healthy poles. This was however the case of freely hanging poles or poles stuck into the concrete.

Unfortunately, this is not the case in real life since most of the poles in Czech Republic are fixed in tamped sand/stone ballast only. This kind of fixing effectively damps most of eigen modes in diverse range so no meaningful conclusion can be made from it (see Fig. 1).

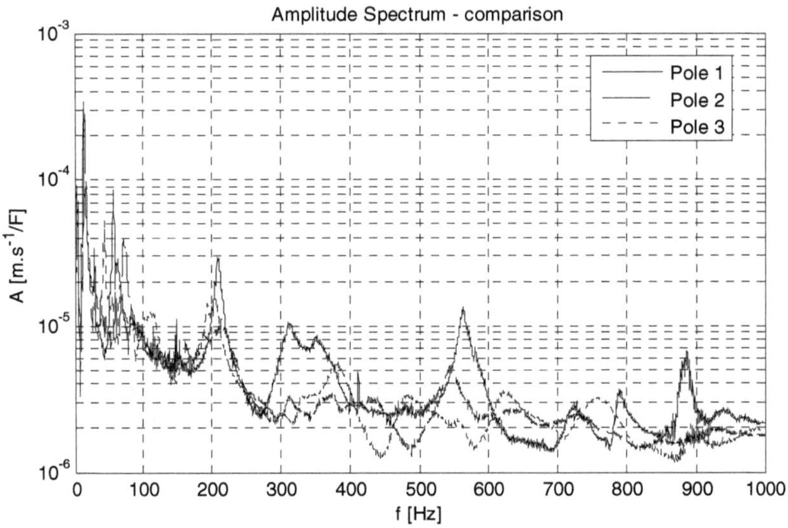

Fig. 1. Amplitude transfer functions of tested poles in different health

Very interesting results were obtained by measurement of pole stiffness by classical approach, i.e. loading the pole by known force at certain height and measurement of pole deflection by positional sensor.

The realization of such approach in field conditions is awkward and very time consuming due to necessity to install heavy construction for force source and second construction for positional sensors.

Another approach was thus used to measure pole stiffness for broader field experiments with success. It is based on low frequency harmonic actuation of the pole. As the method turned out to be successful, mechatronic measurement device was proposed and simulated.

2 Measurement Method

The method of measurement is based on approach used once for machine tools ([1]). The pole is actuated by harmonic force acting successively in two perpendicular directions (see Fig. 2). The frequency of harmonic force is chosen much (at least 5-10x) lower than first pole eigen frequency. The motion of the pole can be considered as quasi-static. Even the first eigen mode's motion can be neglected.

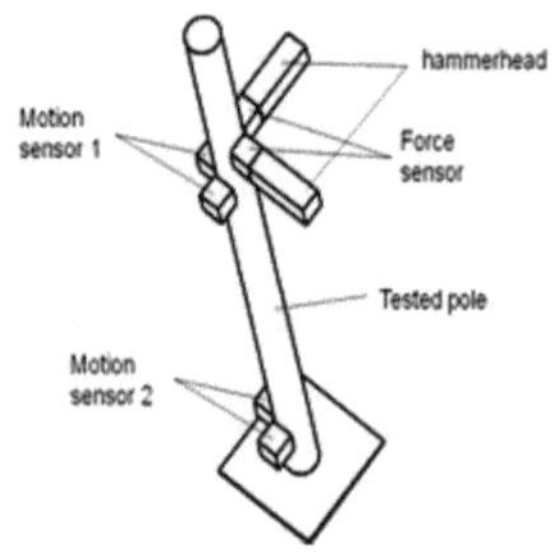

Fig. 2. Concept of the pole measurement

The pole is thus loaded by quasi-static force, which can be applied without stiff base, e.g. by hanging electromagnetic actuator. The actual acting force is measured by force sensor as time series. The response of the pole is also harmonic and can be measured on the level of either position or velocity or acceleration by appropriate sensor. The amplitude of both acting force and response is determined by fitting measured time series on appropriate harmonic function

$$f(t) = A_f \sin(\omega t + \psi) + d,\ a(t) = A\omega^2 \sin(\omega t + \varphi),$$
$$v(t) = A\omega \sin(\omega t + \varphi),\ x(t) = A\sin(\omega t + \varphi) + c \tag{1}$$

by least square method ([2]).

Parameters A_f and A are used to determine pole stiffness. Further, the force level A_f is changed several times so linearity of the pole can also be checked.

In addition, the pole deflection is measured at two points, one near to the acting force and second one on the lowest pole part. The so called relative stiffness of the pole between these points is evaluated. The influence of pole fixing into ground is thus excluded from the result.

3 Experimental Verification

The described method was applied to several groups of lighting poles with different parameters. Each group contained healthy and ill poles. In all cases, the pole state corresponds to the differences of pole stiffness measured by described method. One group of poles is illustrated on the Fig. 3.

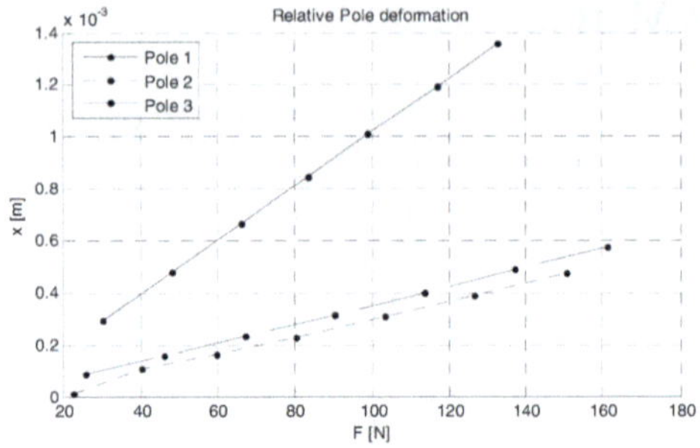

Fig. 3. Relative deflection of tested poles in different health

4 Measurement Device Design and Control

The principal idea of the measurement device is sketched in the figure 4. Based on the measurement principle described above, the device must ensure low frequency harmonic force excitation at certain height while no massive supporting construction is needed.

The supporting construction only holds vertical (gravity) forces. This can be e.g. lifting platform. The hammerhead is connected to excited pole via spring with adjustable stiffness. The mechanical system can be so tuned to have its eigen frequency near the required frequency of harmonic actuating force. Initial motion is invoked by linear actuator inside the hammerhead.

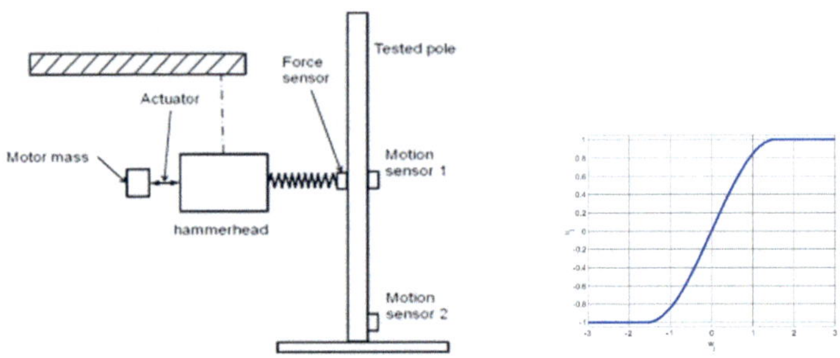

Fig. 4. Measurement device concept and force actuator limitation

The mechanical part of proposed device is expected to behave as linear, so it could be modelled by classical matrix equation, where **z** is vector of the positions.

$$M\ddot{z} + B\dot{z} + Kz = f \quad (2)$$

The friction between device support and hammerhead is neglected. The actuator is however limited in its motion range and achievable force. It can be modelled by additional fast state equation

$$\dot{u}_j = \frac{1}{T}\left[f_u\left(x, u_j\right) + w_j\right] \quad (3)$$

Inverse function f_u is smooth limiting function

$$u_j = sat\sin(m, s, w_j) = \begin{cases} m & pro \;\; x > s.\frac{\pi}{2} \\ m.\sin(s.w_j) & pro \; -s.\frac{\pi}{2} \le x \le s.\frac{\pi}{2} \\ -m & pro \;\; x < s.\frac{\pi}{2} \end{cases} \tag{4}$$

The originally linear system model is so transformed into nonlinear form, where vector **x** consist of positions **z** and their time derivatives d**z**/dt, **u** is input of the system.

$$\dot{x} = A(x).x + g(x).u \tag{5}$$

Control law itself is conveniently designed by NQR approach ([3]) in the state dependent feedback form

$$u = K(x).x \; . \tag{6}$$

NQR design procedure minimizes quadratic optimality criterion

$$J = \int_0^\infty (x^T.Q(x).x + u^T.R(x).u)dt \; . \tag{7}$$

The stiffness of the adjustable spring is calculated to tune the device consisting of said spring and hammerhead with actuator into resonance on required frequency f_{req}

$$k_{ps} = (2\pi f_{req})^2 (m_h + m_a) \; . \tag{8}$$

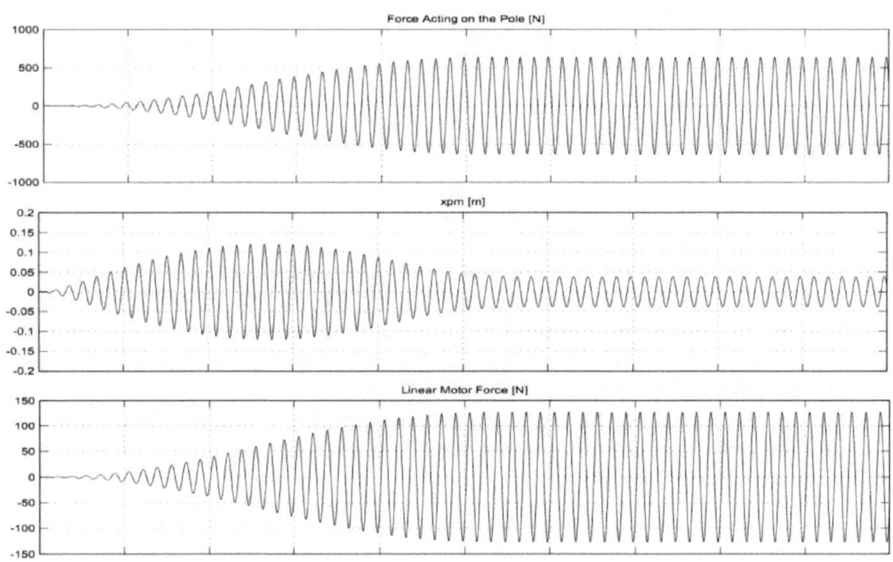

Fig. 5. Illustration of achieved force acting to the pole and required actuator motion and force

The other mechanical parameters of the device, i.e. mainly hammerhead mass, actuator mass were optimized to achieve highest harmonic force acting to the lighting pole through the spring connection, while maintaining force from the actuator on low level.

To reduce required actuator force and motion range, the required acting force was shaped by already mentioned satsin function (4).

Figure 5 shows simulation results which demonstrate possible amplification of acting force. This is achieved by tuning the system to resonance frequency and optimal control.

5 Conclusions

The problem of street lighting pole health monitoring has been addressed. Several methods of pole deterioration detection were proposed and tested based on extensive field experiments. Measurement of the pole stiffness is of particular importance because it provides composite information about pole health.

Based on this experience, the testing device, which does not require time consuming installation nor is heavy and bulky, was proposed. The device is based on excitation of the construction by low frequency harmonic force. The deflection can be thus measured either by deformation, velocity or even acceleration sensors and no stiff point or construction is needed. Due to the nature of harmonic force, it can be developed by mechanical oscillator which again does not require neither bulky stiff supporting construction nor powerful force source. The need of Archimedes's fixed point or Valasek's flexible point ([4]) is removed in this particular case due to nice properties of harmonic force actuation.

The concept of measurement was tested by field experiments. The testing device concept was evaluated by mechatronic simulation model and its realization is being prepared.

References

[1] Prokopec, M.: Dynamic method of static deformation measurement. Strojírenství 12, 832–836 (1962) (in Czech)

[2] Kozanek, J.: A comment to problem of least square method in identification. In: Proc. of Colloquium Dynamics of Machines 95, UT AV CR, pp. 55–60 (1995) (in Czech)

[3] Valasek, M., Steinbauer, P.: Nonlinear Control of Multibody Systems. In: Ambrosio, J., Schiehlen, W. (eds.) Proc. of Euromech Colloquium 404, Advances in Computational Multibody Dynamics, Lisabon, pp. 437–444 (1999)

[4] Valasek, M., Necas, M.: Innovative concept of increasing dynamic stiffness of compliant structures by mechatronic approach, CTU in Prague, Prague. Bulletin of Applied Mechanics, vol. 4(16) (2008)

Neural Networks: Off-Line Diagnostic Tools of High-Voltage Electric Machines

P. Latina, J. Pavlík, and M. Hammer

Brno University of Technology, Faculty of Mechanical Engineering, Institute of Production Machines, Systems and Robotics, Technicka 2896/2, Brno, Czech Republic
{hammer,latina,pavlik.jo}@fme.vutbr.cz

Abstract. This article describes neural networks used in the off-line diagnostics of high voltage generators. Such an artificial intelligence tool is used for the prediction of diagnostic quantity selected and used in the machines tested. The diagnostic quantities are obtained by non-destructive measurement during planned revisions and shutdowns in the machines monitored. Furthermore, the results achieved in the diagnostics and the methods of the neural network application as a possible additional tool of diagnostic methods are presented in this paper.

1 Introduction

Machines diagnostics has become an important branch of science to describe monitored machine condition. Industry uses rotary electrical machines mostly to drive machine tools, conveyors, etc. Rotary electric machines are widely used in power industry where they operate as generators to produce electricity. The diagnostics of such a machine may divide into the diagnostics of mechanical parts (such as bearings) and the diagnostics of electric parts that include electric machine insulation system. The diagnostics of the later part, in the form of OFF-LINE diagnostics, consists of electric diagnostic quantities measurements and the assessment of insulation system operability through criterion values, for example. This article describes the options of enhancing the insulation system diagnostics with the modern OFF-LINE diagnostics approach that utilises artificial intelligence, namely neural networks. The research carried out previously [1] verified the option of neural network utilisation in the insulation system diagnostics of electric machines with a sample piece of material. The material sample had been heat aged and then electrically broken down resulting in obtained breakdown voltage value that explicitly ends material functionality. Breakdown voltage was determined the value the neural network modelled and/or predicted. However, practical measurement of breakdown voltage makes troubles as it results in turning the entire machine inoperable. Therefore, another suitable quantity should be found that bears the information of machine operable condition.

2 Measured Quantities Description

As mentioned above, the quantities that will be used for the modelling and predicting of the quantity that would characterise insulation system condition were obtained through non-destructive measurement on electricity producing machines that were three-phase generators. Data shown in this article were measured on a generator that had been in operation for 23 years. Measurements were done during times down and inspections planned in 1, 2 or 3 years cycles. The neural network needs the input of the quantities that would contain sufficient number of measurements so that it is able to adequately model machine conditions. With respect to carrying out each measurement, there were 17 items of measured quantities in each set in average. For the above-mentioned reason, the following quantities were selected:

- Insulation resistance R_{iz15} of stator winding measured fifteen seconds after the measuring voltage had been brought,
- Insulation resistance R_{iz60} of stator winding measured one minute after the measuring voltage had been brought,
- Stator insulation resistance R_{iz600} in stator winding measured ten minutes after the measuring voltage had been brought,
- Charging current I_{60} measured one minute after the measuring voltage had been brought,
- Charging current I_{600} measured ten minutes after the measuring voltage had been brought,
- Leakage factor $tg\delta$.

3 Output Quantity Selection

The output quantity is the quantity the neural network should learn to remember and also to predict. The basis of that quantity selection included also correlation and regression analysis. The result of correlation analysis was the calculated correlation selective coefficient r_{XY} and Spearman correlation coefficient r_s. The correlation analysis results are presented in Tables 1 and 2.

Table 1. Correlation selective coefficients rXY

Measured quantity	I_{60}	I_{600}	R_{iz15}	R_{iz60}	R_{iz600}	$tg\delta$
I_{60}	1	**0,75**	0,17	0,05	**0,53**	**0,66**
I_{600}	0,75	1	-0,13	0,14	0,11	**0,78**
R_{iz15}	0,17	-0,13	1	-0,15	**0,60**	0,01
R_{iz60}	0,05	0,14	-0,15	1	0,04	-0,03
R_{iz600}	0,53	0,11	0,60	0,04	1	0,34
$tg\delta$	0,66	0,78	0,01	-0,03	0,34	1

Table 2. Spearman correlation coefficients rs

Measured quantity	I_{60}	I_{600}	R_{iz15}	R_{iz60}	R_{iz600}	tgδ
I_{60}	1	0,48	-0,07	0,001	**0,70**	**0,68**
I_{600}	0,48	1	-0,18	0,01	0,10	**0,54**
R_{iz15}	-0,07	-0,18	1	-0,34	0,15	-0,14
R_{iz60}	0,001	0,01	-0,34	1	0,03	0,08
R_{iz600}	0,70	0,10	0,15	0,03	1	**0,62**
tgδ	0,68	0,54	-0,14	0,08	0,62	1

Tables 1 and 2 highlight in bold face the correlation coefficient values, for which the calculated significance level was less than 0,05. The regression analysis results show that most of the quantities depend on time more or less. Out of the above-mentioned results the I_{60} or $tg\delta$ quantities seem to fit output quantity for the neural network. The experimental calculation done showed that the neural network more predicts $tg\delta$ quantity more precisely; therefore $tg\delta$ quantity was selected for the neural network to model and predict. In case of prediction, the neural network answer is based on current state values and the values measured previously. So that the predicted output value may be verified, the number of both input and output quantities was reduced by several values while the remaining values disregarded in the learning phase were used to verify the prediction. The below mentioned results include 3 values of output quantity that were predicted.

4 Software Used in the Experiment

The modelling and prediction experiment was carried out in MATLAB® 7 software environment and its Neural Network Toolbox that created the control environment for the operator to choose and combine the input quantities and modify neural network architecture by changing the number of hidden layers and the number of neurons in those layers. The control environment is shown in Figure 1.

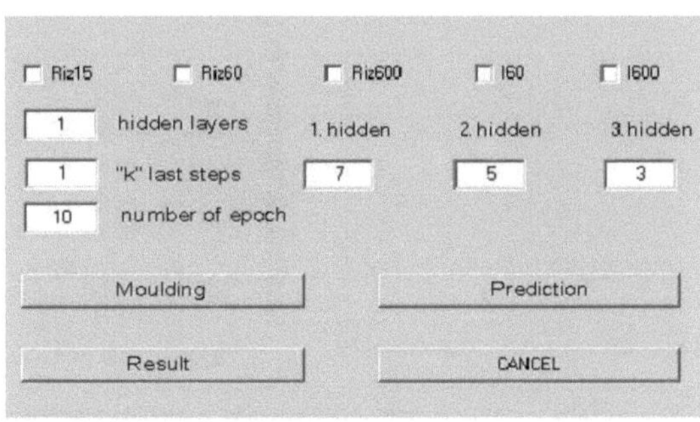

Fig. 1. Control environment

5 Experiment Results

The following selection of modelling and prediction results present measured values of *tgδ* shown in black dash line and neural network answer $tg\delta_m$ in continuous line. The next diagram shows neural network answer absolute error. Absolute error is absolute value of difference between measured quantity and value of neural network's answer. Each result is preceded by a description of chosen input quantities and neural network architecture. Modelling utilised all measured values of applied quantities, i.e. 17 elements in this case. The prediction result omits learning phase showing 3 predicted elements as described above.

5.1 Result 1 (modelling, shown in Fig. 2)

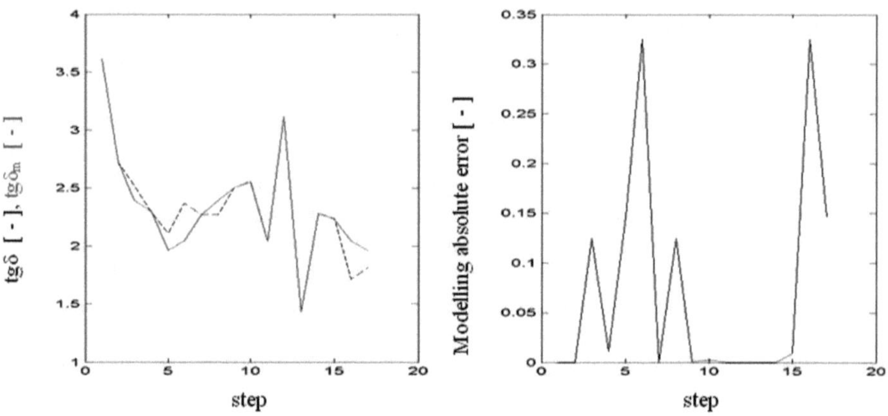

Fig. 2. Output quantity modelling diagram for chosen architecture and absolute error (Result 1)

- Input quantity: R_{IZ60}, output quantity: *tgδ*,
- Number of neurons in the first hidden layer: *7*
- Maximum absolute error of neural network answer: *0,325 [-]*

5.2 Result 2 (prediction, shown in Fig. 3)

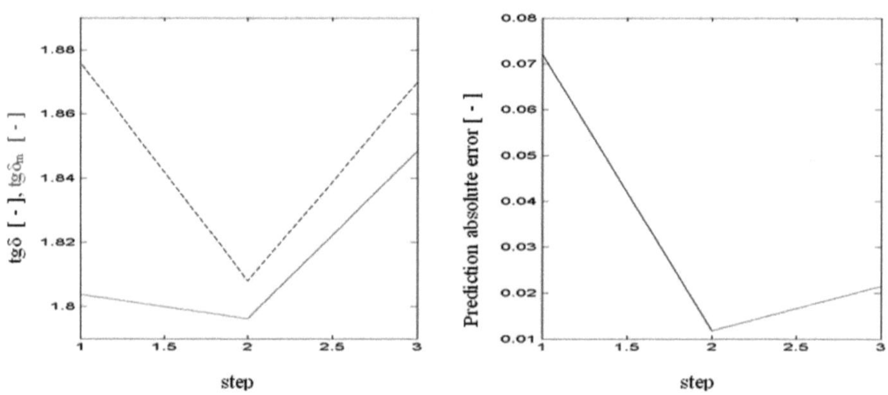

Fig. 3. Output quantity modelling diagram for chosen architecture and absolute error (Result 2)

- Input quantities: I_{60}, I_{600}, output quantity: $tg\delta$
- Number of neurons in the hidden layers: *7-5-2*
- Maximum absolute error of neural network answer: *0,072 [-]*

6 Conclusion

The modelling and prediction experiments performed so far have brought up several important observations. The first of them is that a neural network is able to adapt to the fed data making thus the basis for prediction result subsequently. This is done in modelling mode. If one quantity enters the neural network, the adaptability increases together with growing number of neurons in the hidden layers. This solution is time demanding in terms of calculation. More than one quantity brought to the neural network input considerably reduces the time demands of calculation as well as neural network answer error. The more significant learnt fact for real application is the ability of prediction of quantity value in future as shown in the results of performed predictions (Fig. 3). The observations demonstrated there are similar relations here as in the modelling mode in the terms of chosen architecture and the number of quantities brought to the neural network input. The prediction results let imply that the procedure is feasible for an enhancement tool of OFF-LINE diagnostics, for electric rotary machines in this case. The advantage of artificial intelligence application is in obtaining the information of the future value of the monitored quantity that characterises insulation system condition so that it is possible to avoid unplanned time down of the machine or discover an imminent breakdown or similar resulting thus in fund saving in the end.

Acknowledgments. This article was written within the research project of GAČR 102/08/1118 Intelligent diagnostics of electrical machines.

References

[1] Hammer, M.: Metody umělé inteligence v diagnostice elektrických strojů (Artificial Intelligence Methods in Electric Machine Diagnostics), 1st edn., p. 400. BEN – technická literatura, Praha (2009)

[2] Hammer, M.: Fuzzy modely a umělá inteligence v diagnostice izolačních materiálů (Fuzzy Models and Artificial Intelligence in Insulation Material Diagnostics), 1st edn., Vysoké učení technické v Brně, Fakulta strojního inženýrství, p. 120 (2003) ISBN 80-214-2331-5

[3] Hammer, M.: Diagnostics of Glass-Epoxy Insulating Materials for Electrical Machines, 1st edn., Vysoké učení technické v Brně, Fakulta strojního inženýrství, p. 60 (2003)

[4] Kozlovský, T.: Aplikace umělé inteligence v diagnostice izolačních systémů (Artificial Intelligence Applied to Insulation Systems Diagnostics), Ph.D. Thesis, VUT Brno (2004)

[5] Mařík, V., Štěpánková, O.: Umělá inteligence (4) (Artificial Intelligence 4). ACADEMICA Praha (2003)

Artificial Intelligence in Diagnostics of Electric Machines

M. Hammer, M. Šimková, and M. Ministr

Brno University of Technology, Faculty of Mechanical Engineering,
Institute of Production Machines, Systems and Robotics, Dept. of Electrical Engineering,
Technicka 2896/2, Brno, Czech Republic
hammer@fme.vutbr.cz, {ysimko00,yminis00}@stud.fme.vutbr.cz

Abstract. The article is focused on the applied artificial intelligence in the diagnostics of electric machines. Attention is paid to some developmental trends of artificial intelligence, for example, neural networks, fuzzy systems, genetic algorithms and expert systems. Having in mind the intended future usage in electric machine diagnostics the accent is placed on the analysis of topology and behaviour of an artificial intelligence method – fuzzy systems. The obtained results are evaluated in detail.

1 Introduction to the Issue

A crucial attribute of electric devices is reliability and lifetime. Our team has been working on that issue for several years, particularly in the relation of electric drives. An important element in every electric drive is an electrical machine. Much has been written of electrical machines in the terms of lifetime. However, our experience says that electrical machines' reliability and lifetime largely depend on their insulation systems. Therefore, we have devoted quite a time to electrical machines' insulation system properties. Especially important feature is insulation systems technical diagnostics focused on the selection of suitable diagnostic methods that would result in the evaluation of insulation system current condition or in the prognosis of future lifetime. Currently, the issue is called Electrical machine insulation system residual lifetime assessment. From the reliability point of view, the insulation system is a link of series chain whose condition affects the reliability of the entire electrical machines. Our team has gained considerable experience in diagnostic methods application to electrical machines' insulation systems. Because economic aspects are important to consider also in electrical machine diagnostics, our team has concentrated, particularly recently, on electrical machines in operation at energy companies. The reason is that an unexpected breakdown of every electrical machine also results in secondary losses that may grow to tens million of Euro.

Electrical machines diagnostics is based on various engineering and mathematical processes. The effect of modern approaches that definitely include artificial intelligence has also become to reflect in this field. And artificial intelligence application to electrical machines diagnostics is the topic of this article.

2 Artificial Intelligence Development Trends

Recently, artificial intelligence application has become spread in engineering diagnostics, too. It is positive that artificial intelligence has appeared in electrical machines diagnostics, in the relation to their reliability and lifetime. Implied in the references, such as [1], artificial intelligence has been elaborated in detail in current and prognostic electrical machine diagnostics. It is possible to see multiple development trends there. These particularly include neuron networks, fuzzy systems, genetic algorithms and expert systems. All of the above-mentioned is described in [1], for example.

3 Artificial Intelligence Application to Current and Prognostic Diagnostics

The application of artificial intelligence to electrical machines diagnostics issues from the block diagram shown in Fig. 1.

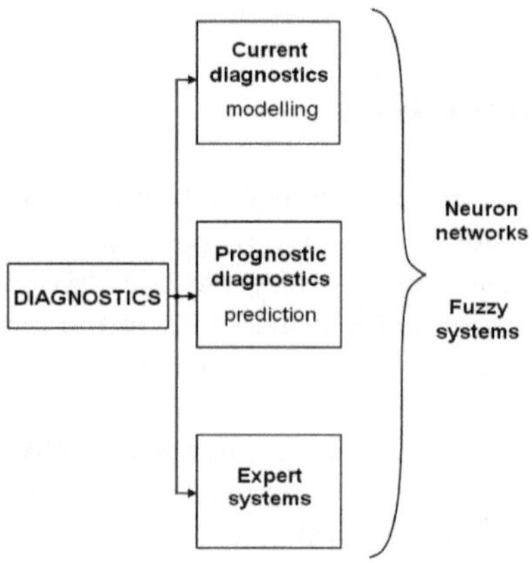

Fig. 1. A block diagram of Artificial intelligence application to electrical machines diagnostics

In this case, diagnostics is seen in three ways. The attention concentrates on current diagnostics (modelling) that is the diagnostics of current insulating condition of electrical machines. The block diagram also includes prognostic diagnostics (prediction) that allows formulating a prognosis of electrical machine insulation condition. Accent is also placed on expert systems. The artificial intelligence domain offers a whole number of powerful application tools. Based on previous experience our team always used the two most powerful representatives of artificial intelligence that were neuron networks and fuzzy systems. They may be considered universal approximators, then. It is the approximation ability that makes them suitable for considered issues solution.

4 An Example of Artificial Intelligence Application to Electrical Machine Diagnostics

This article makes use of knowledge and data in [1]. All of the artificial intelligence methods mentioned in the block diagram in Fig. 1 are described there. Therefore, this article below only analyses an example of the application of one of the artificial intelligence methods, which is Sugeno fuzzy predictor. Besides, it uses common diagnostic quantities (I_{60}, I_{600}, R_{iz15}, R_{iz60}, R_{iz600} and $tg\delta$) that are monitored for the off-line diagnostics of electrical machine in operation of energy industry. I_{60}, I_{600}, R_{iz15}, R_{iz60}, R_{iz600} quantities and their various combinations were selected for input, while $tg\delta$ quantity for output. The data gained from the mentioned diagnostic quantities measurements during the preventive inspections of the generator that is in operation at a Czech water power plant were used for the data file. Those data were interleaved with a linear function or a second order polynomial for the purpose of fuzzy predictor behaviour verification.

The block diagram of Sugeno fuzzy predictor (instrument for realization of fuzzy modelling) application to prognostic diagnostics of electrical rotary machines insulation condition is shown in Fig. 2.

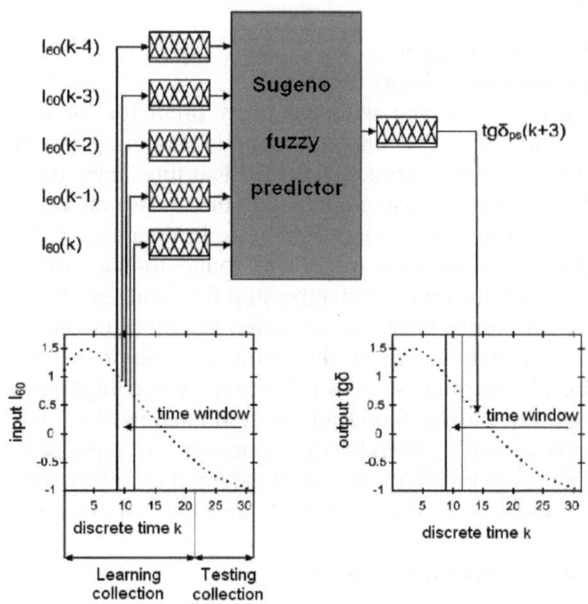

Fig. 2. Sugeno fuzzy predictor in prognostic diagnostics

The designed Sugeno fuzzy predictor contains as many input universes, as input quantities and their previous values are to be processed. The used input and output values of I_{60}, I_{600}, R_{iz15}, R_{iz60}, R_{iz600} and tgδ quantities were always normalised within <-1; +1> interval. The range of input and output universes was thus defined by that interval. For Sugeno fuzzy predictor verification, a various number of membership functions was chosen in each universe, in the forms of *trapmf* or *trimf*. Within the experimental tests of Sugeno fuzzy predictor in prognostic diagnostics, the condition of the maximum absolute and relative error at the model output was observed depending on the number of input universes (i.e. number of input quantities and their previous values), forms and number of membership functions and the number of learning steps).

5 Result Evaluation of Sugeno Fuzzy Predictor Application in Prognostic Diagnostics

Sugeno fuzzy predictor in electrical rotary machine insulation condition prognostic diagnostics is described with several important parameters that influence fuzzy predictor behaviour and its quality. Those parameters were thoroughly analysed on real data basis with the aim of gaining information for the suitable topology and setting of concerned fuzzy predicator. The following paragraphs include a list of the important parameters together with their effect on predicted value of tgδ leakage factor.

The best model:

Membership function: trapmf

1. Data interleaved with a linear function

The best results were achieved with the fuzzy predictor for one input quantity Riz60 (1 input universe). The universe contained 2 membership functions, 30 learning steps and 3 previous values. After 30 learning steps, the fuzzy model answered the input unknown measured data with the maximum absolute error 1.64E-02 and the maximum relative error 8.50E-01%. In the case of three input quantities (3 input universes), the best result was found for the combination of R_{iz15}, R_{iz60}, R_{iz600} input quantities with 2 membership functions per input universe and 2 previous values. The fuzzy predictor answered the input unknown measured data with the maximum absolute error 4.00E-02 and the maximum relative error 2.12E+00% after 10 learning steps. In the case of two input quantities (2 input universes), the best result was found for the combination of I_{60}, I_{600}; R_{iz600}, I_{600} and R_{iz15}, R_{iz600} quantities with 2 membership functions per input universe and 2 previous values. The fuzzy predictor answered the input unknown measured data with the maximum absolute error 4.00E-02 and the maximum relative error 2.06E+00%.

2. Data interleaved with a polynomial

The best result was achieved with the fuzzy predictor for one input quantity: R_{iz15} (1 input universe). The universe contained 2 membership functions, 5 learning steps and 2 previous values. After 5 learning steps, the fuzzy model answered the input unknown measured data with the maximum absolute error 4.11E-02 and the maximum relative error 2.12E+00%. In the case of two input quantities (2 input universes), the best result was found for the combination of R_{iz15}, I_{60} with 2 membership functions per input universe, 5 learning steps and 3 previous values. The fuzzy predictor answered the input unknown measured data with the maximum absolute error 4.65E-02 and the maximum relative error 2.40E+00%. In the case of three input quantities (3 input universes), the best result was found for the combination of R_{iz15}, R_{iz60}, R_{iz600} and 2 membership functions per input universe and 2 previous values. The fuzzy predictor answered the input unknown measured data, after 2 learning steps with the maximum absolute error 5.07E-02 and the maximum relative error 2.62E+00%.

Membership function: trimf

1. Data interleaved with a linear function

The best result was achieved with the fuzzy predictor with one input quantity R_{iz60} (1 input universe). The universe contained 3 membership functions, 30 learning steps and 3 previous values. After 30 learning steps, the fuzzy model answered the input unknown measured data with the maximum absolute error 1.69E-02 and the maximum relative error 8.74E-01%. In the case of two input quantities (2 input universes), the best result was found for the combination of R_{iz60}, I_{60} with 2 membership functions and 3 previous values. The fuzzy predictor answered the input unknown measured data after 5 learning steps with the maximum absolute error 1.93E-02 and the maximum relative error 1.02E+00%.

2. Data interleaved with a polynomial

The best result was achieved with the fuzzy predictor with two input quantities: R_{iz15}, R_{iz600} (2 input universes). The universe contained 2 membership functions, 5 learning steps and 3 previous values. After 5 learning steps, the fuzzy model answered the input unknown measured data with the maximum absolute error 4.64E-02 and the maximum relative error 2.39E+00%.

The above mentioned implies the following conclusions:

1. The presented results of trapmf membership function and linearly interleaved data imply the ideal results may be achieved for one quantity. One input quantity R_{iz60} with 2 membership functions per universe may be considered the optimum setting of the fuzzy predictor.
2. The presented results of trapmf membership function and data interleaved with a polynomial imply the ideal results may be achieved for one input quantity I_{60} or R_{iz15} with 2 membership functions applied.
3. The ideal results for the trimf membership function and data interleaved linearly may be achieved for one or two input quantities: R_{iz60}, I_{60}; R_{iz60}, I_{60}; I_{600}, I_{60} and 3 or 2 membership functions and a bigger number of previous values.
4. The presented results of trimf membership function and data interleaved with a polynomial imply the ideal results may be achieved for one or two input quantities. For a single input universe the input quantity R_{iz15} is suitable. For two input universes, the combination of input quantities R_{iz15}, R_{iz600} is suitable.

6 Conclusion

The obtained results of Sugeno fuzzy predictors usage under the specified conditions prove the assumed appropriateness of their application to electrical rotary machine insulation condition prognostic diagnostics. As the number of input universes and the number of membership functions grow, as predictor creation time demand increases. More conclusions as well as the results for other artificial intelligence methods are mentioned in [1], [2] and [3] for example. The obtained conclusion will be of service to arrangement of models for expert system. This system will be use in actual and prognostic diagnostics of state electrical machines.

Acknowledgments. This article was written within the research project of MPO ČR FI-IM5/173.

References

[1] Hammer, M.: Metody umělé inteligence v diagnostice elektrických strojů (Artificial Intelligence Methods in Electric Machine Diagnostics), p. 400 (2009) ISBN 978-80-7300-231-2
[2] Hammer, M.: Fuzzy modely a umělá inteligence v diagnostice izolačních materiálů (Fuzzy Models and Artificial Intelligence in Insulation Material Diagnostics), 1st edn., p. 120 (2003) ISBN 80-214-2331-5
[3] Kozlovský, T.: Aplikace umělé inteligence v diagnostice izolačních systémů (Artificial Intelligence Applied to Insulation Systems Diagnostics), Ph.D. Thesis, VUT Brno (2004)

Expert Systems in Transformer Diagnostics

M. Šimková, M. Ministr, and M. Hammer

Brno University of Technology, Faculty of Mechanical Engineering,
Institute of Production Machines, Systems and Robotics, Dept. of Electrical Engineering,
Technicka 2896/2, Brno, Czech Republic
{ysimko00,yminis00}@stud.fme.vutbr.cz, hammer@fme.vutbr.cz

Abstract. The contribution is focused on the diagnostics of transformers in power engineering practice. First, attention is paid to the brief description of the general expert systems. Furthermore, the classification of expert systems and the analysis of the diagnostic expert system without vagueness are presented. In this contribution, emphasis is laid on the description of the rule expert system which is applied in the diagnostics of transformers. The composition of this expert system and the specific example of the results obtained are analysed in detail.

1 Introduction to the Issue

Our team has been active in the research of artificial intelligence application to electric machines diagnostics. An application of the most successful ones of artificial intelligence to real life is in expert systems. Much has been written of expert systems, there are existing examples of operating expert systems [1] that were assembled and has been used by industry. This article tries to contribute to expert systems development and verification. It presents the results achieved by our team in this matter as an example. It resumes the research of 2005 of the expert system designed and constructed by Brno University of Technology, Faculty of Mechanical Engineering for the purpose of electric rotary machine – generators diagnostics [5], [1]. This entirely operating expert system was successfully tested and has been used by industry. Our team exploits the gained experience in developing another expert system that should in this case apply to transformer practical diagnostics in energy and mechanical engineering.

2 Expert Systems

The expert systems seen today often lack problem oriented knowledge application. Such expert systems are called empty expert systems. Only if such an empty expert system is completed with a knowledge data base the expert systems orients to the concerned issues becoming a problem-oriented system. The added data base is the key to solve a particular case, then. As far as solved problem nature is concerned, the currently existing expert systems may divide into multiple groups out of which we are interested in "diagnostic expert systems". Diagnostic expert systems are intended to efficiently interpret data so that it could be possible to determine which hypothesis of explored system behaviour best corresponds with actual data that refer to the case concerned. The problem solution procedure takes the form of partial hypotheses sequence evaluation within the fix determined model of the problem under exploration established by an expert. A diagnostic expert system consists of several basic parts, such as the knowledge base (principles), data base, inference (control) mechanism or interpreting module. A block diagram of a typical diagnostic expert system is shown in Fig.1.

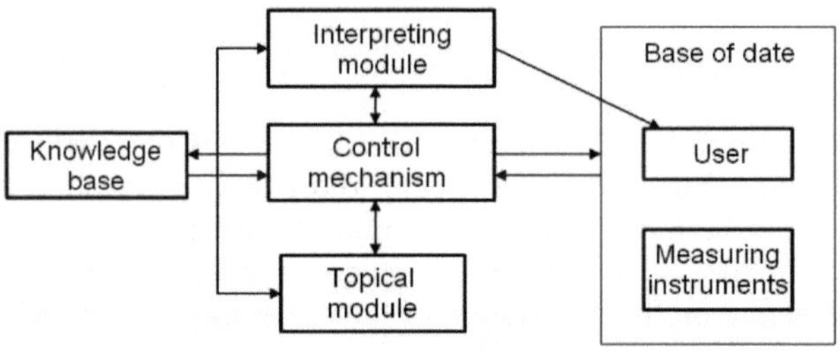

Fig. 1. A block diagram of a typical diagnostic expert system

Of the currently existing techniques of knowledge representing in expert system, the most widely used one is representing knowledge through principles that represent knowledge by determining the relations of the input and output information; thus, the latter becomes known once the former is known. The principles are defined in the following form:

IF assumption_1 AND(OR) assumption_2 THEN conclusion_1.

Assumption part | Conclusion part

That definition may be generally called an assumption → conclusion form. That form includes the principles in the knowledge base forming the basis for decision. Each of the principles is compared with input assumptions to make the conclusion on conformity basis but additional procedures and functions may be called and the products of them reflect in the final expertise. The inference process stops only that there is no more principles that might conform the assumption. The principle-based knowledge representation advantage is it is schematic and understandable. On the other hand of disadvantage, there is a risk of infinite chains, additional new inconsequent knowledge or non-transparency. In the case of large knowledge bases with a complex internal structure, an oriented diagram called interference network is suitable for clear demonstration of internal relations. The diagram assigns every sentence to a nod, every principle to an oriented edge leading from and to a sentence. The inference network generally represents a state space where the nods represent the statuses of environment and the oriented edges represent the transitions from one status to the other. The control (inference) mechanism is to solve the task of finding the optimum way in oriented diagram (state space), which means it determines the sequence of principles of the principle base being applied. The methods of searching in state space include particularly backward chaining. The backward chaining method issues from the determined goal analysing back the assumptions of its achieving, while it solves conflict situations in every step making multiple choices from sentences confirming the selected hypothesis. The backward chaining inference mechanism issues from the result part of the principles and checks for met assumption parts while the proofs are obtained from the user through expert system dialogs. Knowledge and information commonly used by people are of various complexity and/or abstraction. The best accessible and applicable data are saved explicitly. This way of new knowledge extrapolation is called deduction representing one of the most powerful methods of reasoning that allows deduction of new sentences from the essential axioms and derived sentences. Deduction is implemented in expert systems through logical propositions executed by logic programming. The principal tool for working in logic is currently the Prolog programming language or LISP, etc.

3 Diagnostic Expert Systems

This article tells about the development of the real above-mentioned and briefly theoretically described principle-based diagnostic expert system that is intended to analyse and describe actual transformer condition.

In the beginning phase of our diagnostic expert system building we had to choose suitable tools for its development and implementation. Based on previous experience we chose an empty ESTA expert system, whose core had been developed in Visual Prolog 6 programming environment. That environment houses the development of the actual knowledge base that turns the empty expert system into a problem-oriented expert system. It is also the environment of user and expert system communication. The environment appearance looks like a typical Windows operation system application environment. Every knowledge base contains principles related to a particular field. In case of ES the knowledge base contains the principles of each diagnostic test or the principles of diagnostic quantities criterion value evaluation. The main construction elements of the knowledge base are sections and parameters. The sections comprehend the actual principles that guide the user through the knowledge base to the point the user is provided with the consultation final result. The sources used in knowledge base construction include mainly the following: Czech national and international engineering standards, company internal standards of leading energy companies, technical literature and the results of consultations with experts. From the very principle of expert systems (it particularly applies to principle-based expert systems), it is obvious that deduced results' quality first of all depends on the quality of knowledge encoded in the knowledge base by the knowledge engineer.

The knowledge base classifies the transformers by operation voltage level: voltage level 400 and 220 kV, voltage level 110 kV, voltage level of high-voltage and power ≥ 1.6 MVA – distribution and voltage level of high-voltage and power ≥ 1.6 MVA – self-demand. A standard or a manual defines for each operation condition (new, in operation, after thorough inspection, spare instrument) a collection of specified diagnostic methods, an example of that is in Fig. 2. The user and expert system dialog is done in the manner mentioned further. It is first necessary to select the voltage level that is concerned in the consultation. Then, it is necessary to choose the operation condition of the machine and enter whether it is the condition, of a new machine for example, before commissioning or at the end of guarantee period, etc. Once the required data are entered the user selects, in a dialog box, the desired manner of transformer condition diagnostics. After the final expertise has been generated, the evaluated parameters are stored in a file, so that the consultation may be repeated any time on the basis (see further).

The very operation of the expert system under development of our team is easy in fact. After the input data are loaded into the expert system (from a form or entered by the user during consultation), the values are compared to criterion values following the principles. If a value complies with the criterion, the principle consequent stores to the knowledge base and the procedure continues. If the value fails to comply with the criterion, the principle is ignored and the next principle in sequence is used. The final recommendations generated by the expert system on the basis of input data are presented to the user in the form of graphic windows in the expert system environment and also saved to a text file.

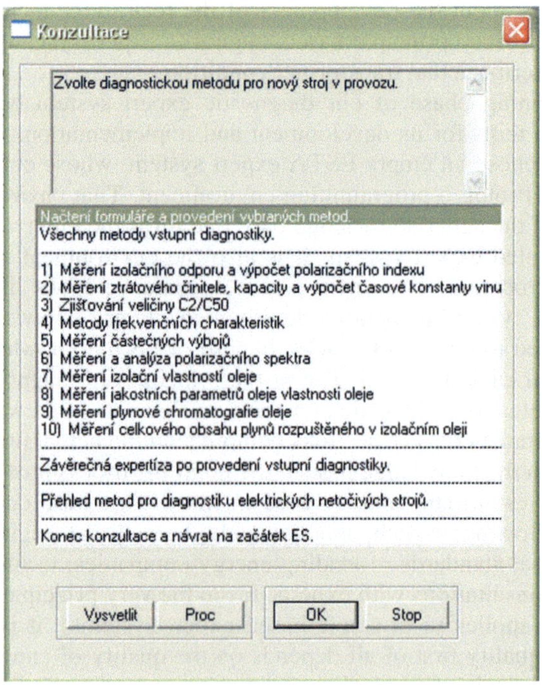

Fig. 2. The dialog box of a new transformer diagnostic method selection

The expert system can generate two types of final recommendations: 1. Final recommendations for each of the diagnostic methods. 2. Final recommendations for the entire set of prescribed diagnostic methods. They closely relate to one another, as the comprehensive recommendation for the entire set of diagnostic methods is generated from individual methods recommendations. A recommendation example is shown in Fig. 3.

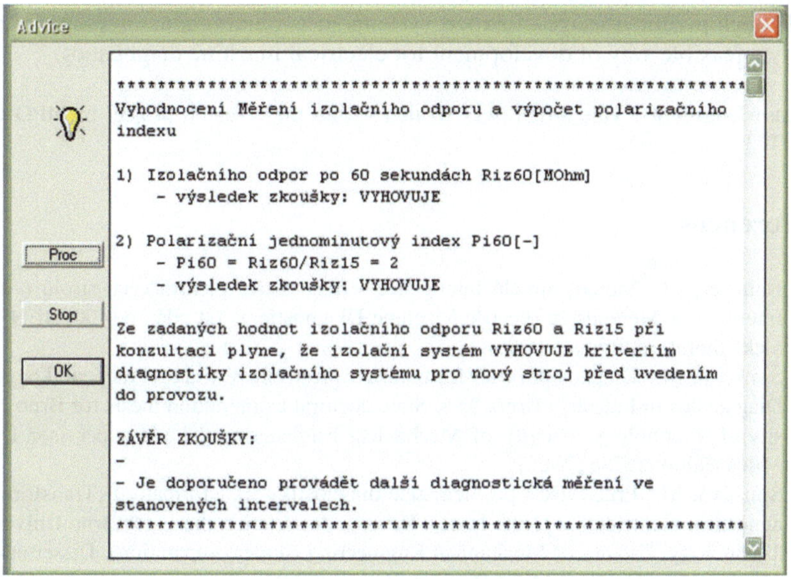

Fig. 3. The window with the final recommendation for the diagnostic method of transformer winding insulation resistance measuring

As it is obvious in Fig. 3, there are three buttons in the window with expert system conclusion:

"Proč" (Why) – is to get explanation of the manner the final recommendation was achieved,
"OK" – is to confirm the dialog and to continue in consultation,
"Stop" – interrupts consultation.

The actual content of final recommendation is organised in the following structure: evaluated diagnostic method name; diagnostic method parameters list and the results of comparison with criterion values, verbal evaluation of diagnostic method parameters compared to criterion values, final recommendation concerning the condition of insulation system or oil charge. Obviously, a requirement of the expert system is backward reproducibility and performed consultation backup. The expert system offers two ways of problem solving: 1. to save consultation process to a text file, 2. to save consultation process to a file.

4 Conclusion

The expert system described in this article has been developed for the diagnostic investigation of transformers in operation in energy industry. For the moment, it is a simple principle-based expert system that is going to be improved and enhanced with implemented additional modern elements on artificial intelligence basis, such as further lifetime prognosis and other.

Though practical usage of expert systems meets ambiguous opinions, it is definitely a feasible way of development for electrical machine diagnostics.

Acknowledgments. This article was written within the research project of MPO ČR FI-IM5/173.

References

[1] Hammer, M.: Metody umělé inteligence v diagnostice elektrických strojů (Artificial Intelligence Methods in Electric Machine Diagnostics), 1st edn., p. 400. BEN – technická literatura, Praha (2009)

[2] Šimková, M.: Diagnostika transformátorů v průmyslové praxi (Practical Transformer Diagnostics in Industry) Brno, 34 s. State doctoral examination thesis for Brno University of Technology, Faculty of Mechanical Engineering. Thesis supervisor Doc.Ing. Miloš Hammer,CSc (2008)

[3] Šimková, M.: Příspěvek k problematice diagnostiky transformátorů (Transformer Diagnostics: Contribution to the Issue). Brno, A dissertation thesis for Brno University of Technology, Faculty of Mechanical Engineering (under preparation). Dissertation thesis supervisor Doc.Ing. Miloš Hammer, CSc. (2009)

[4] Hammer, M.: Fuzzy modely a umělá inteligence v diagnostice izolačních materiálů (Fuzzy Models and Artificial Intelligence in Insulation Material Diagnostics), 1st edn., p. 120 (2003) ISBN 80-214-2331-5

[5] Szabó, R.: Expertní systémy v diagnostice izolačních systémů elektrických točivých strojů (Expert Systems in Electrical Rotary Machine Insulation System Diagnostics), Brno, p. 114. Dissertation thesis for Brno University of Technology, Faculty of Mechanical Engineering. Dissertation thesis supervisor Doc.Ing. Miloš Hammer, CSc (2005)

N-link Inverted Pendulum Modeling

A. Gmiterko[1] and M. Grossman[2]

[1] Technical University of Kosice, Faculty of Mechanical Engineering, Institute of Special Technical Science, Department of Applied Mechanics and Mechatronics, Letná 9, 04200 Košice, Slovakia
alexander.gmiterko@tuke.sk

[2] Ricardo Prague s.r.o. Palac Karlin, Thamova 11-13,186 00 Praha 8, Czech Republic
martin.grossman@ricardo.com, martin.grossman@gmail.com

Abstract. In this article is the automatized procedure for deriving of n-link inverted pendulum motion equations presented. Example of 2 link inverted pendulum is included. The LQR algorithm using Maple input equations is proposed. Also the comparison between SimMechanics and Simulink is included.

1 Introduction

The inverted pendulum has become a standard benchmarking problem. At the department of Applied Mechanics and Mechatronics, we have decided to build single, double and later triple inverted pendulum. In most of the publications [1], [2], [3], [4], we see that authors use already derived equations or derive them for each case of a single, double or triple pendulum system.

Nevertheless, there is a lack of general equation for the n-link pendulum. For this purpose, we decided to derive and automatize the process of generation of this general equation. It turned out that it was an invaluable help for us, in time of building the inverted pendulum model in Matlab and Simulink. This paper is divided into three parts. In the first part, the geometric model is shown. In the second part will be introduced the derivation of general equation using Maple and an example will be presented. In the last part, we discuss the further possibilities, using Maple and other symbolic mathematical software. By its mean, hopefully one gets better understanding of behavior these systems (non-minimal phase type system).

1.1 Pendulum's Geometry

The basic assumptions are:Pendulum movement is in E^2(plane). Cart and links are rigid homogeneous bodies (RBD rigid body dynamics). Joints with or without friction, but no compliance.Links are cylinders with variable lengths and masses.

In the Table.1 are listed all variables and constants used in the model. From the illustration at the (Fig.1), we see that the n-link inverted pendulum consists of n links connected with the revolute joints. The cart motion is allowable only in x direction by translational joint. We decided not to picture the forces, since their implementation will be very straightforward using Lagrange-Euler formula. Instead, the angles between the links (which are actually measured by sensors), the angles between link and its rotation angle (u_i) from vertical position were chosen. This is just for more clear derivation.

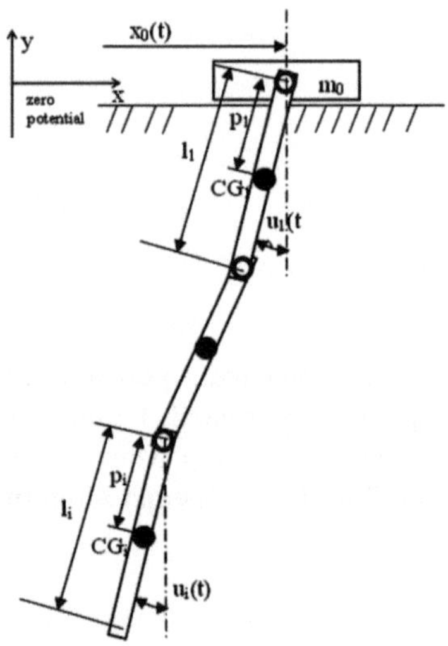

Fig. 1. N-link inverted pendulum, down position

1.2 Dynamic Behaviour and Equations

The standard method of deriving dynamical equations of multi rigid body systems is using Euler-Lagrange formula. Its usage is relatively very easy, and for our purpose, it is especially adequate. This method does involve only derivatives of time, speed and position and therefore is very suitable for manual and of course half-manual (by software usage) procedure. We have though remember, that we are losing information about inner forces in joints. Other type of dynamical formulation will later obtain these. The most important and only little more difficult part of the procedure is obtaining the kinetic and potential energy of the whole n-link pendulum. By inspection, we find that the kinetic energy and potential energy of the system consisting of n links and one cart is *Ek(n)* (1) and *U(n)* (2), respectively

Variable	description	dimension
$x_0(t)$	x position of the cart	m
$x_{1...n}(t)$	x position of the i-th link	m
$y_{1...n}(t)$	y position of the i-th link	m
$u_{1...n}(t)$	angle of i-th link with y axis	rad
$Icg_{1...n}$	moment of intertia of i-th link to its center of gravity	$kg.m^2$
n	number of links	-
m_0	mass of the cart	kg
$m_{1...n}$	mass of the i-th link	kg
$l_{1...n}$	length of the i-th link	m
$p_{1...n}$	distance of the gravity center of the i-th link (CG_i) from its beginning	m
CG_i	center of gravity of the i-th link	-

Fig. 2.

$$Ek(n)=m_0\left(\frac{d}{dt}x(t)\right)^2+\sum_{k=1}^{n}\left(\frac{1}{2}m_k\left(\left(\frac{d}{dt}x(t)-\sum_{i=1}^{k-1}l_i\cos(u_i(t))\frac{d}{dt}u_i(t)-p_k\cos(u_k(t))\frac{d}{dt}u_k(t)\right)^2+\left(\sum_{i=1}^{k-1}-l_i\sin(u_i(t))\frac{d}{dt}u_i(t)-p_k\sin(u_k(t))\frac{d}{dt}u_k(t)\right)^2\right)\right)+\sum_{k=1}^{n}\frac{1}{2}Icg_k\left(\frac{d}{dt}u_k(t)\right)^2 \tag{1}$$

$$U(n)=-g\sum_{k=1}^{n}\left(m_k\left(\sum_{i=1}^{k-1}l_i\cos\left(u_i(t)\right)+p_k\cos\left(u_k(t)\right)\right)\right) \tag{2}$$

After deriving, the equation (1) and (2) we can put these in suitable form into any software allowing symbolic mathematical computation like Mathematica, Maple,

Maxima or even Matlab with symbolic toolbox (Maple 3.0) etc (fig. 2). The next easy step is computing the Lagrangian L (3) and motion equations (4),

$$L = Ek(n) - U(n) \tag{3}$$

$$\frac{\partial}{\partial t}\left(\frac{\partial}{\partial \dot{\xi}} L\right) - \frac{\partial}{\partial \xi}(L) = F^{\xi} \tag{4}$$

where ξ is generalized coordinate (in our case $x(t)$ or $u(t)$) and $\dot{\xi}$ is its time derivative. F^{ξ} is generalized non-conservative force acting on generalized coordinate.

As an example, let us consider inverted pendulum consisting of 2 links. Then the L using (3) becomes $L(2)$ and the equations of motion are given by (4) with $L(2)$.

Therefore, with this procedure we can derive all n equations of motion.

Next step would be solving these equations. The equations are nonlinear differential equations and the Matlab was used to find numerical solution. Already the double inverted pendulum could be considered as the chaotic system, and thus it is important to bear in mind this behavior during the process of finding solution.

2 Symbolic Computation

In our experience Maple have many advantages but also some disadvantages. It has relatively very simple interface allowing even the beginners quick and impromptu problem solving.

During our work we have found out, that the version 11, have problems with handling more then 3 inner sums with other then latin-2 (not Greek) letters and subscripts. It is also considerably slower then the Mathematica 5.2 (tried the same problem solving). In addition, graphics quality of implicit equation is not satisfactory.

2.1 Lqr Controler Design, for Stabilizing the Double Inverted Pendulum in Upright Position

The motion equations of DIP were derived and linearized in upright position using Maple.

After deriving state matrices using Maple, the LQR K matrix coefficients were calculated within Matlab.

These coefficients are already able to stabilize DIP but with this approach we don't include other objectives and constraint as time, maximal travel in x direction or other. Therefore for the first and rough estimate of K constants we used optimization toolbox in Simulink to decrease travel of cart in x direction. This was achieved by constraining the maximal travel in x direction and optimizing all 6 constants in K matrix.

After this parametric optimization using gradient method, we were able to decrease the travel by 50% and keep the power consumption under acceptable level.

2.2 Simulink vs. SimMechanics, Control Loop Simulation

The control loop was implemented with 2 different approaches. First, was using Simulink. We translated nonlinear equations to Simulink by means of F(u) function blocks. Then the computed K matrix was implemented and optimization of K matrix parameters was performed.

The second approach, we modeled the inverted pendulum using SimMechanics, again used the generated K matrix and perform optimization. After successful stabilization in both cases, we implemented noise disturbance in joints, wind noise force air friction and cart friction and damping.

In (Fig.3) and (Fig.4), we can see decrease of cart travel by using optimization of K coefficients.

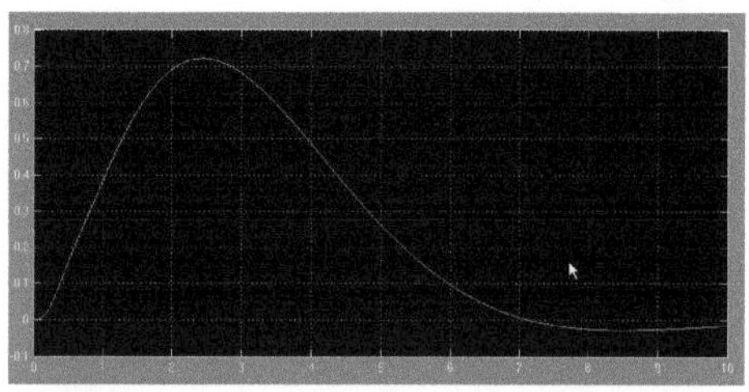

Fig. 3. Cart travel with standard K matrix

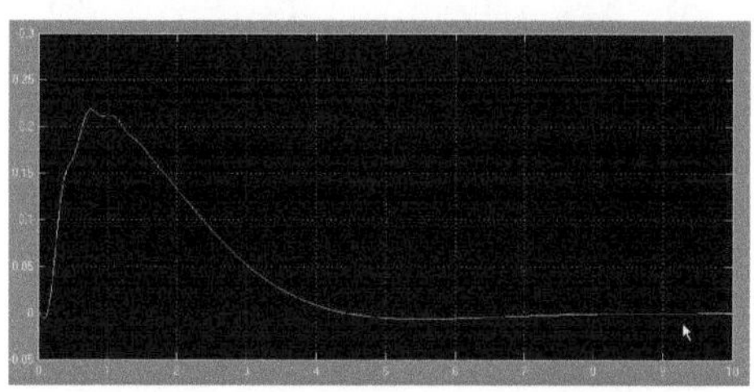

Fig. 4. Cart travel, K optimized

3 Conclusion

Symbolic computation software is already few years becoming standard tool, helping to solve engineering problems. The typical commercial products are Maple and Mathematica.

After implementation of LQR algorithm using Simulink and SimMechanics we can conclude that SimMechanics gave exactly same results as standard block implementation in Simulink. In fact the SimMechanics building procedure was very fast and straightforward. By using SimMechanics we gained option of Animation.

On the other hand, the simulations in SimMechanics were about 10 times slower. Beside that, the equations of DIP are still need for computing the LQR K matrix. After our experience, the SimMechanics is a good choice for very complicated models, were deriving of equations could take long time, in other cases we would suggest using the Simulink or Matlab itself.

Acknowledgments. The authors would like to thank to Slovak Grant Agency -project VEGA 1/0454/09 "Research on mechatronic systems imitating snake locomotion in confined and variable area" and VEGA 1/0201/08 "Research of structures and behavior of the mechatronics mobile engineering system on the organ level and construction item level in order to properties improvement of the mobile engineering system".

References

[1] Spong, M.W.: The Swing Up Control Problem for the Acrobot. IEEE Control Systems, 49–55 (February 1995)
[2] Stimac, A.K.: Standup and Stabilization of the Inverted Pendulum, MIT master thesis (1999)
[3] Zhong, W., Rock, H.: Energy and Passivity Based Control of the Double Inverted Pendulum on a Cart. In: Proceedings of the 2001 IEEE International Conference on Control Applications, México City, México, September 5-7 (2001)

Human Pilot Behaviour Model during of Flight Control

R. Jalovecký and P. Janů

University of Defence, Faculty of Military Technology,
Department of Aerospace Electrical Systems, Kounicova 65, 662 10 Brno, Czech Republic
{rudolf.jalovecky,Premysl.Janu}@unob.cz

Abstract. The paper represents analysis of human properties during aircraft flight control. Mathematic approach will be shown to create a replacement behaviour model of human - pilot in the creation of a replacement model from view of automatic control. There is not much information in literature how to access to analysis of human properties from the view of automatic control. Simulation of human "response" on input impulse represents an essential building block in the design of auto flight control circuits of an aircraft. Particular attention is devoted to simulations of human pilot - aircraft set during selected flight parameters control of an aircraft.

1 Introduction

Human behaviour represents very difficult procedure, whose description by suitable mathematic functions is more complicated in this case. During getting leastwise approximate replacement block scheme of regulation circuit of human behaviour is supposed his behaviour in occurrence case of exactly defined stimulations.

Orientation reflections about the possible structure of a pilot appear very sporadically in the literature [1]. With development of computer technology advanced simulation systems are available. Since that this consideration are beginning to be concretized [2], [3]. Their focus is however directed at the area of driving cars in automobile conveyance. Questions of human behaviour - the pilot and his definition from the perspective of the automatic regulation with usage of simulation tools dedicates article [4].

2 Simulation Human Characteristic Model

Authors of this paper have already published in [5] and [6] initial approaches to creation of models simulating human characteristic. Complete information about possible behaviour of human during different situations would present greatly extensive model, which would have been changing according to actual abilities of human. Human behaviour model is approached to a lot of simplifications during its creating.

Defined types of a human-pilots are used from mentioned papers [5] and [6] . For other simulated cases human-pilots type "A" and "C" are taken into account. There are linear models with transport delay characterized by transmission functions.
Pilot type "A"

$$F_{(p)} = \frac{Y_{(p)}}{X_{(p)}} = K \frac{(T_3 p + 1)}{(T_1 p + 1)(T_2 p + 1)} e^{-\tau p} \tag{2.1}$$

where:

K - Increasing of force on the steers in relation to their deflection (from 1 to 100)

T_1 - Reaction time constant, i.e. reaction ability to rate of change of input signal (5 to 20s) (prediction time constant)

T_2 - Dynamics properties of the pilot power members components (0.1 to 0.2s) (neuromuscular time constant)

T_3 - Integrating time constant, i.e. pilot's ability to realize varying activity (0.2 to 1s)

τ - Transmission delay (0.1 to 0.4) (time of pilot reaction).

Pilot type "C"

$$F_{(p)} = \frac{Y_{(p)}}{X_{(p)}} = K \frac{1}{(T_2 p + 1)} e^{-\tau p} \tag{2.2}$$

With mentioned "types" models of human pilot's dynamics, the simulation of responses to single input impulse according to Fig. 1 has been provided.

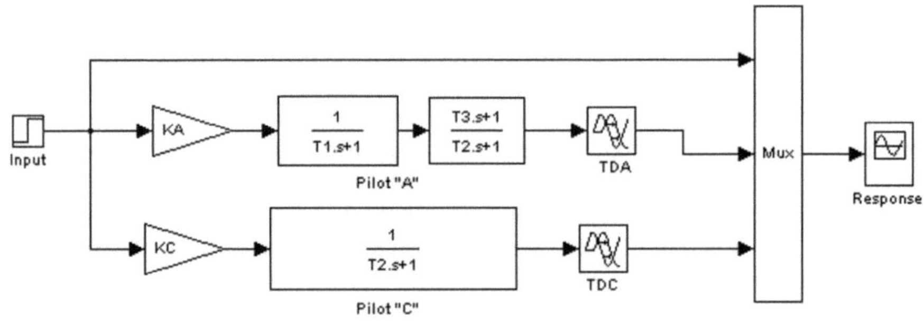

Fig. 1. Block scheme of man characteristics simulation

Values of time constants for instance of simulations are mentioned at Tab.1 [7].

Table 1. Ranges of time constants for definition of human pilot characteristic

T1 = 5 ÷ 20 s	KA = 2	TD_A = 0.15 s
T2 = 0.1 ÷ 0.2 s	KC = 1	TD_C = 0.40 s
T3 = 0.2 ÷ 1 s		

For this paper was chosen other way of imaging waveforms of simulated quantities. Imaging of waveforms at 2D graphs presents conventional standard with sufficient survey of required waveforms. Interesting and more clearly is imaging of the waveforms at 3D graphs, when alternation of chosen simulation parameter is carried out on the third axis. In MATLAB – SIMULINK programme command MESHGRID(time, time constant) is used for creating grid of 3D graph. Command MEŠH(time, time constant, response) is used for depiction of 3D graph. Data before depiction values of response must be saved to two dimensional matrix (time, time constant), with matches to third axis of 3D graph (response).

All sequentially presented graphs use this feature. Specified third parameter is alternation of time constant of one inertial member describing variation of human characteristic.

Example of typical response onto "unit" input pulse is on Fig. 2. Third simulation parameter (T1) here represents reaction pilot's time variation (prediction time constant). The smaller is this time constant, the faster is human in reaction on stimulus – change, indeed at the expense of extensive overrange of required deviation. In consequence it means regulation value overshoot.

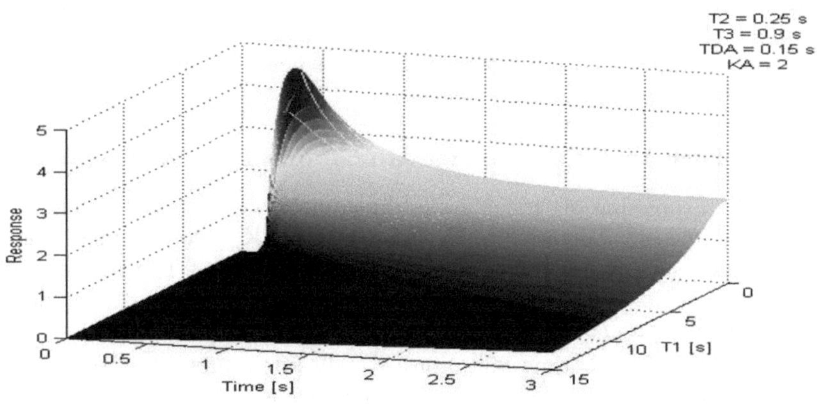

Fig. 2. Response of dynamic models of human - pilot "A"

3 Pilot Behaviour during Flight Control

Article [6] mainly solves behaviour of human-pilot integrated into loop for fast oscillations dumping on an aircraft. Article conclusions are possible to generalize on known fact, that human is not able to dump fast oscillation incurred during aircraft movement in the space effectively enough. Quite different situation indeed

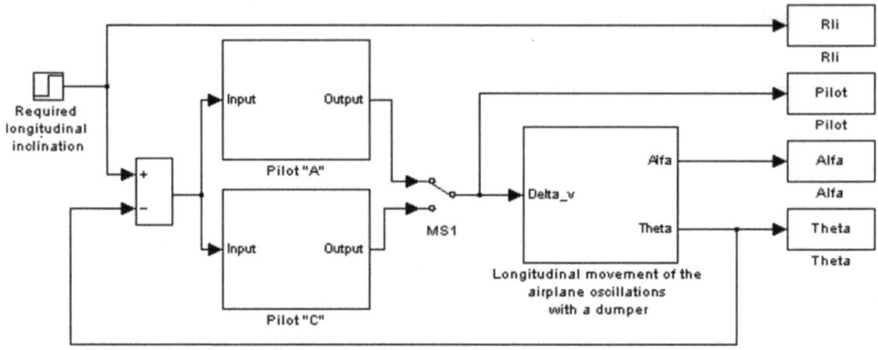

Fig. 3. Block scheme for simulation of body pitch angle control of aircraft by pilot

occurs, if aircraft is controlled at selected dimensions, at selected altitude, under selected course and certain speed.

Simulation model of aircraft movement at longitudinal axis with oscillation dumper is used for identification of regulation loop pilot-aircraft characteristics during body pitch angle control [7]. Aircraft is represented here by simplify model with relatively big rate of instability (coefficient nij [7], used at this model characterize combat well controllable aircraft) and used negative feedback ensures sufficient longitudinal movement stability. According to block scheme on the Fig. 3 required body pitch angle is controlled by pilot type "A" or "C" (s. symbol in [5]). Switch MS1 in the scheme enables switching pilot type which is needed.

Individual behaviour "responses" of pilot-aircraft set, let us say only it's body pitch angle with oscillations dumper then can indicate possibilities of human behaviour body pitch angle aircraft control.

3.1 Body Pitch Angle Control of the Aircraft with Type "A" Pilot Integrated

Mathematic expression of "transmission function" for pilot type "A" (Eq. 2-1) is transformed into block diagram MATLAB-SIMULINK system. Human characteristic - "Parameters of individual block" were mentioned at Tab. 1, the transport delay can be realized by block "Transport delay". The value of "Time delay" is put into this block as a coefficient TDA.

Simulation result for time variation of body pitch angle ϑ is introduced on Fig. 4. Variation of prediction time constant T1 value is included in the third axis of depicted waveform. At the time t=1s requirement on body pitch angle change on value 0.05 comes. Now pilot begins to control the aircraft. Owing to human characteristic-transmission delay, really begins to control body pitch angle with time delay. Waveform shows, that time constant T1 variation at this human type dramatically doesn't change possibility of fast control of aircraft body pitch angle. The lower is this time constant the faster is ability of human pilot to react on changes of stimulus. It is quite clear, that ideal zero value is not possible to range at human. On the contrary if this time constant will be high behaviour of human (from the view point of automatic regulation) as "low-pass" is described.

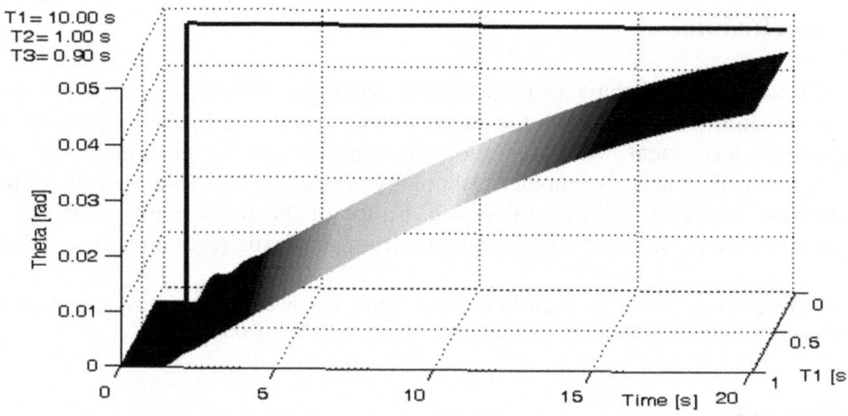

Fig. 4. Simulation results during body pitch angle control by human-pilot type "A"

3.2 Body Pitch Angle Control of the Aircraft with Type "C" Pilot Integrated

Mathematic expression of "transmission function" of the pilot type "C" (Eq. 2-2) is transformed into a block of automatic regulation with parameters mentioned at Tab.1. Transport delay is contained in coefficient TDC.

Simulation result for time variation of body pitch angle ϑ is introduced on Fig. 5. Input stimulus on body pitch angle change is the same as in previous case. Here too pilot begins to react on change after certain duration but with mentioned "regulation characteristic" proves expressively faster to reach nearly required value of body pitch angle. Indeed time constant T_1 variation has unpleasant influence thus, the lower (that is required by suitable training), additional "vibrating" of body pitch angle instantaneous value comes. But this oscillation is continuously dumped by suitable feedback of oscillation dumper.

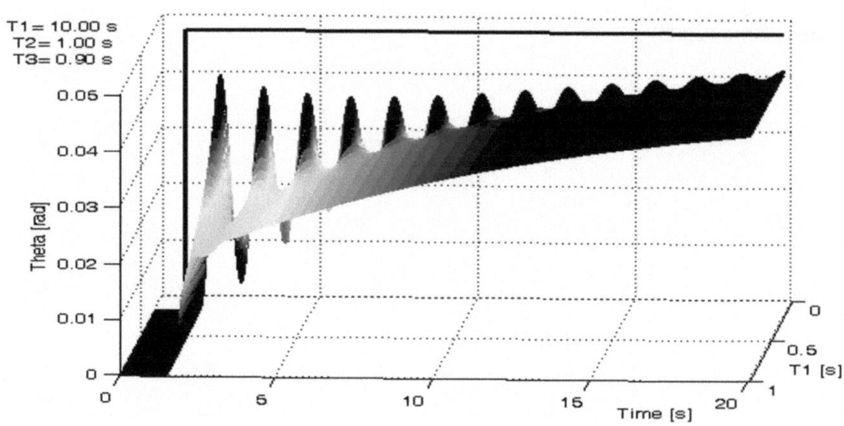

Fig. 5. Simulation results during body pitch angle control by human-pilot type "C"

4 Conclusions

Possible way of pilot behaviour modelling during his trial of body pitch angle control of the aircraft was shown in the paper. It was represented by simplify model of longitudinal movement with oscillation dumper. By orientation simulation was shown, that with usage of computer technology is possible to realise analysis of behaviour and supposed pilot characteristic. Range of the paper can't cover whole problem of human characteristics simulation from automatic regulation view point.

Acknowledgments. The work presented in this paper has been supported by the Ministry of Defence of the Czech Republic (Research Plan No. MO0FVT 0000403).

References

[1] Zihla, Z.: Automatic control of aerospace, Parts 1 (in Czech). Provisional textbook, U-1220/I, p. 395. Military academy, Brno (1987)

[2] Havlikova, M.: Models of driver's behaviour (in Czech). Elektrorevue, vol. 22 (2004), `http://www.elektrorevue.cz/clanky/04022/index.html` (cited 2009-1-20)

[3] Havlikova, M.: Human reliability and hazard of his failure in systems MMS (in Czech). Automatization 49(7-8), 461–462

[4] Jalovecky, R.: Man in the Aircraft Flight Control System (in Czech). In: Measurement, Diagnostics and Dependability of Aircraft Systems, pp. 67–74. University of Defence, Brno (2008)

[5] Jalovecky, R.: Man in the Aircraft Flight Control System, Advances in Military Technology, Brno, University of Defence, June 2009, vol. 4(1) (2009)

[6] Jalovecky, R., Janu, P.: The Features Model Of A Pilot During Of Dumping The Rapid Oscillations In An Aircraft. In: Jalovecky, R., Stefek, A. (eds.) Proceedings of the International Conference on Military Technologies 2009, pp. 333–344. University of Defence, Brno (2009)

[7] Jalovecky, R.: Onboard systems flight control I (in Czech), p. 82. University of Defence, Brno (2008)

Servocontroller for a Class of Nonlinear Continuous-Time System

J.E. Kurek

Institute of Automatic Control and Robotics, Warsaw University of Technology
Boboli 8, 02-525 Warszawa, Poland

Abstract. Servocontroller for a class of nonlinear time-varying continuous-time systems with unmeasurable bounded disturbances is presented. The controller can be also used for control of uncertain nonlinear systems.

1 Introduction

Nonlinear systems we meet everywhere. Usually they are controlled using linear models in the neighborhood of the operating point. In this paper we present an adaptive algorithm for asymptotic control for a class of nonlinear time-varying continuous-time systems with unmeasurable bounded disturbances based on nonlinear model of the system. In the design of controller we use the second Lyapunov stability theorem. The algorithm is based on control concepts presented in papers devoted to control of chaotic systems [1, 2, 3].

In section 2 we formulate the considered problem and in section 3 we present controller algorithm – solution to the problem. Finally concluding remarks are given.

2 Problem Formulation

Consider system described by the following nonlinear time-varying model

$$\dot{x}(t) = f(t,x) + \sum_{i=1}^{l} h_j(t,x) z_j(t) + u(t) \tag{1}$$

where $x \in R^n$ is a state vector, $u \in R^n$ is a control input, $z_j \in R^{n_j}$ are vectors of unknown disturbances, and $f(t,x)$ and $h_j(t,x)$ are maps $f: R^n \to R^n$, and $h_j:\ R^{n_j} \to R^n$. It is assumed that disturbances z_j are bounded

$$\left\| z_j(t) \right\| \le \zeta_j \tag{2}$$

where $\|\cdot\|$ denotes the Euclidean vector norm: $\|x\| = \sqrt{\sum_{i=1}^{n} x_i^2}$, $x \in R^n$.

The problem can be now stated as follows: given system (1) and (2), and the reference state $x_r \in R^n$, $x_r \in C^1$, find a controller such that

$$x(t) \to x_r(t) \quad \text{for } t \to \infty \tag{3}$$

3 Controller Design

Consider the following control rule

$$u(t) = \dot{x}_r(t) - f(t,x) - \alpha e(t) - \sum_{j=1}^{l} \frac{h_j(t,x) h_j^T(t,x) e(t)}{\left\| h_j^T(t,x) e(t) \right\|} \zeta_{uj}(t)$$
$$\zeta_{uj}(t) = \left\| h_j^T(t,x) e(t) \right\|, \quad j = 1,\ldots,l \tag{4}$$

where e is control error, $e(t) = x(t) - x_r(t)$, $\zeta_{uj}(t) \in R$, $j=1,\ldots,l$, are adaptive gains and α is a tuning parameter, $\alpha > 0$.

Then, the following theorem can be proven.

Theorem 1. The control law (4) guarantee for system (1) satisfaction of requirement (3).

Proof. From (1) and (4) we obtain

$$\begin{aligned}
\dot{e}(t) = \dot{x}(t) - \dot{x}_r(t) &= f(t,x) + \sum_{i=1}^{l} h_j(t,x) z_j(t) + u(t) - \dot{x}_r(t) \\
&= f(t,x) + \sum_{i=1}^{l} h_j(t,x) z_j(t) \\
&\quad + \dot{x}_r(t) - f(t,x) - \alpha e(t) - \sum_{j=1}^{l} \frac{h_j(t,x) h_j^T(t,x) e(t)}{\left\| h_j^T(t,x) e(t) \right\|} \zeta_{uj}(t) - \dot{x}_r(t) \\
&= \sum_{i=1}^{l} h_j(t,x) z_j(t) - \alpha e(t) - \sum_{j=1}^{l} \frac{h_j(t,x) h_j^T(t,x) e(t)}{\left\| h_j^T(t,x) e(t) \right\|} \zeta_{uj}(t)
\end{aligned} \tag{5}$$

Then, consider the following Lyapunow type function

$$V(t,x) = \frac{1}{2} \left(e^T(t) e(t) + \sum_{j=1}^{l} e_{\zeta j}^2(t) \right) \tag{6}$$

where $e_{\zeta j}(t) = \zeta_{uj}(t) - \zeta_j$.

Time derivative of the function V along trajectory of the closed-loop system (1) and (4) can be calculated as follows

$$
\begin{aligned}
\dot{V}(t,x) &= e^T(t)\dot{e}(t) + \sum_{j=1}^{l} e_{\zeta j}(t)\dot{e}_{\zeta j}(t) \\
&= e^T(t)\sum_{i=1}^{l} h_j(t,x)z_j(t) - \alpha e^T(t)e(t) - e^T(t)\sum_{j=1}^{l}\frac{h_j(t,x)h_j^T(t,x)e(t)}{\left\|h_j^T(t,x)e(t)\right\|}\zeta_{uj}(t) \\
&+ \sum_{j=1}^{l}[\zeta_{uj}(t) - \zeta_j]\left\|h_j^T(t,x)e(t)\right\| \\
&= e^T(t)\sum_{i=1}^{l} h_j(t,x)z_j(t) - \alpha e^T(t)e(t) - \sum_{j=1}^{l}\frac{\left\|h_j^T(t,x)e(t)\right\|^2}{\left\|h_j^T(t,x)e(t)\right\|}\zeta_{uj}(t) \\
&+ \sum_{j=1}^{l}\zeta_{uj}(t)\left\|h_j^T(t,x)e(t)\right\| - \sum_{j=1}^{l}\zeta_j\left\|h_j^T(t,x)e(t)\right\| \\
&= e^T(t)\sum_{i=1}^{l} h_j(t,x)z_j(t) - \alpha e^T(t)e(t) - \sum_{j=1}^{l}\zeta_j\left\|h_j^T(t,x)e(t)\right\|
\end{aligned}
\tag{7}
$$

Using (2) one simply finds

$$
e^T(t)h_j(t,x)z_j(t) \le \left\|e^T(t)h_j(t,x)\right\|\zeta_j \tag{8}
$$

Thus, we obtain from (7) and (8)

$$
\dot{V}(t,x) \le \sum_{j=1}^{l}\left\|h_j^T(t,x)e(t)\right\|\zeta_j - \alpha e^T(t)e(t) - \sum_{j=1}^{l}\zeta_j\left\|h_j^T(t,x)e(t)\right\| = -\alpha e^T(t)e(t)
$$

Hence

$$
\dot{V}(t,x) < 0 \quad \text{for } e(t) \neq 0
$$

Henceforth, along trajectory of the closed-loop system (1)-(4) the function V tends to zero. This, however, implies that

$$
e(t) \to 0 \quad \text{and} \quad e_{\zeta}(t) \to 0 \quad \text{for } t \to \infty
$$

i.e.

$$
x(t) \to x_r(t) \quad \text{and} \quad \zeta_{uj}(t) \to \zeta_j \quad \text{for } t \to \infty
$$

for the closed-loop system.

Remarks

1. Adjusting the tuning parameter α one can change velocity of the control error convergence to zero.
2. Assuming $h_1(t,x)$ is a map $R^n \to R^n$ and $z_1 \in R^n$ one can control uncertain time-varying system

$$\dot{x}(t) = f(t,x) + \Delta_f(t,x) + \sum_{i=2}^{l} h_j(t,x)z_j(t) + u(t)$$

where $\Delta_f(t,x) = h_1(t,x)z_1(t)$ and $\|z_1(t)\| \le \zeta_1$.

3. For reference state generated by dynamic system, e.g.

$$\dot{x}_r(t) = f_r(t,x_r) + h_r(t)u_r(t)$$

where $u_r \in R^n$ is bounded, piecewise continuous reference input, $f_r(t,x_r)$ and $h_r(t)$ are maps $f_r : R^n \to R^n$ and $h_r : R^n \to R^n$, with known initial condition $x_r(0) = x_{r0}$ one can use instead of (4) the following control input

$$u(t) = f_r(t,x_r) + h_r(t)u_r(t) - f(t,x) - \alpha e(t) - \sum_{j=1}^{l} \frac{h_j(t,x)h_j^T(t,x)e(t)}{\left\|h_j^T(t,x)e(t)\right\|} \zeta_{uj}(t)$$

4 Concluding Remarks

Algorithm for design of servocontroller for a class of nonlinear time-varying continuous-time systems with unmeasurable bounded disturbances has been presented. The algorithm is rather simple. A disadvantage of the proposed controller is that the considered class of control systems is rather limited. An additional problem is also how to estimate unknown state vector x of the control system in the case that only some of them can be measured. However, the presented approach can be considered in the future research for the design of nonlinear control systems.

References

[1] Estrada, J.L., Duarte-Mermoud, M.A., Travieso-Torres, J.C., Beltrán, N.H.: Simplified robust adaptive control of a class of time-varying chaotic system. Compel. 27(2), 511–519 (2008)

[2] Li, Z., Chen, G., Shi, S., Han, C.: Robust adaptive tracking control for a class of uncertain chaotic systems. Physics Letters A, 40–43 (2003)

[3] Estrada, J., Duarte-Mermoud, M.A.: Simplification of a control methodology for a class of uncertain chaotic systems. WSEAS Transactions on Electronics, 347–352 (2004)

Mechatronic Stiffness of MIMO Compliant Branched Structures by Active Control from Auxiliary Structure

M. Nečas and M. Valášek

Czech Technical University in Prague, Faculty of Mechanical Engineering,
Department of Mechanics, Biomechanics and Mechatronics,
Karlovo nám. 13, Praha 2, 121 35, Czech Republic
{martin.necas,michael.valasek}@fs.cvut.cz

Abstract. This paper deals with a concept of mechatronic stiffness applied to branched mechanical structures. It is based on the introduction of auxiliary structure that functions as a support for embedded actuators. Feedback control enables significant increase of a dynamic stiffness of the original mechanical structure around its eigenfrequencies. The system is controlled by state space and state derivative controllers synthesized by LQR approach. Presented concept aims at machine tools where mechanical compliance is a major obstacle.

1 Introduction

Dynamic stiffness of mechanical structures largely predetermines their behavior when subjected to external time varying loads. The limiting factor here is typically the first eigenfrequency and therefore methods for its shifting or complete attenuation are of a great importance.

The first (classical) approach tries to increase the rigidity of the machine tool parts through the application of extensive optimalizations and structural modifications of their existing structures such as introduction of composite materials [1].

The second approach is based on the concept of mechatronic active stiffness, where active elements are being embedded within structures in order to actively affect vibration damping or deformation [2],[3]. The structural stiffness in this case is influenced indirectly. Gao and Cheng [3] analyzed a similar concept, however, their motivation was focused on vibration isolation where the force actuation and the measurement point are not collocated. This paper analyzes a system where an auxiliary support structure is brought into a required location and by control feedback effectively increases the stiffness of the structure at the required location.

2 Concept of Mechatronic Stiffness

Fig. 1 illustrates the basic idea behind the patented concept [4]. Next to the structure 2 (performing a technological operation) another parallel auxiliary structure 3 is built. Both structures are connected by a force actuator 4. This auxiliary structure is also compliant but functions mainly as a source of a support for force actuator 4 just like in the case of famous Archimedes quote: "Give me a place to stand, and I will move the Earth." with the only difference that this "place" does not exhibit an infinite stiffness.

The existing control mechanisms typically act in relative coordinates or are eventually brought in the structure from the outside. In this case in order to achieve collocated control the position of both attaching points of the force actuator must be measured by sensors 6 with respect to the fixed reference frame 1.

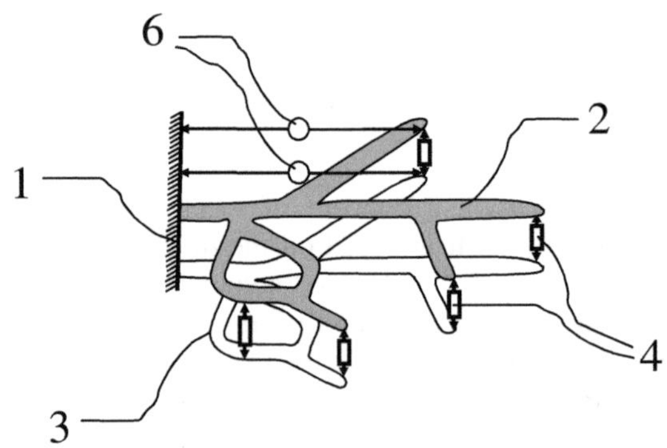

Fig. 1. Mechatronics stiffness from support structure

3 Proposed Structure and Its Controllability

Structures previously studied by authors [2], [5] were limited to simple two-mass models or 2D planar structures. In all cases it was found that under specific conditions the system completely looses its controllability [2]. The most trivial case is when the primary and secondary structures have the identical modal characteristics. It is therefore necessary to make sure that both structures exhibit "different" modal behavior. Conditions of controllability for systems with more degrees of freedoms parameterized by structure parameters are complicated. It must be stressed that some intuitively sound structural setups may perform poorly or cannot be used at all.

In this paper we propose a 3D spatial structure as illustrated in Fig.2. It consists of two branched beam structures 1 and 2 mutually interconnected by three

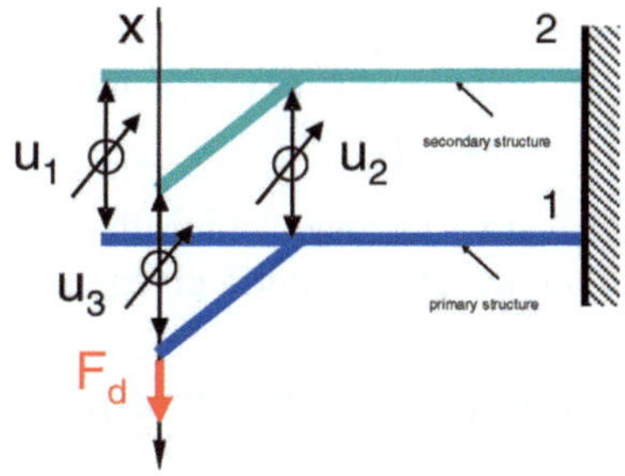

Fig. 2. Simulated branched mechanical structure

piezoactuators u_i. Structure 2 is designed to be stiffer then structure 1 to prevent to loss of system's controllability.

4 System Modeling and Control Strategies

4.1 System Modeling

The structure was discretized by FEM using 3D beam elements yielding standard dynamic equations in the linear form

$$M\ddot{x} + C\dot{x} + Kx = Fu + EF_d, \tag{1}$$

where F_d represents disturbance force acting as shown in Fig.2. Transformation of the system into space description yields standard state space description.

$$\dot{x} = Ax + Bu + EF_d. \tag{2}$$

Individual piezoactuators were modeled as springs with controllable free length according to [6]

$$u_i = k_s(kU_i - \Delta\tilde{x}), \tag{3}$$

where $\Delta\tilde{x}$ represents relative nodal displacement, k_s represents stiffness of the piezoactuator stack, U_i is the piezostack control voltage and k is proportionality constant k. Realistic values of parameters in the model were used.

4.2 Control Strategies

Two types of controllers were derived using LQR optimization strategy: State space ($u = -Kx$) and state derivative feedback ($u = -K\dot{x}$) controller. State space controller was derived using standard form of performance criteria

$$J = \int_0^\infty (x^T Q x + u^T R u)\,dt\,. \tag{4}$$

Non-standard state derivative feedback controller was derived using methodology developed in [7] with optimization criteria given as

$$J = \int_0^\infty (x^T A^T Q A x + u^T (R + B^T Q B) u + 2 x^T A^T Q B u)\,dt\,. \tag{5}$$

Both types of controller were set up to give similar magnitudes of control action within allowable limits of considered piezoactuators with the main objective to increase the $F_d(\omega)/x(\omega)$ dynamic stiffness around the system's eigenfrequencies.

5 Simulation Results

Simulation results in Fig. 3 verify the proposed concept of mechatronic stiffness. It can be observed that both controllers achieve significant attenuation of resonant frequencies of passive system and create a system with homogenized dynamic stiffness across a wide range of frequencies.

State derivative feedback enables global increase of dynamic stiffness while suppressing structure eigenfrequencies. State derivative feedback only suppresses the eigenfrequencies but offers an additional benefit of using possibly only acceleration sensors with velocity obtained by integration. Fig. 4 shows a relative comparison between controlled and passive system.

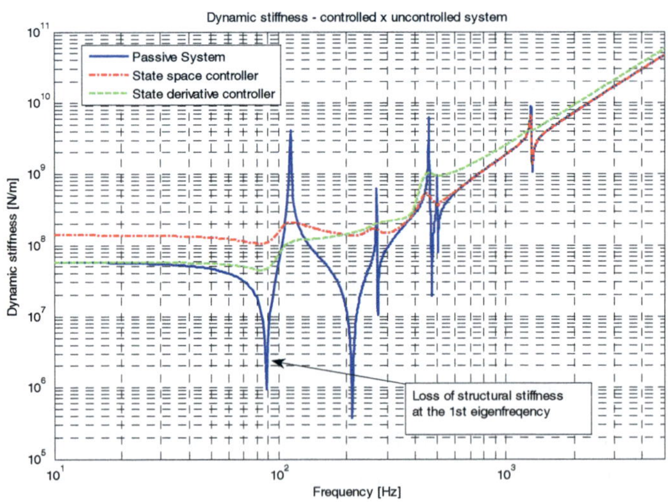

Fig. 3. Dynamic stiffness of passive and controlled system

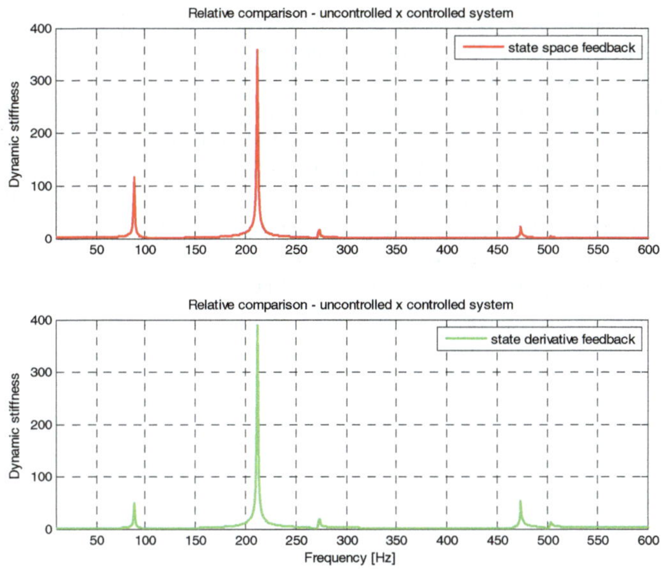

Fig. 4. Relative comparison of controlled and passive system

6 Conclusions

Applicability of the proposed concept of mechatronic stiffness from auxiliary structure was successfully extended to a branched spatial structure. State space and state derivative controllers both achieved significant increase of dynamic stiffness of uncontrolled structure around the eigenfrequencies of passive system and virtually eliminated their effect. The achieved homogenization of dynamic stiffness across a wide span of frequencies has a large practical potential, especially in the area of primarily targeted machine tools. It should be noted, that the potential use of state derivative feedback enables a construction of practical controllers free of position measurement, possibly using only accelerometers as primary detectors.

Acknowledgments. This work was supported by the GA ČR under Grant 101/08/P633, authors highly acknowledge provided financial support.

References

[1] Lee, A.G., Suh, J.G.: Design and Manufacture of Composite High Speed Machine Tool Structures. Composites Science and Technology (64), 1523–1530 (2004)
[2] Nečas, M.: Increase of dynamic stiffness of constructions by mechatronic approach, Dissertation Thesis, Czech Technical University in Prague, Prague (2008) (in Czech)

[3] Gao, J.X., Cheng, L.: Modeling and control of power flow In a double-beam system. Thin-Walled Structures, vol. 43. Elsevier, Amsterdam (2005)
[4] Valášek, M.: Method and Device for Change of Stiffness of Mechanical Constructions, PV2006-123, Patent application (2006) (in Czech)
[5] Nečas, M., Valášek, M.: Increasing dynamic stiffness of MIMO compliant structures by active control from auxiliary structure. In: SMART 2009 Thematic Conference on Smart Structures and Materials, Porto, Portugal (2009)
[6] Šika, Z.: Active and Semi-active Vibration Suppression of Machines, Habilitation Thesis, CzechTechnical University in Prague, Prague (2004) (in Czech)
[7] Abdelaziz, T.H.S., Valášek, M.: State derivative feedback by LQR for linear time-invariant systems. In: Proc. of 16th IFAC World Congress (2005)

An Active Control of the Two Dimensional Mechanical Systems in Resonance

P. Šolek and M. Horínek

Slovak University of Technology, Faculty of Mechanical Engineering,
Institute of Applied Mechanics and Mechatronics, Námestie slobody 17,
Bratislava, Slovak Republic
{peter.solek,martin.horinek}@stuba.sk

Abstract. This article deals with the investigation of optimal actuator and sensor placement. The approach proposed in this article is based on the evaluation of the H_2 and H_∞ norms. The optimal actuator and sensor placement satisfied the conditions of controllability, observability and spillover prevention. The flexible structure is defined as a finite dimensional, controllable and observable linear system. The results of the optimal placement are valuate for two dimensional system.

1 Introduction

For the target of improving the performance control of flexible structures, it is useful to investigate various sensor and actuator locations. The first purpose of the investigation is to determine the minimal number of actuators and sensors and to meet requirements of controllability, observability and spillover prevention. Second purpose is that the minimal subset of actuators and sensors has the same controllability and observability properties as the original set.

The importance of actuator and sensor placement is supported in many investigations and contributions. The articles [1]-[2] used the norms H_2 , H_∞ and Hankel singular values for the actuator and sensor placement. The contributions [3]-[4] use observability and controllability grammians for the actuator and sensor placement. Next big group of articles use the various formulations of optimization problems [5]-[6] for the solution of the actuator and sensor placement the flexible structures.

2 Norms

System norms serve as a measure of intensity of its response to standard excitations, such as unit impulse or white noise of unit standard deviations. The standardized response allows comparing different systems.

When (A, B, C) is a system state-space representation of a linear system and let $G(\omega) = C(j\omega I-A)^{-1}B$ is the transfer function. The H_2 norm is formulated in form

$$\|G\|_2^2 = \frac{1}{2\pi}\int_{-\infty}^{\infty} tr(G^*(\omega)G(\omega)d\omega \quad \text{or} \quad \|G\|_2 = \sqrt{tr(C^T C Wc)} = \sqrt{tr(BB^T Wo)} \tag{2.1}$$

where Wc and Wo are the controllability and observability grammians.

The H∞ norm is formulated as

$$\|G\|_\infty = \sup \frac{\|y(t)\|_2}{\|u(t)\|_2} \tag{2.2}$$

where y(t) is the system output and u(t) is the system input.

3 Actuator and Sensor Placement

Actuator and sensor placement are solved independently and both procedures are similar. Indicate by G the transfer function with all S candidate actuators. The index of placement σ_{2ki} that evaluates the k-th actuator at the i-th mode in terms of the H_2 norm is defined with respect to all the modes and all admissible actuators

$$\sigma_{2ki} = w_{ki}\frac{\|G_{ki}\|_2}{\|G\|_2} \qquad k = 1,\ldots,S \qquad i = 1, \ldots,n \tag{3.1}$$

where $w_{ki} \geq 0$ is the weight assigned to the k-th actuator and the i-th mode
n is the number of modes
G_{ki} is the transfer function of the i-th mode and k-th actuator

The weight reflects the importance the mode and the actuator in applications, and reflects the dimensions of the input. The weight can be determined from a measurement or from a computation.

The index of placement $\sigma_{\infty ki}$ evaluates the k-th sensor at the i-th mode in terms of the H_∞ norm. This index is defined for all modes and all admissible actuators.

$$\sigma_{\infty ki} = w_{ki}\frac{\|G_{ki}\|_\infty}{\|G\|_\infty} \qquad k = 1,\ldots,S \qquad i = 1,\ldots,n \tag{3.2}$$

The matrix of placement gives an insight into the placement properties of each actuator (sensor) because the index of placement of the k-th actuator (sensor) is defined as the rms sum of the k-th column. In case of the H_2 norm, it is the rms sum of the k-th actuator indexes over all modes

$$\sigma_{sk} = \sqrt{\sum_{i=1}^{n}\sigma_{ik}^2} \qquad k = 1 ,\ldots,S \tag{3.3}$$

And in the case of H_∞ and Hankel norms it is

$$\sigma_{sk} = \max_{i}(\sigma_{ik}) \qquad i = 1,...,n \qquad k = 1, ...,S \tag{3.4}$$

4 Example

Using H_2 and H_∞ norms for determining optimal sensor placement is presented in the following example dealing with a plane plate clamped at its both ends (fig. 1). Calculation of natural frequencies and modes of the plate was done using finite element methods in program Ansys. Analyzed model of plate has six degrees of freedom in each node: displacements in directions x, y, z and rotations around these directions. The length of the plate is 50 centimetres, width is 40 centimetres and its thickness is 2,5 milimetres.

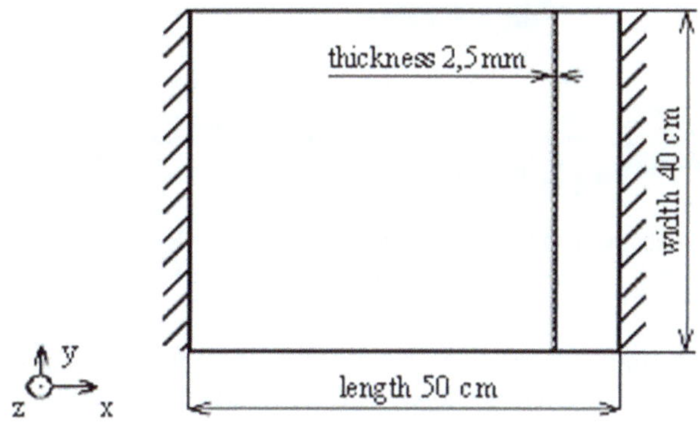

Fig. 1. Scheme of the plate clamped at two edges

Using the above presented H_∞ norm placement technique finds the best place for actuators functioning in z direction to control the first, second, third, and fourth own mode and to control simultaneously the first two modes then first three and four modes. Own modes of the plate is shown in figure 2.

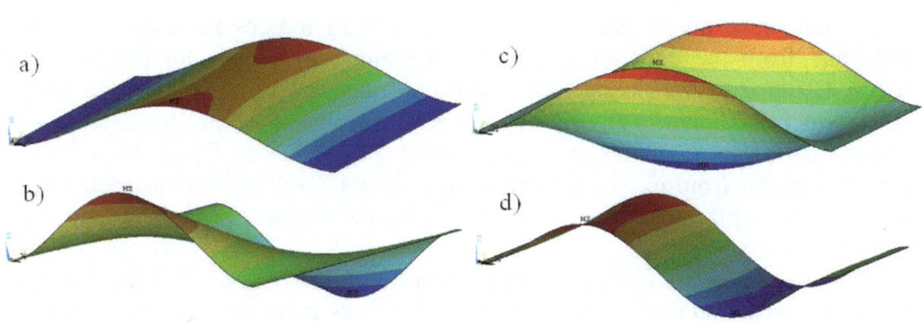

Fig. 2. Own modes of the plate presentation. a) first mode, b) second mode, c) third mode, d) fourth mode

We obtain H_∞ norm $\|G_{ki}\|_\infty$ for the kth mode (k = 1, 2, 3, 4) and ith actuator location. From these norms we obtain the sensor placement indices for each mode from (3.2), using weight such that $\max_i (\sigma_{\infty ki}) = 1$. The plots of $\sigma_{\infty ki}$ are shown in figure 3. The plot of the actuator placement indices for the first mode in figure 3 a) shows the maximum in the middle of two longer edges, and indicates that an actuator shall be placed at that place. In figure 3 b) and 3 c) indices reach their maximal values in the middle of two longer edges of the plate, as it was in the case of the first mode, although the second and third modes are different. Actuator placement indices for the fourth mode, reach two maxima on the both longer edges of the plate what is shown in figure 3 d).

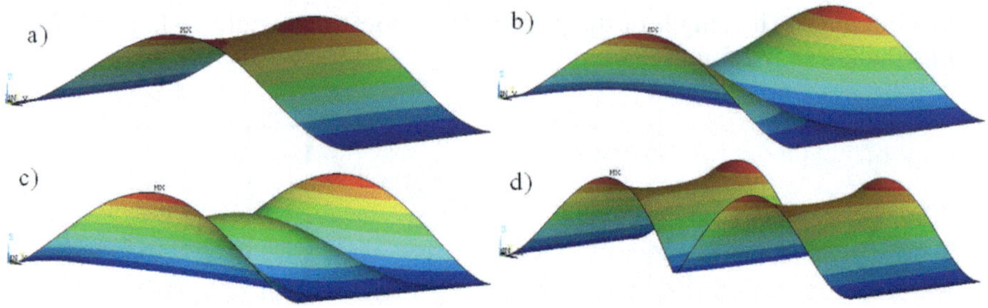

Fig. 3. Actuator placement indices as a function of actuator locations: a) for the first mode, b) for the second mode, c) for the third mode, d) for the fourth mode

Next, we determine the indices for the first two modes, using (3.4), namely $\sigma_{\infty 12i} = \max(\sigma_{\infty 1i}, \sigma_{\infty 2i})$, which is shown in figure 4 a). This maximum is in the middle of the longer edge of the plate, as individually for the first mode and the second mode the index reaches its maximum at the same place. Also for the first three modes we obtain the same result according to $\sigma_{\infty 123i} = \max(\sigma_{\infty 1i}, \sigma_{\infty 2i}, \sigma_{\infty 3i})$, which is shown in figure 4 b). For actuator placing for controlling first four modes according to $\sigma_{\infty 1234i} = \max(\sigma_{\infty 1i}, \sigma_{\infty 2i}, \sigma_{\infty 3i}, \sigma_{\infty 4i})$, we obtain three maxima on the both longer edges of the plate, which is shown in figure 4 c).

For determining the best actuator placement, the H_2 norm $\|G_{ki}\|_2$ for the kth mode (k = 1, 2, 3, 4) and ith actuator location was used for solving this example as well. From these norms and using (3.1) we obtain indices for each individual mode and from the equation (3.3) we determine the indices for the first two modes. Using $\sigma_{2,12i}$ we obtain index with one maximal value in the both longer edges, which is shown in figure 4 d) and a similar result we get for determining indices for first three modes from $\sigma_{3,123i}$ what we can see in figure 4 e). For the first four modes from $\sigma_{4,1234i}$ we obtain indices with two maxima at each longer edge, and these places are the most suitable for the actuator placement, shown in figure 4 e).

Influence of sensors and actuators placement was tested by simulation of control with external an excitation force. The force was assumed harmonic with frequency varying from zero to fourth natural frequency. In our case where the plate

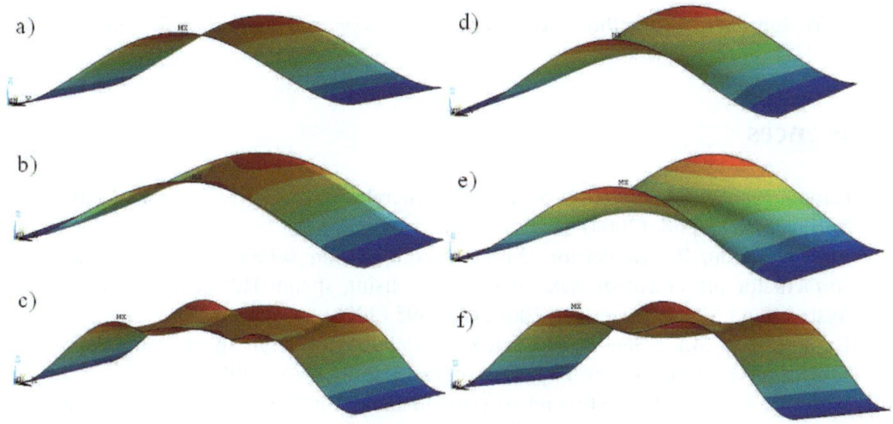

Fig. 4. Actuator placement H_∞ / H_2 indices as a function of actuator location. a) for the first two modes of norm H_∞, c) for the first three modes of norm H_∞, d) for the first four modes of norm H_∞, d) for the first two modes of norm H_2, e) for the first three modes of norm H_2, f) for the first four modes of norm H_2.

was clamped at its both shorter edges, the output of uncontrolled structure and controlled structure during that excitation is shown in figure 5.

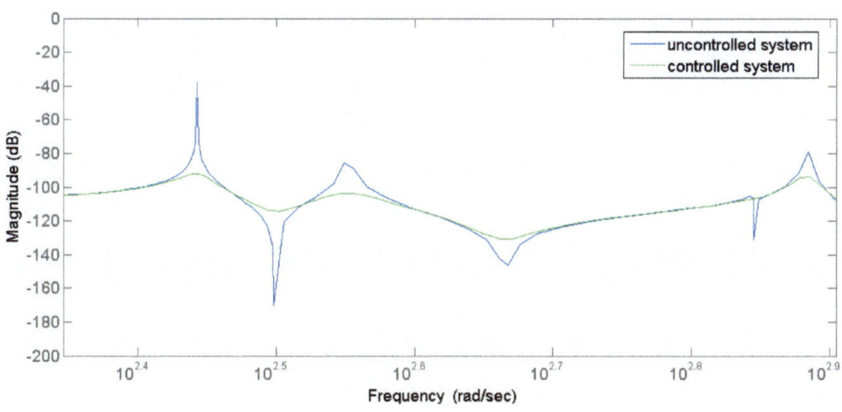

Fig. 5. Output of excitation of the uncontrolled structure and controlled structure

5 Conclusion

When sensors and actuators are placed according to indices of the norm H_∞ for the first four modes, which is shown in figure 4 c), than there are six actuators on the plate. In case of sensors and actuators placement according to indices of the norm H_2 for the first four modes, as it is shown in figure 4 f), then we have only four actuators but the input controlling force has to be 1,255 times greater than in the previous case for reach the same output.

Acknowledgments. The authors acknowledge the support by the Slovak Grant Agency VEGA -1/4128/07.

References

[1] Gawronski, W.K.: Advanced Structural Dynamics and Active Control of structures. Springer, New York (2004)
[2] Liu, W., Hou, Z., Demetrion, M.A.: A computation scheme for the optimal sensor/actuator placement of flexible structures using spatial H2 measures. Mechanical Systems and Signal processing 20(4), 881–895 (2006)
[3] De Lorenzo, M.L.: Sensor and actuator selection for large space structure control. Journal of Guidance, Control and Dynamics 13, 249–257 (1990)
[4] Kim, Y., Junkins, J.L.: Measure of controllability for actuator placement. Journal of Guidance, Control and Dynamics 14, 895–902 (1991)
[5] Gregory Jr., C.Z.: Reduction of large flexible spacecraft models using integral balancing theory. Journal of Guidance, Control and Dynamics 7, 725–732 (1984)
[6] Yaowen, Y., Zhanli, J., Chee, K.S.: Integrated optimal design of vibration control system for smart beams using genetic algorithms. Journal of Sound and Vibration 282, 1293–1307 (2005)

Control Loop Performance Monitoring of Electrical Servo-Drives

R. Schönherr, M. Rehm, and H. Schlegel

Chemnitz University of Technology, Faculty of Mechanical Engineering,
Institute for Machine Tools and Production Processes,
Mechatronics, Reichenhainer Str. 70, Chemnitz, Germany
{ruben.schoenherr,matthias.rehm}@mb.tu-chemnitz.de
holger.schlegel@mb.tu-chemnitz.de

Abstract. In production industry modern machines include a great number of feedback controlled drives. The reliable operation of these mechatronic systems including the requested function of the control loops are essential requirements for failure-free processes. In current systems, the observation of the control-loops is mostly restricted to basic monitoring functions, such as the supervision of limits. Yet, in the process- and facility-automation various methods of control loop performance monitoring (CLPM) have successfully been applied. The implementation and adaptation of these methods into the field of controlling electrical drives is considered to be a great contribution to a higher process reliability and efficiency of machines in production industry. Detecting oscillations automatically in the inner circuits of cascaded control loops is of great importance concerning process reliability and efficiency. The objective of this study is to monitor all major signals within the cascaded position control loop during operation in order to detect disadvantageous oscillations. As a first step, procedures for detecting oscillations were applied to the speed controller of electrical servo-drives.

1 Introduction

Beginning with Harris' approach, in the last twenty years diverse methods for monitoring and assessing control loops have been developed [2, 4-8]. These methods allow the identification of an insufficient operation of the control loops. Current monitoring systems, for instance in the field of chemical engineering, include such identification methods so that they are classified as state of the art for those applications. The existing monitoring procedures can be categorized in two groups. Some of the approaches, like the Harris-index [2] provide a single value which characterizes the performance of the control loop. Others analyze the behavior of the feedback control in order to detect specific malfunctions such as oscillations inside the loop.

1.1 Formulation of Problem

In current electrical drive-systems the signals inside the cascaded control loop are exclusively used for the control, but not for an observation of the control loop. Typically, monitoring functions are restricted to observe signal limits e.g. of the following error. Adapting and implementing CLPM methods in drive control loops would therefore result in a better utilization of the information, provided by the available signals.

For monitoring the drive by analyzing the feedback controls internal signals only, no supplemental sensors are required. As a result, these monitoring methods provide an interesting alternative for applications, where additional condition monitoring systems are too unprofitable. In addition, due to the development towards the automatical tuning and adaptation of controllers, there is an increasing demand of diagnose and monitoring functions to ensure stable system behaviour.

Because of raising energy costs, common direct mains-operated motors like pumps and fans are increasingly replaced by efficient feedback controlled drives. As a result, these applications offer access for the described methods of control loop performance monitoring.

1.2 Challenge

Control loops of electrical servo drives and controllers applied in the field of chemical engineering differ in various aspects. These differences, such as in the dimensions of time constants and cycle times, or different sensor signals demand specific control loop behavior. Most chemical plant controls base on regulator systems with a fixed set point combined with an optimized disturbance response. In contrast, mechatronic drives have to follow dynamic reference variable behaviour. Hence, the known methods from the chemical industry can not be easily applied to the cascaded drive control loops.

However, the approach to use already available controller signals to monitor the process remains of importance for drive controls.

2 Oscillation Detecting Methods

Several oscillation detecting methods base on the evaluation of spectra by a Fast Fourier Transformation to obtain relevant information about occurring vibrations. The automatical interpretation of these information is not a trivial task [3]. Hägglund as well as Forsman and Stattin suggested oscillation detecting methods based on the analysis of signals in the time domain [1, 5]. The less demanding calculation and interpretation of the results is one main advantage of these techniques. Thus, both methods are online applicable even in dynamic applications.

The analysis of the integral of absolute error (IAE) is the focus of both approaches.

$$IAE = \int_{t_{i-1}}^{t_i} |e(t)| dt \tag{1}$$

The variable e represents the control error while t_i expresses the moments of zero-crossing. The main difference between the two methods is the criterion to detect oscillations. According to Hägglund an oscillation is present if a certain number of areas exceed a critical value within a defined period. The supervision time as well as the limit of IAE are defined depending on the natural frequency of the control loop. In contrast, the method of Forsman and Stattin is based on estimating the similarity of successive areas, each above and below zero. One criterion for the similarity is the size of the area (A_i; B_i), the other one is the time between zero-crossings (ε_i; δ_i) (Fig. 1).

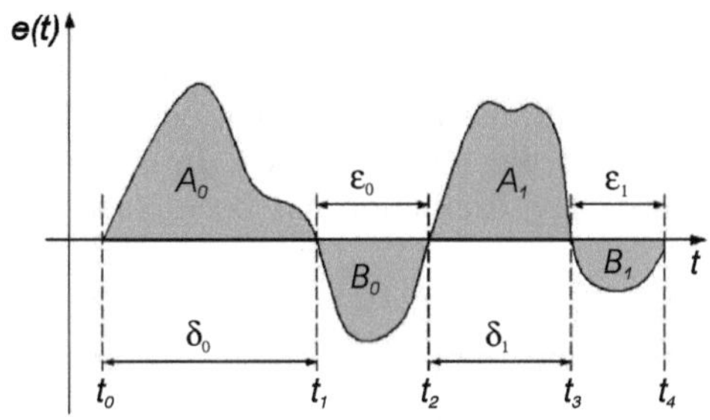

Fig. 1. Values for calculating the oscillation-index according to Forsman und Stattin. [5]

In order to obtain significant results, a specific filtering is recommended, especially in respect of the oscillation-index of Forsman and Stattin. In addition, the application of a filter allows to match the monitoring procedure to concrete frequency ranges.

According to the original area of application, both oscillation detecting methods purpose the identification of enduring oscillations: Hägglund recommends a supervision time about 50 times greater than the cycle time at natural frequency and Forsman and Stattin suggest averaging the oscillation index over at least 20 periods of the oscillation.

3 Methods of Resolution

The introduced methods are applicable to drive control loops with diverse objectives. One option is the application to detect enduring oscillations. Alternatively, the methods can be implemented to identify aggressive controller tuning. For this purpose an appropriate adaptation is required. With the following description both function modes are considered in detail.

3.1 Detection of Sustained Oscillations

Different mechanical problems such as worn-out bearings are displayed by periodic disturbances. These lead to periodic control error values in the affected control loop. In case the axis is moving with constant command velocity, the cycle duration is constant as well. Because they can be derived from the mechanical construction, the relevant frequencies are assumed to be known. Thus, the essential requirements for monitoring mechanical issues with both oscillation detection methods are met.

Both procedures were successfully applied to different servo drives. Mechanical oscillations with various causes have been detected, within different frequency ranges. The effect of a discontinuously working magnetic brake on the drive's velocity is shown in Fig.2. On the left, the resulting oscillation was detected by Hägglund's method. The regularly appearing exceeding of the limit of IAE is demonstrated. Values of $n > 10$ indicate a permanent oscillation. For this purpose, the procedure was adjusted to the rotation speed of the axis. On the right in Fig.2, the results of applying the method of Forsman and Stattin are shown. The results displayed with the solid line were achieved by using the unfiltered velocity signal. This caused the procedure to fail detecting the oscillation (oscillation index $h < 0.4$). The results displayed with dashed lines were obtained using a second order filter, which was adjusted to the rotation speed. In contrast to the experiment, where no filters were used, the oscillation has been clearly detected (oscillation index $h = 1 > 0.8$).

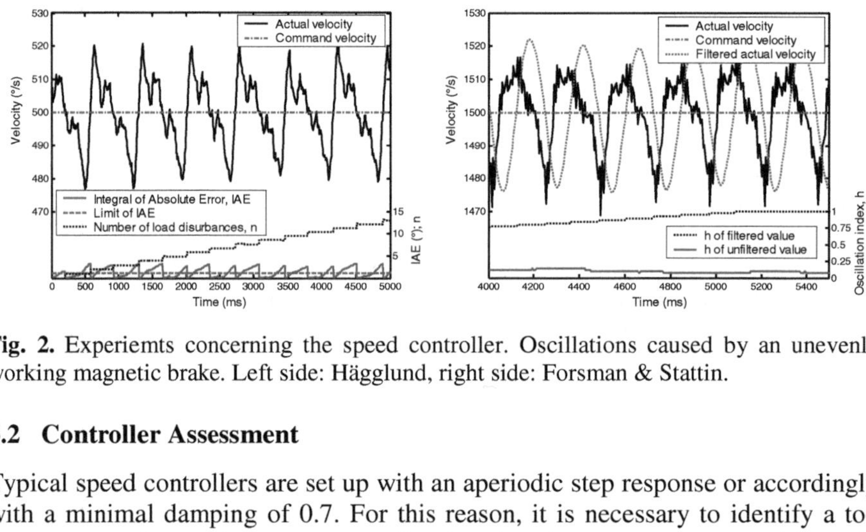

Fig. 2. Experiemts concerning the speed controller. Oscillations caused by an unevenly working magnetic brake. Left side: Hägglund, right side: Forsman & Stattin.

3.2 Controller Assessment

Typical speed controllers are set up with an aperiodic step response or accordingly with a minimal damping of 0.7. For this reason, it is necessary to identify a too aggressive tuning of the speed controller already when step responses with single or multiple overshoots occur and thus even before enduring oscillatory behaviour. The information required about the behaviour of the speed controller can only be distinguished from existing noise and mechanical influences, if a sufficient excitation of the loop is given.

The method of Hägglund has been adjusted according to the new requirements. Thereto, the analysis was limited to a short period after a defined excitation. Currently, the excitation is realized by an impulse or step function, but the further aim is to identify and use phases of sufficient excitation during normal operation. Fig.3 demonstrates, how an impulse of the command variable is used to excite the control loop. The behavior of the control error is schematically shown for three different controller settings (Fig.3, a), b), c)).

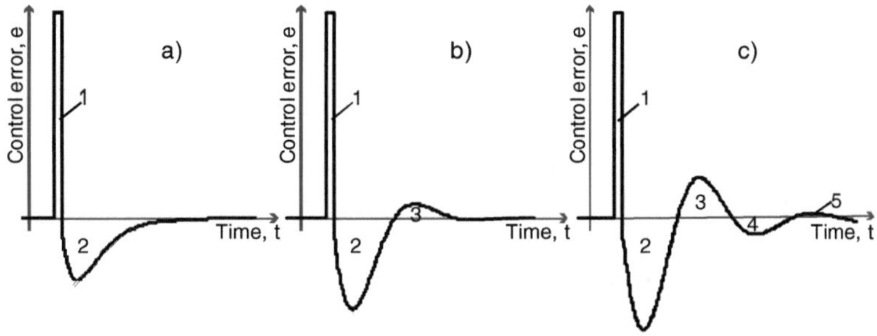

Fig. 3. Course of control error after impulse excitation.Detectable areas are numbered. The first area is caused by the impulse of the command variable. a) aperiodic setting, b) damping about 0.7, c) Damping about 0.3.

The graphs in Fig.3 illustrate, that

- 2 areas are detected, if an aperiodic setting is used,
- 3 areas are detected, if a single overshoot is present,
- 4 or more areas are detected, if the controller is tuned to aggressively.

If a single overshoot is allowed, the implemented exponential weighting factor λ (cf. [1]) leads to the following results: Values of $n > 3$ indicate a tuning of the controller, which is to tight. Values between 1 and 2 refer to an aperiodic course of the response and imply sluggish controller settings.

The results achieved by applying the method in the speed control loop of a real drive are shown in Fig.4. The position loop control was closed during the experiments. Setting 2 was identified to be to aggressive ($n > 3$). In addition, the aperiodic characteristic of setting 1 is represented by n as well.

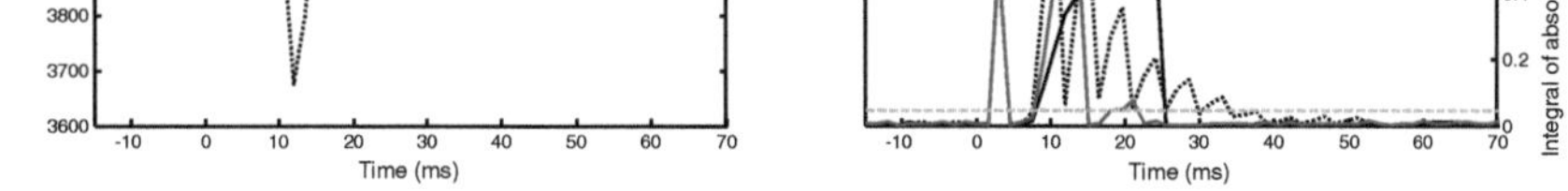

Fig. 4. Different speed controller settings excited by an impulse. (setting 1 sluggish, setting 2 aggressive, setting 3 normal). Left hand side: courses of speed, lower right hand side: calculated and limit values of IAE, upper right hand side: detected "load disturbances" (cf. [1]) or detected areas (λ respected).

4 Summary

In respect of monitoring drive control loops the paper has demonstrated an approach how to utilize the signals available more extensive.

Exemplified by applying the methods of Hägglund as well as Forsmann and Stattin to a speed controller, it has been shown that diverse CLPM methods are applicable to feedback controls in mechatronic systems. Implementing these modified techniques in regular industrial drive systems is considered to be a great contribution to a higher process reliability and efficiency of machines in production industry.

References

[1] Hägglund, T.: A control-loop performance monitor. Control Eng. Practice 3(11), 1543–1551 (1995)
[2] Harris, T.J.: Assessment of control loop performance. Canadian Journal of chemical Engineering 67, 856–861 (1989)
[3] Horch, A.: Oscillation Diagnosis in Control Loops –Stiction and other Causes. In: Proceedings of the American Control Conference Minneapolis, USA (2006)
[4] Ko, B., Edgar, T.F.: Assessment of Achievable PI Control Performance for Linear Processes with Dead Time. In: Proceedings of the American Control Conference Philadelphia, USA (1998)
[5] Jelali, M.: Regelkreisüberwachung in der Metallindustrie Teil 1, vol. 54(1), pp. 36–46 (2006)
[6] Jelali, M.: An overview of control performance assessment technology and industrial applications. Control Eng. Practice 14, 441–466 (2005)
[7] Ordys, A.W., Uduehi, D., Johnson, M.A.: Process Control Performance Assessment. Springer, London (2007)
[8] Visioli, A.: Practical PID Control. Springer, London (2006)

High Level Software Architecture for Autonomous Mobile Robot

J. Krejsa, S. Věchet, J. Hrbáček, and P. Schreiber

Brno University of Technology, Faculty of Mechanical Engineering,
Technická 2896/2, Brno, 616 69, Czech Republic
krejsa@fme.vutbr.cz

Abstract. The paper deals with the high level software architecture for autonomous mobile robot Bender II, medium size wheeled nonholonomic mobile robot with Ackerman chassis for both indoor/outdoor environment. High level software uses hybrid approach combining common layered architecture with reactive control. Local navigation including static/dynamic obstacle detection/avoidance is based on laser range finder data. Global navigation in indoor environment is based on fusion of odometry based position estimation with precomputed scan matching localization technique. Architecture description is accompanied by tests results obtained in real environment.

1 Introduction

Autonomous mobile robots exhibit rapid progress in past decade, as illustrated for example by results of DARPA Grand Challenge competition, as while in 2004 no robot was able to successfully finish the test track, in 2005 five robots finished the track and 22 of 23 gained better results than the best robot from the year before [1]. Together with improving the quality of sensors and other elements accompanied with falling prices of the components the control software is the key issue for successful and robust performance of mobile robot. There exist a number of high level software schemes, usually divided into three categories: 1. hierarchical (vertical) arrangement in which system cycles through perception-processing-action cycle, 2. reactive control which creates a set of behaviors directly connecting perception and action (e.g. [2]) and 3. hybrid architectures, which combines both [3]. The latest one achieves the greatest focus around researchers [1].

This paper deals with the universal high level architecture developed for both indoor and outdoor missions, as even there are differences with respect to the environment, the basic scheme is general. First the overview of the architecture is given, followed by detailed view on key modules, implementation details and test results.

2 High Level Architecture

The key idea of the architecture is task decomposition. Contrary to commonly used architectures we divide robot tasks into simple subtasks that are inserted into hierarchical tree during robot mission according to current situation observed by robot sensors. As robot is supposed to operate both indoor and outdoor and perform various types of missions, there are two levels in the architecture, first one deals with the mission type, second one with currently performed mission, as denoted on figure 1. Mission level is controlled by Mission manager (MM) – the top control object in the architecture that receives the initial global goal, sets appropriate mission and processes outer requests. Missions are of certain types, regardless the environment. Commonly used mission types might be e.g. Motion from location A to location B on free space, Motion from on the path, Wandering on free space, Mapping - motion in unknown environment, Manual control, etc.

During robots operation there is only one active mission at any time; MM processes the outer requests by forming mission tree or queue according to the request. For example, initial mission is wandering on the parking lot in front of the building. MM creates appropriate mission, sets it as current and starts. When outer request arrives, e.g. drive into nearby building to the office, queue of missions is created, firstly motion on free space (parking lot) to the road entrance, secondly motion on the road to the building, finally motion on free space inside.

The decomposition inside each mission is similar; however instead of mission queue the hierarchical tree of tasks is created. MM is replaced by Task manager (TM), structure existing only once (globally), assigned to currently performed mission. TM holds and services the tree of tasks. Each task represents a simple goal, commonly used tasks are e.g. Move on path, Move on free space, Localize, Avoid obstacle, Solve trap, Return to base, etc.

The task is performed until it is successfully finished or fails. Depending on failure reason the appropriate new task is assigned by TM and inserted into the tasks tree (unsuccessfully finished task remains open). Failure reasons do not necessarily have to be of high level causes. It could be hardware failure, suddenly appearing obstacle, etc. Reactive model is used for low level, e.g. robot performs

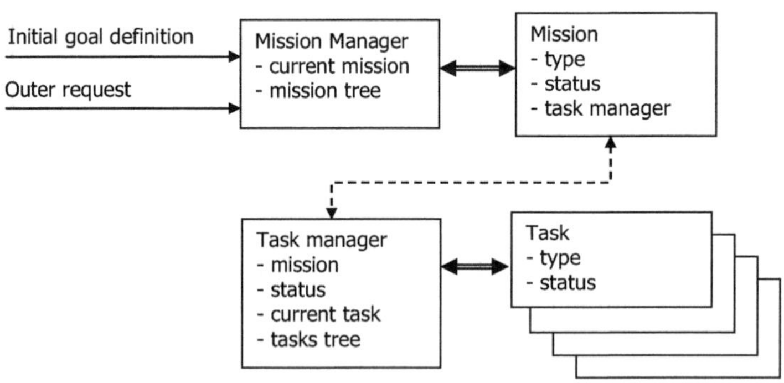

Fig. 1. Top level decomposition

emergency stop maneuver when confronted with suddenly appearing obstacle, and message is sent to current task, causing it to fail. The flow can be illustrated on figure 2, where example with obstacle avoidance is given. The robot is moving on the path (for now we do not care how the path is recognized, how robot steers on the path, as it is what the task deals with) and task fails as there is obstacle detected (figure 2a). Due to the failure the TM inserts appropriate task into the tree. Avoid obstacle task fails as there is no way how obstacle can be avoided (figure 2b). TM now inserts new task into the tree, the Solve trap task (figure 2c). Once the trap is solved, the task ends and flow returns to upper task (figure 2d) and this mechanism continues until the motion on the path is restored (figure 2f).

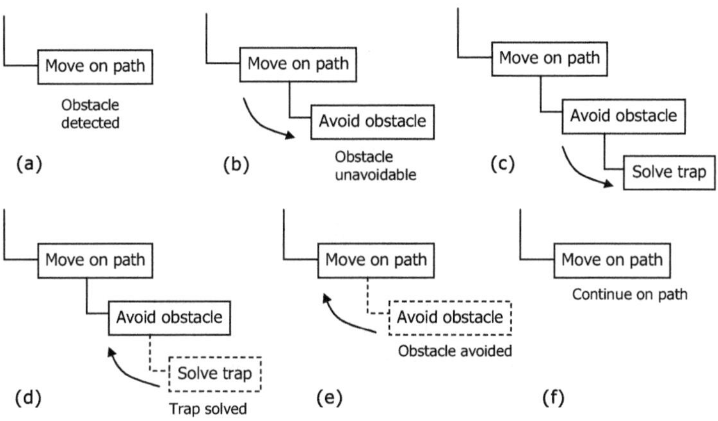

Fig. 2. Task tree example

The limited space does not allow us to go into greater detail in all the tasks, let us take a look at least into Avoid obstacle task. The flow is shown on figure 3. The

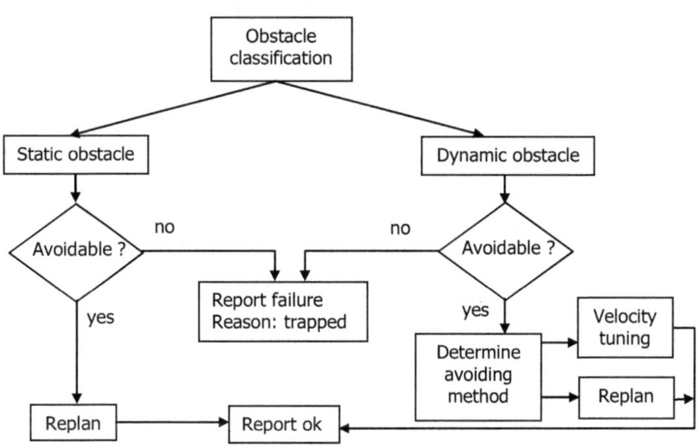

Fig. 3. Avoid obstacle task

task first determines whether the obstacle is static or dynamic. In case of static obstacle the task determines if it can be avoided, if yes the planning algorithm replans the trajectory and task returns success. In case of dynamic obstacle the task has to resolve if it can be avoided, and if yes it must decide whether to re-plan the trajectory or use velocity tuning algorithm ("wait" for the obstacle to move away).

There are several ways how to build the core of TM, the decision mechanism which selects appropriate tasks for insertion in the case of current task failure. Currently we are using simple set of rules that acts as primitive expert system, with low level functions "hard wired" (e.g. in the case of hardware component failure the low level functions independently stop the robot immediately, report the HW failure to current task in operation, task fails and whole mission is aborted). However, the simple set of rules can be replaced with e.g. Bayesian based decision maker, etc.

The architecture is modular, which can further be demonstrated on indoor/outdoor perception module, see Figure 4. All the sensors (apart from internal robot HW state check) can be divided into two groups, regardless the environment: global location data and local environment data. Local information is directly used for obstacle avoidance and indirectly for determining the global location estimate via localization. Therefore the architecture can use generalized perception modules, use it for the given tasks as information sources and modules themselves update the environment model depending on sensors available and environment type the robot is currently in. This way the indoor/outdoor fuses simply into environment.

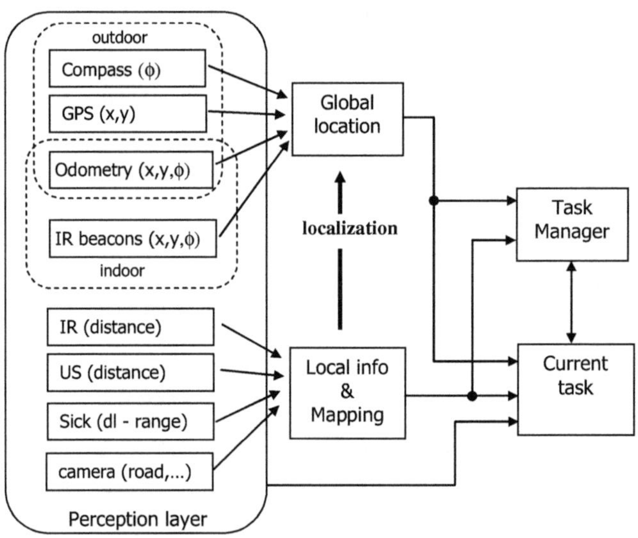

Fig. 4. Perception modules

3 Implementation and Tests

The architecture was implemented and tested on autonomous mobile platform Bender II: a medium size wheeled robot with Ackerman steering, for both indoor/outdoor environment, equipped with odometry IRC sensors, compass, GPS module, CCD camera and laser range finder. Rear wheels are independently driven by separate Maxon RE40 DC motors with planetary gears 43:1 and chain drives. Steering is controlled with Hitec HS-5745MG servo. Ground plan dimensions are 600x325mm, total weight is 20kg. Batteries used are LA 2x12V, 7.2Ah, allowing about 90 minutes operating range. The control software runs on Intel Mini-ITX board, which uses a simple bus driver board connected through a USB-to-UART converter to drive the onboard RS-485 bus. All hardware devices are interconnected by this bus (except the SICK laser scanner connected directly to the computers COM port). Every bus device type is controlled by a dedicated low level software layer class. Due to independently driven rear wheels the robot is equipped with a software differential, implemented by the middle-level software that also provides modules exploiting the bus devices. For example, there is an odometry module that periodically asks for data from both encoders (sampling period 0.6 s) and performs a simple data fusion. On top of the SW differential and odometry modules is based the algorithm caring for the robots ride.

High level software is based on previously described architecture. It is implemented in C# with time critical routines written in various languages. Localization routines use PCSM algorithm [4], planning algorithms are based on RRT [5] with path smoothness module based on limited radius circular arc fitting and optimization module. Dynamic obstacle recognition is based on laser range finder data analysis (indoor) together with image processing (outdoor).

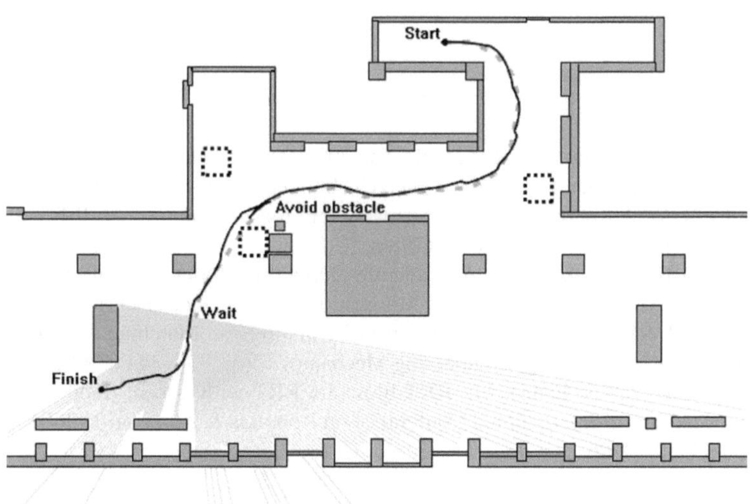

Fig. 5. Robot track example

Bender II was tested in range of environments on a number of missions. Due to limited space only single test example is described. The mission goal was to travel in indoor environment from defined start location to defined goal location. During the mission a number of static obstacles not included in the map and precomputed localization data were positioned both on expected track and around. At the same time the dynamic obstacles (people) were wandering at various speed and distances in the environment. Figure 5 shows planned track of the robot together with its estimated position. The gaps in location estimation are caused by improving the estimate precision via PCSM localization. As you can see, globally planned track goes through static obstacles, as those were not incorporated in the map and therefore the planning algorithm was not aware of those. As the steering of the robot is limited, reversing is included in obstacle avoidance for obstacles detected too late for direct avoidance.

4 Conclusion

Presented high level software architecture represents general framework for both indoor and outdoor environment autonomous mobile robots. The architecture is sensor independent and can be extended when robot hardware equipment changes / improves or when new robot missions are defined. The number of tests performed with Bender II autonomous robot that uses this architecture proved that it is usable and flexible. Further improvements can be obtained by replacing the currently used simple rules mechanism in Task manager module by more advanced modules, e.g. based on Bayesian approach.

Acknowledgments. Published results were acquired with the support of the Ministry of Education, Youth and Sports of the Czech Republic, research plan MSM 0021630518 "Simulation modeling of mechatronic systems".

References

[1] Buehler, M., Iagnemma, K., Singh, S.: The 2005 DARPA Grand Challenge: The Great Robot Race (Springer Tracts in Advanced Robotics). Springer, Heidelberg (2007)
[2] Brooks, R.A.: A Robots Layered Control System for a Mobile Robot. IEEE Journal of Robotics and Automation 2, 14–23 (1986)
[3] Connell, J.H.: SSS: A Hybrid Architecture Applied to Robot Navigation. In: Proc. of the IEEE Int. Conf. on Robotics and Automation, Nice, France (1992)
[4] Krejsa, J., Věchet, S.: Dealing with Sensor Errors in Scan Matching for Simultaneous Localization and Mapping. Engineering Mechanics 15(5), 337–344 (2008)
[5] Kalisiak, M., van de Panne, M.: RRT-blossom: RRT with a local flood-fill behavior. In: Proceedings of International Conference on Robotics & Automation 2006, Orlando, Florida, USA (2006)

Real Time Maneuver Optimization in General Environment

J. Mazal

Faculty of Economics and Management,
University of Defence, Kounicova 65, 61200, Brno, Czech Republic
jan.mazal@unob.cz

Abstract. The paper deals with the issue of optimum maneuver in general environment on local and global level and with effective exploitation of traffic segment overload, considering traffic density dependence and dynamic condition change on communications or terrain areas. Solution of this issue is dependent on time density expectation (extrapolation) of partial traffic segments computed based on optimum path solution and modification of weighted graph model for next element process iteration. Integration of exploited solution with C4I systems is viewed as a partial aspect of advanced decision support application in the area of military unit distribution and maneuver in combat or non-combat operations. It enables to identify critical configurations in traffic network in operations.

1 Introduction

Development of contemporary C4ISTAR systems is connected with substantial increase in potential and effectiveness of controlling troops equipped with these systems. Nevertheless, these aspects usually remain on theoretical level if some of the control processes are not automated. Due to current capabilities of communication systems enabling instant transfer of bulk data informing of the actual situation on the battlefield, capabilities of even the very best commanders to deal with them effectively exceeded. Therefore, without implementing advanced methods of support and optimization of routine decision-making, it is impossible to control troops effectively in the environment of the 21st century battlefield (what was sufficient or acceptable in the 50´s or 60´s of the last century does not have to necessarily suffice today). Another aspect is also the ability of today´s systems to automate implementation of commander´s decision, in our case ensure control of a maneuver of a unit in real time under given tactical conditions. Regarding the large number of processes that have to be implemented in this area, again it is unfeasible that any human element would be able to perform them in such an extent and quality that can be achieved using today´s computing systems.

2 Implementation

Real time path optimization in general environment is able to solve in two layers, higher and lower layer, where high layer is independent to level of automation of local element control and determine rough (in terms of high model resolution) path configuration on digital terrain model. Which is connected to element ability to move in particular resolution step and closely divided from level of detail of basic virtual terrain model, where is path optimized? Meanwhile low level solution is trying to deal with local motion problem between particular points delivered by high layer. Base information used for local optimization problem solution is 3D near environment configuration scan, including destination point, from which is collision information extracted, leading to actual environment collision characteristic graph, where is applied search algorithm for optimal path solution. Algorithm solving this type of problem issuing from critical path solution in non-oriented graph, optimized for real time application. Solution of both problems has the same base principles, but with different data models and process of its construction. The key point of solution is particular ability of parallel execution of that solution and implementation of GPGPU concept to element processing. In these areas there are certain possibilities, one of the small complication is fact, that there is not possible to avoid serialization of particular phases, because iteration steps of algorithm is previous result dependant, nevertheless process solution of each phase is possible to parallelize. Here is another small complication of global memory share of particular threads. General effectiveness is dependent on parallel process count, what is possible to execute in each phase of iteration process, this number is variable to calculation process and in general point of view it is trapezoid or triangle trend, as is illustrated on a graph.

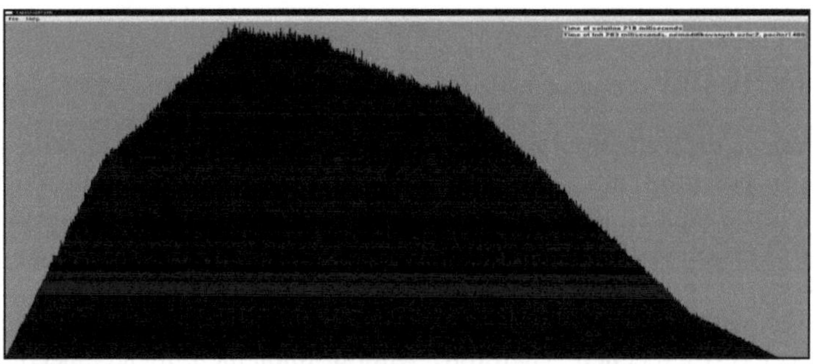

Fig. 1. Graph of number of executed threads in each phase of solution, total count of phases is 1400 in that case

There is example of model containing 1 million of elements and 4 millions of connections, where final solution was achieved in 1400 iterations phases, where in each phase was possible to execute 1 to 1286 parallel threads operating over elements as is shown on a graph. Parallel solution in that case could speed up the final solution more than 100 times.

3 Maneuver Optimization of Large Number of Elements in the Time Resulting in Effective Exploitation of Distribution Network

This is a relatively new phenomenon, remaining in the initial phase of its development, which a number of those involved in telematics try to deal with. The main issue in the commercial sector is the need to equip end elements with adequate communication and imaging technology. However, the situation differs in the military. The trend of gradual implementation of C4I systems even at lowest levels of command frankly calls for this application. At the beginning, it is very difficult to estimate the extent of effectiveness enhancement. Nevertheless, intuitively it is possible to assume that it will be substantial, particularly with increasing density of traffic and growing incidence of critical events. Corresponding detailed method dealing with given problem mathematically would substantially exceed the scope of this article and take attention from the general primary principle to computing details of the algorithm. The text bellow describes its generalized variant including explanation of key moments of its solution process where the starting structure is represented by a chart of a traffic network with initial weight configuration of individual nodes that in the basic version determine traffic-carrying capacity of given segment. In this respect, it is possible to consider also quality of given road, which eventually affects maximum speed in given segment and therefore its capacity. Relation between these variables can be derived from results gained in experiments. General relations can be often resolvable only with substantial difficulties and therefore, it is necessary to determine given parameters empirically and individually. Influence of individual elements in real time present in given segment of traffic network are to be taken into account. Again, for efficient optimization calculation applied to real-life situation, these factors have to be analyzed in detail based on data resulting from experiments. Fine-tuning of selected criteria approximates iterative process of a solution, where some parameters randomly oscillate in certain intervals around their mean value, which is based on statistic assessment of predicted actual states continuously refined to bring extrapolated position configurations of most elements into consonance with actual situation.

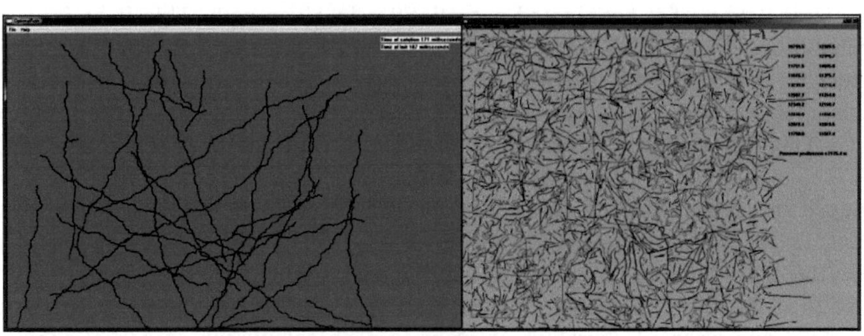

Fig. 2. Sample of Optimum Path Output

General principles of the solution can be characterized as follows:
As already mentioned, starting model is represented by a mathematical chart of a traffic network, which represents initiatory data model for solving given tasks. During the solution, the initial data model is being gradually modified based on influence of individual elements and their anticipated (calculated) path, where the time of appearance in individual (calculated) segments of optimum path is extrapolated. In fact, each of these paths for each of the elements is calculated in a different model - model, which has been from different prospective influenced by previous elements. However, it is necessary to maintain sequence of individual elements in each iterative cycle for any change in this sequence would change initial conditions for each of the elements and as a result it would cause unpredictable changes in the course of optimum path in time, thus chaotic guidance of given element throughout the model. This fact is intuitive and its disruption would be equivalent to changes in starting times in the course of any race, which would result in continuous changes of placings of individual racers even if there would be no real change. Another separate part is effective searching for optimum path of individual element. The key element in this part is effectiveness of the algorithm finding optimum (usually the shortest) path in a large (millions of nodes and dozens of millions of links) not oriented weighted chart quickly enough to enable the solution "in the real time" for a large number of moving elements. In this case, the situation if partially complicated by the fact that it is impossible to calculate optimization of movements of individual elements simultaneously for modification of the initial data model for each element depends o previous optimum path calculation (for the previous element). Therefore, regarding implementation of the parallel calculations architecture, it is necessary to concentrate on the process of calculation of optimum path of an individual element with adaptive modification of spatiotemporal attribute of the basic segment (link). Today, authors have achieved the solution times around 0.1 s per element with complexity of 1M nodes and 8M links – without implementation of parallel architectures, as for instance CUDA (that again reduce the time needed for solution significantly...). After computing the optimum (minimum, maximum) path of the element, times of transits though individual segments (of the optimum path) by given element are estimated by extrapolation of it anticipated potentialities in given path. This is an issue of probability, which is also considered in effect on the time dimension of the model, where we can anticipate that with growing time the probability of appearance of given element in given segment in given time reduces (the question is to what extent). This fact must be considered to modify the traffic network (chart data model), i.e. individual links according to anticipated time transit of given element. Longer time reduces the probability of appearance of given element in given segment, i.e. the value of modification of the original weight coefficient for given time interval decreases. In a certain way, this new dimension increases the complexity of the basic algorithm for calculation of the shortest (longest) path in weighted non-oriented chart, which prior to each calculation of partial sums of weight for given node must modify these weights according to spatiotemporal

appearance of previous elements[1]. Otherwise, further elements can input the iterative process during the calculation. However, as already mentioned, the sequence of already solved elements cannot change. Consequently, any element can be in relations between iterative segments excluded and new elements can be included at the end of the line (not to affect the sequence of the previous ones). This generally outlined procedure has to iteratively performed for the entire set of elements where in the following iterative phase, information on position of individual elements (expectedly based on GPS data) should be updated in data model and changes in capacity (quality) of individual traffic segments recorded. This requires on-line digital communication of a central computing server and individual elements in order to assure real-time responds to changes and optimization of movements of a large number of elements in spatiotemporal context. This is another aspect confirming the necessity to employ C4I (C4ISTAR) systems at all levels of command structures, especially for mobile elements operating under conditions of the 21st century battlefield.

4 General Algorithm Description

Regarding the fact that a detailed description of the algorithm or extract of is program code would exceed the scope of this article, bellow is outlined a sequence of individual steps that aggregate individual processes ensuring implementation of mainly trivial and routine sub-processes:

1. Input or creation of initiatory data mode (chart) of a traffic network, especially initiation of weight coefficients, for links of individual nodes.
2. Initiation of starting positions of individual elements and determination of destinations for each of the elements.
3. Sequencing elements in the queue for solution.
4. Calculation of optimum path for the first element to its destination.
5. Extrapolation of timetable of path being solved and determination of probability coefficients for given element in given time and given segment.
6. Inserting this calculation in a table of paths and inclusion of an indicator in transit segments of the chart for this item of the table of paths.
7. Solution of path optimization for the next element (in the queue). Prior to partial summation of individual links starting from given node, expected time of arrival to the node (and middle of the link) is being calculated as well as spatiotemporal component of each link from the table of paths colliding with anticipated time of (the center of) the link is being integrated (separate algorithm).
8. Until the end of the queue of elements is reached, the algorithm continues with the step 5, after that it continues with step 9.

[1] Using today's assets, it is impossible to solve the task without effective identification and elimination of regressive cycles.

9. Should any of the elements reach its destination, it is excluded from the queue; should there be a requirement to add a new element, it is included at the end of the queue.
10. Update of positions of all elements in the model, setting of given element indicator at the beginning of the queue and continuation with the step 4.

5 Conclusion

In conclusion, considered problematic is tightly closed together mostly the common approach of algorithmic solution, which is relatively universal and search of optimal way is similar to light emition, where, light is dispersed in every direction and each point of light approach is behave as original emitter with another all directional light spread (rather simplified description, but apposite). Path search automation and optimization is nowadays relatively actual affairs and in future its importance will be on increase. And wide implementation in fully or semi autonomous vehicles is undisputable, mostly in military applications, same as time optimization of traffic overload on communications, what could start in area with appropriate conditions. Besides military applications we can note that according to current development, build-up of new communications will be probably unable to compensate the increase in traffic, especially in the Czech Republic. So, it is more than likely that in the near future absence of a similar system will cause substantial problems or even permanent collapse in traffic. Let us hope that before we reach this critical moment, this problem will be solved. Nevertheless, not even implementation of similar system can solve all cases of overload. After a global congestion of the entire network there is no solution but prolongation of load distribution or construction of new segments.

References

[1] Parker, J.R.: Algorithms for Image Processing and Computer Vision, New York, p. 417. Wiley computer publishing, Chichester (1997)
[2] Mazal, J., Kacer, J., Jolly, B.: Tactical path optimization. In: Proceedings of Conference 2nd International Symposium Advances in Mechatronics 2007. Brno: Expertia o.p.s., vol. 8 (2007) ISBN 978-80-7231-314-3
[3] Jablonsky, J.: Operational research. PRAGUE: Professional publishing 323 (2002) ISBN 80-86419-32-1[in Czech]

Geometric Robot Motion Strategies

M. Šeda and T. Březina

Brno University of Technology, Faculty of Mechanical Engineering,
Institute of Automation and Computer Science, Technická 2896/2, Brno, Czech Republic
{seda,brezina}@fme.vutbr.cz

Abstract. In robot motion planning in a space with obstacles, the goal is to find a collision-free path of robot from the starting to the target position. There are many approaches depending on types of obstacles, dimensionality of the space and restrictions for robot movements. Among the most frequently used are roadmap methods (visibility graphs, Voronoi diagrams, rapidly exploring random trees) and methods based on cell decomposition. A common feature of all these methods is the generating of trajectories composed from line segments. In this paper, we will show that generalised Voronoi diagrams can be used for fast generation of smooth paths sufficiently distant from obstacles.

1 Introduction

In recent years a number of new data structures and algorithmic techniques have been developed that have improved and simplified many of the previous approaches used in network optimisation, robot motion planning [3], [5], [11], etc. Geometric data structures defined in computational geometry have a surprising variety of uses [1], [2], [6].

Computational geometry emerged from the field of algorithm design and analysis in the late 1970s. It has many application domains including computer graphics, geographic information systems (GIS), robotics, and others in which geometric algorithms play a fundamental role. Computational geometry deals with specific geometric data structures, the most important ones being Voronoi diagrams, Delaunay triangulation, visibility graph and convex hull.

Before we study examples of their applications, we will introduce them and summarise the basic definitions.

2 Basic Notions

A Voronoi diagram of a set of points (called sites) in the Euclidean plane is a collection of regions that divide up the plane. Each region corresponds to one of the sites and all the points in one region are closer to the site representing the region than to any other site. More formally [1], [2], [4], [6]:

Definition 1. Let P be a set of n points in the plane. For two distinct sites p_i, $p_j \in P$, the *dominance* of p_i over p_j is defined as the subset of the plane that is at least as close to p_i as to p_j. Formally,

$$\text{dom}(p_i, p_j)=\{x\in \Re^2 \mid d(x, p_i)\leq d(x, p_j)\}, \tag{1}$$

where d denotes the Euclidean distance.

Definition 2. *Voronoi region* (or *Voronoi polytope*, *Voronoi cell*, *Voronoi face*, *Dirichlet polygon*, *Thiessen polygon*) of a site $p_i \in P$ is a close or open area $V(p_i)$ of points in the plane such that $p_i \in V(p_i)$ for each p_i, and any point $x \in V(p_i)$ is at least as close to p_i as to any other sites in P (i.e. $V(p_i)$ is the area lying in all of the dominances of p_i over the remaining sites in P).

Formally,

$$V(P_i)=\left\{x\in \Re^2 \mid d(x,p_i)\leq d(x,q): \forall q\in (P-\{p_i\})\right\}= \\ = \bigcap_{q\in P-\{p_i\}} \text{dom}(p_i,q) \tag{2}$$

Definition 3. A *Voronoi diagram* (or *Voronoi tessellation*) $V(p)$ for a given set $P=\{p_1, p_2, \dots , p_n\}$ of points (or sites) is a polygonal partition of the plane into Voronoi regions $V(p_1), V(p_2), \dots , V(p_n)$. The vertices of polygons $V(p_i)$ are called the *vertices of the Voronoi diagram*, and their edges are called the *edges of the Voronoi diagram*. A Voronoi diagram is called *degenerate* if four or more of its Voronoi edges have a common endpoint.

Clearly, each edge of the Voronoi diagram belongs to just two Voronoi regions and

$$V(P)=\bigcup_{p_i\in P} V(P_i) \tag{3}$$

Definition 4. Let $P=\{p_1, p_2, \dots , p_n\}$ be a set of n distinct points and $O=\{O_1, O_2, \dots , O_m\}$ be a set of m closed regions that represent a set of obstacles that are neither transparent nor traversable and do not overlap. Two vertices that can see each other are called (*mutually*) *visible*, and the segment connecting them is called a *visibility edge*.

Definition 5. Let $O=\{O_1, O_2, \dots , O_m\}$ be a set of m obstacles, S be a set of their vertices and p_{start} and p_{target} be the starting and target positions. A *visibility graph* is a graph $G=(V,E)$ whose set of vertices V is given by $S \cup p_{\text{start}}$ and p_{target} and the set of edges E is given by the visibility edges on V.

The *rapidly exploring tree* grows from the starting position that initialises the tree. At each step, a point is randomly generated and, by a shortest possible way, connected to the current tree [5].

3 Robot Motion Planning

Assume that R is a robot of a convex shape, $R(x,y)$ denotes its reference point at (x,y) and the obstacles are also convex. (Non-convex obstacles can be easily divided into several convex parts.) The *obstacle region* (or *configuration-space obstacle* or *C-obstacle*) of an obstacle P and the robot R is defined as the set of points in the configuration space such that the corresponding placement of R intersects P. Denote C_P the obstacle region of an obstacle P. Then

$$C_P = \{(x,y) \mid R(x,y) \cap P \neq \varnothing\} \tag{4}$$

It is evident that the drawback of the rapidly exploring trees is a high number of generated edges resulting in a broken trajectory. It can be smoothed, to a certain extent, using splines.

The shortest path between two points in the plane with polygonal obstacles can be easily solved in the corresponding visibility graph by the Dijkstra algorithm. Using a binary heap implementation, its time complexity is given by $O(|E| \log |V|)$, where E is the set of edges and V is the set of vertices. Fig.1 shows the shortest path between the starting and target positions using the visibility graph. The visibility graph of a set of disjoint polygonal obstacles with k edges in total can be computed in $O(k^2 \log k)$ [2].

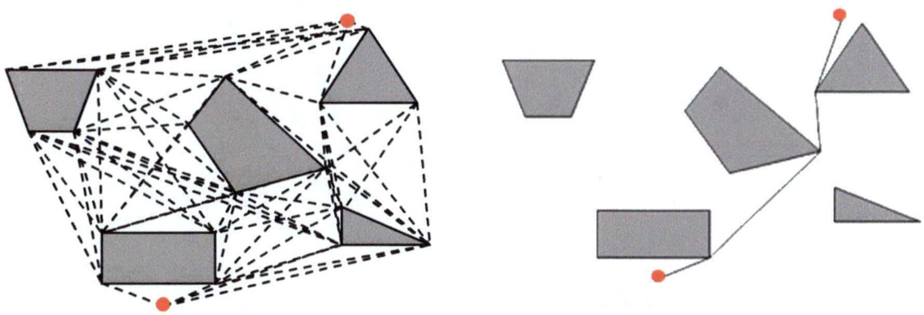

Fig. 1. Visibility graph and the shortest path

This drawback can be removed using Voronoi diagrams.

For reasons of time complexity, it is necessary to know the properties of the Voronoi diagrams and the algorithms of their constructions. We only mention the most substantial properties:

(i) The number of vertices in a Voronoi diagram of a set of n point sites in the plane is at most $2n-5$ and
(ii) the number of edges is at most $3n-6$.

The algorithms used to construct Voronoi diagrams (*divide and conquer*, *incremental* and *plane sweep*) need $O(n \log n)$ time.

If a generator set of a Voronoi diagram represents point obstacles and other obstacles are not present in the plane, then the robot can walk along the edges of the Voronoi diagram of P that define the possible channels that maximise the distance to the obstacles, except for the initial and final segments of the tour. This allows us to reduce the robot motion problem to a graph search problem: we define a subgraph of the Voronoi diagram consisting of the edges that are passable for the robot. However, some of the edges of the Voronoi diagram may be impassable. Then these edges must be omitted from the diagram.

Of course, when we first construct the configuration space for obstacles, it is not necessary to test whether a robot can walk along the edges of the Voronoi diagram.

For scenes with point, straight-line and polygonal obstacles, the simplest way of finding optimal trajectories is to compute ordinary Voronoi diagrams for vertices of obstacles and then remove those of its edges that intersect obstacles. We get more precise solutions by approximating the polygonal edges by line segments and then applying the previous approach [8], [9], [10].

An implementation of this approach is described in [10]. Using this program, we can determine the number of line segments that approximate the edges of polygonal obstacles and compute the final Voronoi diagram with more precise edges. However, the resulting trajectories were not smooth. If we deal with obstacles as sets of their boundaries approximated by lines, then we can precisely compute the bisectors between point and line or between two lines and built a *generalised Voronoi diagram* as follows.

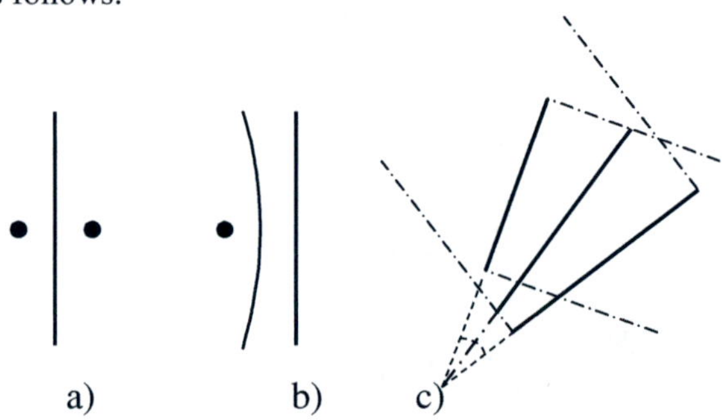

Fig. 2. Edges in a generalised Voronoi diagram

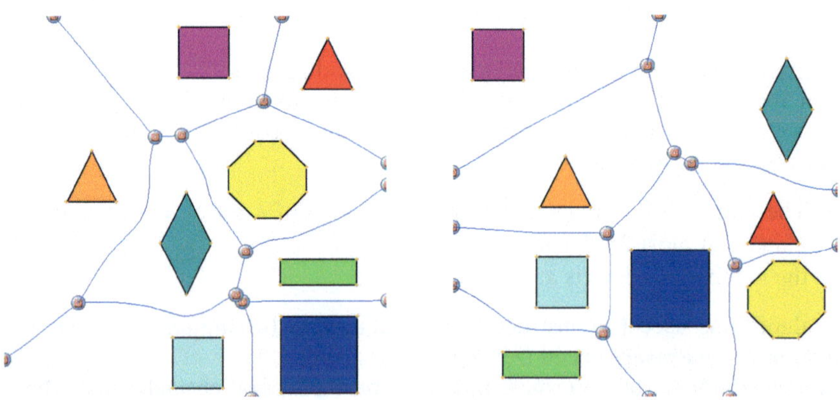

Fig. 3. Generalised Voronoi diagram for a scene with 8 polygonal obstacles in two configurations

Fig. 2 shows that a) bisector for two points is given by an axis perpendicular to the centre of their connection, b) bisector for a point and line is given by a parabolic arc and the point is its locus, and c) bisector of two lines is given by the line dividing the angle between the given lines. Therefore, the edges of the generalised Voronoi diagrams are composed by straight lines and parabolic arcs.

Fig. 3 shows a generalised Voronoi diagram for a scene with 8 polygonal obstacles from the implementation of V. Pich [7]. Since the redrawing of the diagram for moving obstacles runs in real time, we can assume that it could be used for robot motion planning in a dynamic scene.

4 Conclusions

In this paper, we briefly summarised the main geometric data structures and showed their possible use for robot motion planning.

Since traditional robot motion planning methods based on potential fields and decomposition have many drawbacks such as convergence to local minima, and combinatorial explosion, or generating infeasible solutions, we focused on roadmap methods using the generalised Voronoi diagrams to get smooth and safe trajectories in polynomial time. If a scene contains movable obstacles and these obstacles move along continuous curves, then the corresponding generalised Voronoi diagram also changes continuously and, therefore, the position of the robot will be changed continuously.

In the future, we will do more experiments with movable obstacles and try to determine their maximal number for real-time control.

Acknowledgments. This research has been supported by the Czech Science Foundation GA ČR in the frame of GA ČR 102/09/1668 project Control Algorithm Design by Means of Evolutionary Approach and the Czech Ministry of Education in the frame of research plan MSM 0021630518 Simulation Modelling of Mechatronic Systems.

References

[1] Aurenhammer, F.: Voronoi Diagrams – A Survey of a Fundamental Geometric Data Structure. ACM Computing Surveys 23(3), 345–405 (1991)

[2] de Berg, M., van Kreveld, M., Overmars, M., Schwarzkopf, O.: Computational Geometry: Algorithms and Applications. Springer, Berlin (2000)

[3] Choset, H., Lynch, K.M., Hutchinson, S., et al.: Principles of Robot Motion: Theory, Algorithms, and Implementations (Intelligent Robotics and Autonomous Agents). MIT Press, Cambridge (2005)

[4] Fortune, S.: Voronoi Diagrams and Delaunay Triangulations. In: Du, D.A., Hwang, F.K. (eds.) Euclidean Geometry and Computers, pp. 193–233. World Scientific Publishing, Singapore (1992)

[5] LaValle, S.M.: Planning Algorithms. University Press, Cambridge (2006)

[6] Okabe, A., Boots, B., Sugihara, K., Chiu, S.N.: Spatial Tessellations and Applications of Voronoi Diagrams. John Wiley & Sons, New York (2000)

[7] Pich, V.: Application of Voronoi Diagrams in Robot Motion Planning (in Czech), Master Thesis, p. 64, Brno University of Technology (2008)
[8] Šeda, M.: A Comparison of Roadmap and Cell Decomposition Methods in Robot Motion Planning. WSEAS Transactions on Systems and Control 2(2), 101–108 (2007)
[9] Šeda, M.: From Exact Methods to Heuristics. Vědecké spisy VUT v Brně, Edice Habilitační a inaugurační spisy 276, 1–40 (2008)
[10] Švec, P.: Using Methods of Computational Geometry in Robotics. PhD. Thesis, Brno University of Technology, p.135 (2007)
[11] Zilouchian, A., Jamshidi, M.: Intelligent Control Systems Using Soft Computing Methodologies. CRC Press, Boca Raton (2001)

Semi-autonomous Motion Control Layer for UGV-Type Robot

M. Hiiemaa and M. Tamre

Tallinn University of Technology, Faculty of Mechanical Engineering,
Department of Mechatronics, Ehitajate tee 5, Tallinn, Estonia
maido@staff.ttu.ee

Abstract. The concept of proposed Intelligent Motion Control Layer (IMCL) is studied and evaluated on Matlab and Simulink based virtual simulation models. All interfacing between the control model and the virtual UGV physical model simulate the interfacing of a real UGV hardware (latencies, approximation, uncertainty). The virtual physical model of the UGV is tested on virtual environments (ramps, sections of different surface properties, etc) simulating real terrain. Simulation plots clearly show the IMCL capabilities of detecting and compensating different types of disturbances while collecting valuable data about the terrain for later use. Using IMCL capabilities in UGV-type of robots simplifies complicated mission control of the robot where many parameters (velocity, acceleration, jerk) must be kept within limits simultateously. Prolonged use of the intelligent motion control is expected to minimizes energy consumption and prevent mechanical damage.

1 Introduction

There are a number of applications where remotely controlled Unmanned Ground Vehicle (UGV) can be used. Some of the applications involve transporting explosives or delicate equipment and therefore have some additional requirements for motion control parameters. These requirements most likely involve acceleration (a) and jerk (j) to be kept within their mission dependant limits (a_{limit}, j_{limit}) while performing speed commands (v_{cmd}) from the UGV operator. IMCL (Intelligent Motion Control Layer) receives velocity input command and outputs a series of acceleration commands to reach the target speed without violating the predefined limits. To do that, regardless of data delays, uncertainties and external disturbance, several presumptions and calculations are being made inside IMCL. The behavior of IMCL has been written mostly in Matlab and integrated to Simulink by central Level-2 M-file S-Function. Most of the UGV hardware and physical model is implemented as standard Simulink model.

Disturbance is mostly caused by friction and terrain ascents and may therefore change when the direction of velocity changes even when the terrain does not. While disturbance from terrain ascent remains the same, the disturbance from friction is always opposite to the direction of velocity. Splitting disturbance into components on run-time is possible using data from digital gyroscope, which gives IMCL layer information about current terrain ascent angle. Knowing these components is helpful when predicting disturbance magnitude on velocity direction change.

2 Motion Control Simulation

The simulation models describe only part of the UGV platform [1] that IMCL was originally made for. The platform uses 4 wheels equipped with 4 encoders just for driving straight forwards and backwards. All wheels can be repositioned by 4 "legs" which all have individual built-in suspension. Simulation uses greatly simplified model where input is taken only from one encoder and drive commands are expected to drive all wheels equally. Turning, slipping, suspension and any possible airtime events are not being simulated.

2.1 UGV Hardware

The simulation uses long sample time ($t_{smp} = 0.1$ s) to give a good insight of the latency influences to the system performance. Sample time equals in the interest of simplicity main cycle time and simulation time-step. New acceleration command will be generated at the end of every main cycle and it is expected to remain constant until the next command output. IMCL receives position information from rotary encoders and outputs acceleration command in discrete time. The integer multiplier for encoder delay is D_2, for motor controller delay it is D. Simulink models artificially create these delays. Encoder data delay in a simulation is $D_2 \cdot t_{smp}$. Motor controller delay in a simulation is $D_2 \cdot t_{smp}$ giving 0.7 s of total feedback delay. The delays in IMCL as well as Simulink (hardware and physical model) can be changed without IMCL performance degradation as long as changes are being made in pairs. IMCL internal delays must always match the delays created in Simulink model.

Due to encoder limited resolution (only 4096 separate positions per revolution in a simulation) the position data becomes approximated to an integer value before it enters IMCL. This approximation creates noise to all parameters that are being calculated based on position data. The most effective solution to minimize the noise is increasing the encoder resolution.

The IMCL output type is acceleration command, which in some circumstances could be described differently. It can be converted to force by multiplying it with UGV mass (400 kg) or average torque by multiplying the force with UGV wheel radius (0.25 m). This conversion is trivial and can be hard-coded to IMCL easily. As most of the servo controllers use analogue (±10 V) signal for input, another conversion has to be made anyhow. Matching the IMCL output to the specific servo controllers is not being simulated.

2.2 IMCL Concept

The main idea of IMCL delay and disturbance compensation is based on operating with two arrays (a_{cmd} , a_{cmp}), environment parameters (a_{asc} and a_{fri}) and motion parameters calculated based on encoder readings (v , and a). All commands that have been sent to UGV physical model by the Command Planner are kept in a_{cmd} . Commands that were sent to compensate disturbance are kept in a_{cmp} . In practice these two types of commands are added together before being sent out and will be handled by Simulink as one single acceleration command. Both of the arrays (a_{cmd} and a_{cmp}) are being left-shifted just before inserting new values to the rightmost positions at the end of each main cycle. Although the effects of acceleration commands take a while to be propagated back to IMCL by encoder, the commands are still available for calculations in IMCL. The shifting arrays virtually simulate the delay propagation locally to get the best insight of the current and future motion parameters often with sufficient accuracy. Any disturbance is being detected shortly after the fist significantly affected encoder reading arrives to the IMCL. The detection is not momentary because of the encoder data delay. The difference is mostly caused by environment disturbances and is detected by comparing one of previous commands with actual UGV performance that can be verified by encoder feedback.

$$a_{error} = a - a_{command} \tag{1}$$

Variable $a_{command}$ is a sum of components that have been taken from command history (arrays a_{cmd} and a_{cmp}) from the positions that perfectly match the feedback delay. The disturbance (a_{error}) can be separated into two components which act differently when the direction of velocity changes.

$$a_{error} = \sin(\alpha) - \mathrm{sign}(v) \cdot \mu \tag{2}$$

As ascent angle (α) is already acquired by digital gyroscope, friction coefficient (μ) can be calculated.

$$\mu = \frac{\sin(\alpha) - a_{error}}{\mathrm{sign}(v)} \tag{3}$$

To suppress noise propagation, simple filtering is being used.

$$\mu_{filt} = \frac{\mu_{filt} + \mu_{filt}'}{2} \tag{4}$$

The compensation parameter (a_{comp}), which will be added to the acceleration command, is basically the opposite of a_{error} . At the beginning of each main cycle, new velocity (v) and acceleration (a) are being calculated based on most recent encoder reading (x) and few previous readings (x', x'').

$$v = \frac{x - x'}{t_{smp}} \tag{5}$$

$$a = \frac{x - 2x' + x''}{t_{smp}^{2}} \tag{6}$$

These parameters (v and a) described UGV motion well D_2 main cycles ago and are at the moment of their acquirement already obsolete because of the delays. To efficiently calculate new suitable acceleration command it is first necessary to calculate current motion parameters by locally simulating physical UGV behavior. This can be done using previously described data.

$$a_{emn} = a_{cmdn} + a_{cmpn} + a_{error} \tag{7}$$

$$v_{emn} = v_{emn-1} + a_{emn} \cdot t_{smp} \tag{8}$$

These calculations are being repeated for each feedback delay ($D_2 + D$ times). The simulated motion parameters are now virtually one time-step from next command and they can be passed to the Command Planner. Command Planner is a part of IMCL, which calculates next command that is required to reach the requested velocity of UGV without exceeding acceleration and jerk limits and without velocity overshoot. When the UGV acceleration nears the limit, it is being suppressed not to exceed the acceleration limit. Relatively long sample time and high jerk limit could otherwise easily cause the acceleration overshoot.

$$j_{next} = j \cdot j_{limit} \cdot \frac{a_s - a_{emn}}{a_{limit}} \tag{9}$$

Depending on the simulated motion parameters and velocity command, a_s can be equal to $-a_{limit}$, a_{limit} or zero. Keeping a_s only on 3 levels is not ideal because the noise from encoder propagates to the IMCL output. Jerk is being limited by simply trimming the value to its nearest limit if it is not within limits yet. Next Command Planner output (a_{next}) can be calculated by multiplying jerk to main cycle time (t_{smp}) and adding the result to the last acceleration value from local simulation (a_{emn}).

The IMCL also detects the best moment to start decreasing acceleration magnitude to smoothly achieve the velocity from the velocity command while acceleration approaches zero. Locally simulating the acceleration decrease scenario and checking the resulting velocity is used. This primitive method is preferred because it does not require the local simulation algorithm to be changed when Command Planner is being modified.

3 Simulation Results

There is exact command waveform on Figure 1 that was inserted to the simulation model (thick line) to be followed by UMCL and actual velocity (thin line) that was actually performed. Peaks are caused by roughly changing disturbance.

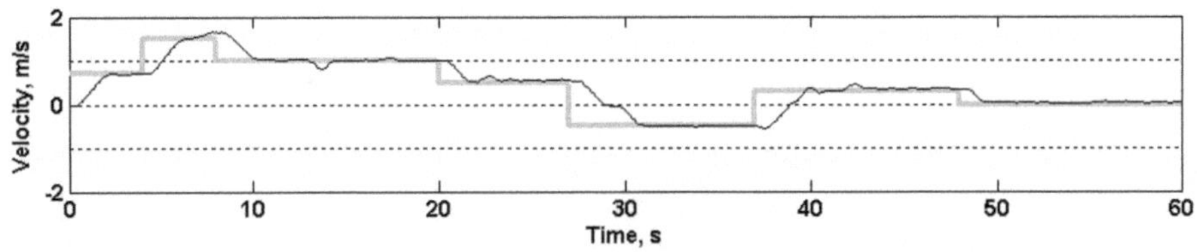

Fig. 1. Velocity Command and Actual UGV Velocity

Figure 2 shows the disturbance caused by friction (thick line) and ascents (thin line). The friction component momentarily changes direction when UGV velocity changes.

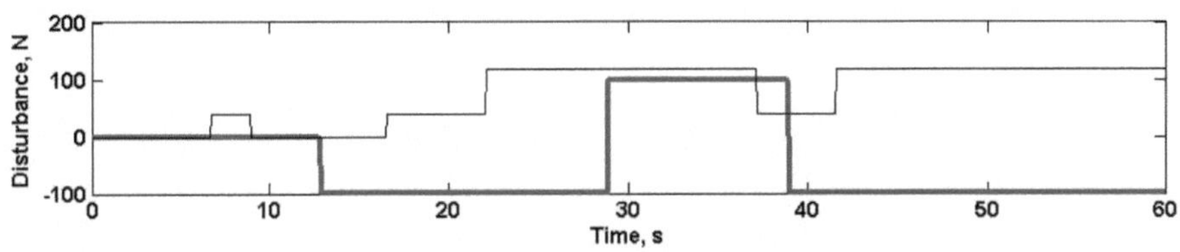

Fig. 2. Components of Disturbance

4 Conclusion

Testing intelligent motion control layer in a simulation gave satisfactory results even in a situation where total feedback delay was extremely long (0.7 s) and sampling rate only (10 Hz). Deployment of previously recorded data (by the same system) is expected to improve compensation effectiveness even more. The tests suggest that the approach could be used to develop advanced control algorithms for UGV type robots and the work in this direction is continuing at TUT.

References

[1] Tallinn University of Technology, Department of Mechatronics, Universal Ground Vehicle, Research project L523, 2005-2008

[2] Hiiemaa, M., Tamre, M.: UGV-type Robot Motion Control Simulation. In: Int. Symposium Topical Problems in the Field of Electrical and Power Engineering (to be published)

[3] Kaiser, D.: Fundamentals of Servo Motion Control. Parker Compumotor, 11 (2001)

[4] Lurie, B., Enright, P.: Classical Feedback Control With Matlab. Control Engineering 6, 448 (2000)

Model Based Controller Design for Automotive Electronic Throttle

R. Grepl[1] and B. Lee[2]

[1] Brno University of Technology, Faculty of Mechanical Engineering,
Institute of Solid Mechanics, Mechatronics and Biomechanics,
Technicka 2896/2, Brno, Czech Republic
grepl@fme.vutbr.cz
[2] Department of Mechanical and Automotive Engineering, Keimyung University, Korea

Abstract. This paper deals with the modelling, parameter estimation and controller design for automotive electronic throttle. The rapid control prototyping methodology and hardware have been used. The properties of simulation model suitable for hardware in the loop experiments are mentioned. Finally the experimental results of proposed controller performance are introduced.

1 Introduction

The electronic throttle is typical mechatronic product which is nowadays a standard part of every automobile. The former mechanical linkage between the driver pedal and throttle is replaced by "drive-by wire" solution – pedal sensor and rotary electromechanical servo system (Fig. 1) connected by means of "wire" [9].

There are very demanding requirements given on the quality of the control (settling time about 0.15s without overshoot, robustness against temperature changes and the wear, etc.) [11]. Moreover, there are two important facts which bring further difficulties: a) the relatively high dry friction as a consequence of low cost mass production; b) "limp home" (LH) mechanical safety feature (in case of failure of electronics or control, the throttle must return to neutral position – this is guaranteed by strongly nonlinear spring) [7]. The linear control tools (PID, LQR) cannot be effectively used in this case.

The typical approach to electronic throttle control is presented in [7], [11]. The key component is friction compensator, usually based on regulation error [5]. Interesting results related to SMC and SMO have been published in [12], [1]. In [11] is presented model based design of controller related to particular Visteon solution.

This paper deals with the controller design based on nonlinear model of the throttle servo system. According to driver and/or traction control commands, there is generated angle reference signal (smooth for slow changes of throttle angle, „stairs" for sudden changes). And at the same time, the friction compensator is switch to appropriate mode (smooth/stairs reference).

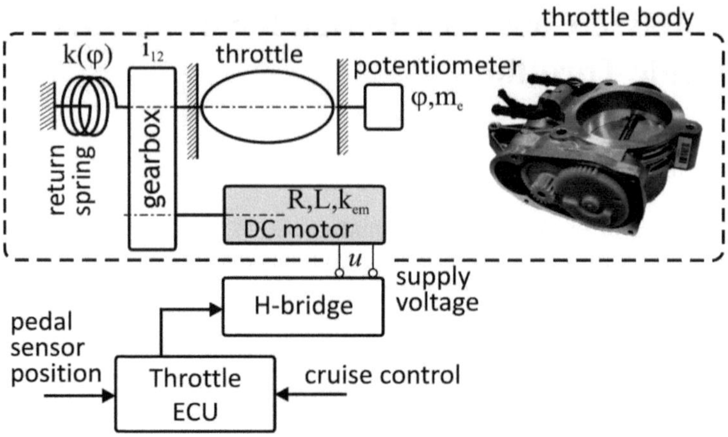

Fig. 1. Schema and photograph of automotive electronic throttle

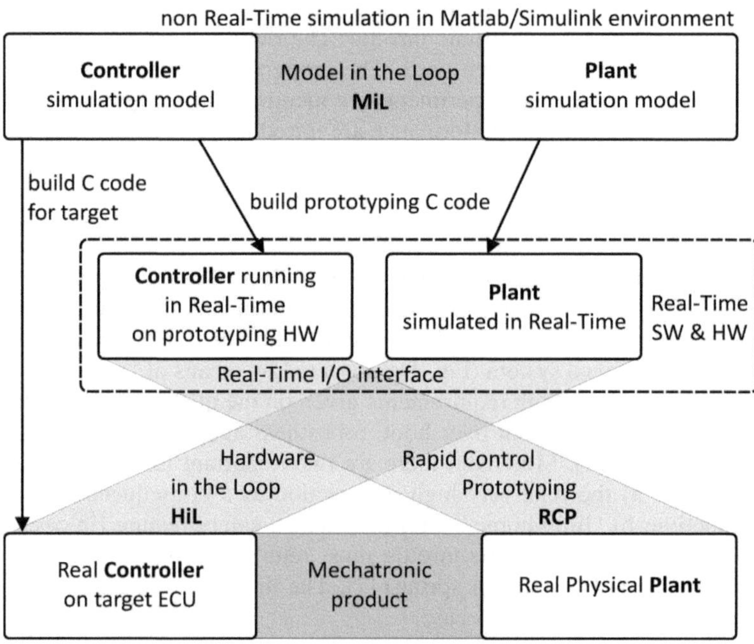

Fig. 2. Overview of Real-Time simulation/computation techniques in the context of model based controller design

Presented results have been obtained using Matlab/Simulink environment with dSPACE modular hardware. Fig. 2 shows the standard development process starting with non Real-Time simulation (MiL – Model in the Loop) leading through the RCP and/or HiL to final mechatronic product.

2 Experimental Identification of Static System Properties

The aim of presented research is the controller design for existing throttle body. Therefore the system identification using RCP hardware was necessary. Fig. 3 a) shows the system response on slow sinusoidal input signal (normalized -1;1) $u = 0.35\sin(0.2t)$.

The hysteresis caused by dry friction is significant as well as nonlinearity of return spring. The opening from neutral (LH) position requires approx. u = 0.2 and throttle is fully opened at u=0.28.

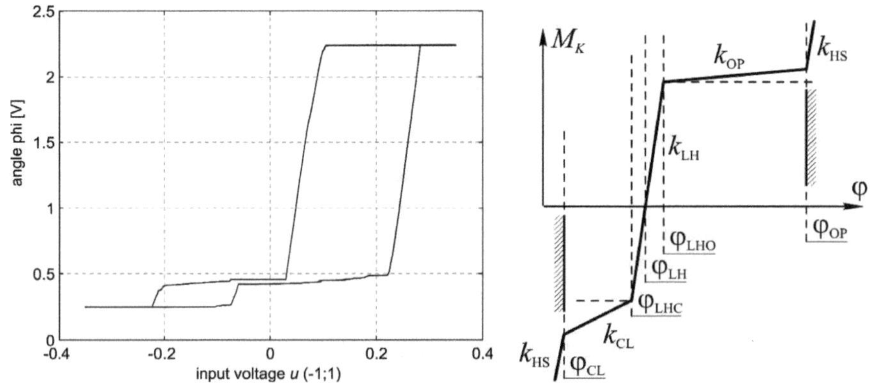

a) b)

Fig. 3. Static characteristic of the system: a) measured quasi-static response; b) spring characteristic [3]

3 Modelling and Parameter Estimation of Electromechanical Plant

If the armature inductance is neglected, the rotary electromechanical servo system with return spring can be simplified to 1 dof mechanical system of form [3]

$$J\ddot{\varphi} = u - b\dot{\varphi} - \tau_S(\varphi) - \tau_F(\dot{\varphi}) + \tau_{HS} \tag{1}$$

where J is inertia, b is viscous damping including back electromechanical force, τ_S is spring force τ_F is friction force, τ_{HS} is hardstop force when the plate hit the limit (see Sec. 4.) and u is DC motor supply voltage controlled by H-bridge. All parameters are normalized relatively to voltage u (-1;1). According to Fig. 3 b), the nonlinear spring is modelled as [3]

$$\tau_S = \begin{cases} k_{LH}(\varphi - \varphi_{LH}), & -\varphi_{LHC} < \varphi < \varphi_{LHO} \\ u_{LHO} + k_{OP}(\varphi_{LHO} - \varphi_{LH}), & \varphi > \varphi_{LHO} \\ -u_{LHC} + k_{CL}(\varphi_{LHC} - \varphi_{LH}), & \varphi < \varphi_{LHC} \end{cases} \tag{2}$$

The appropriate friction modelling is necessary for successful parameter estimation. The Coulomb static model is insufficient and dynamic model must be used instead. The LuGre model is probably the most often used, however the Reset Integrator model has less parameters to be estimated and proves similar behaviour.

$$\dot{p} = \begin{cases} 0 & \text{if } (\dot{\varphi} > 0 \wedge p \geq p_0) \vee (\dot{\varphi} < 0 \wedge p \leq -p_0) \\ v & \text{otherwise} \end{cases} \tag{3}$$

$$\tau_F = \frac{(1 + a(p))\tau_{\mathrm{Fkin}} p}{p_0} + \beta \dot{p}, \quad a(p) = \begin{cases} a & \text{if } |p| < p_0 \\ 0 & \text{otherwise} \end{cases} \tag{4}$$

The model introduces additional state p to the system, which can be understood as the bending of virtual bristle. The parameter τ_{Fkin} is kinematic friction, a defines the static friction increase in low velocities (Stribeck), p_0 determines the stiction range and β is damping coefficient.

Many methods are available for parameter estimation. Many researches successfully used the GA for nonlinear system identification [6]. In this work, the Nelder-Mead simplex search algorithm was used to obtain dynamic and static parameters of the model [2], [3].

4 Plant Model Optimized for HiL

The HiL experiments executed on Real-Time hardware requires simulation model with the limitation of throttle plate movement. In reality, there is very complex contact dynamical behaviour when the plate bounces off the limit. The simplest approach applicable in Simulink ODE solver environment is penalty method based on Hertz impact model. At the end of working range of the plate is inserted linear virtual spring of form [3]

$$\tau_{\mathrm{HS}} = k_{HS}\varphi + |\varphi| b_{\mathrm{HS}}\dot{\varphi} \tag{5}$$

Moreover, the experiments prove that during the impact the armature inductance cannot be neglected. The change of angular velocity of the plate is very sudden and in supply voltage is constant, the current changes very quickly.

The model must also use fixed sample time and thus Zero Crossing Detection feature cannot be applied.

5 Nonlinear Model Based Controller

The nonlinear controller shown in Fig. 4 consists of standard discrete PID supplement with feedforward spring compensator based on eq. 2 and feedback/feedforward friction compensator.

$$u = u_{\mathrm{PID}} + u_{\mathrm{SC}} + u_{\mathrm{FC}} \tag{6}$$

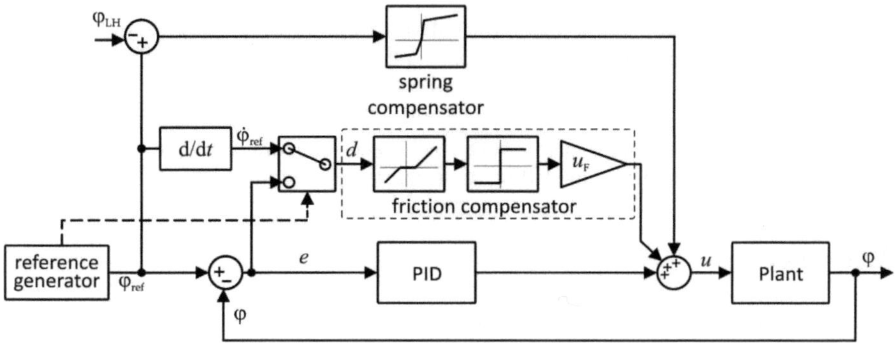

Fig. 4. Schema of nonlinear controller

$$u_{\mathrm{PID}} = k_{\mathrm{P}}(\varphi - \varphi_{\mathrm{ref}}) + k_I \sum_N (\varphi - \varphi_{\mathrm{ref}}) + u_{\mathrm{D}} \tag{7}$$

The derivative action is filtered using derivative impulse area invariant method [3]

$$u_{\mathrm{D}}(z) = \frac{T_{\mathrm{d}}}{T_{\mathrm{s}}}\left(1 - e^{-\frac{GT_{\mathrm{s}}}{T_{\mathrm{d}}}}\right)\frac{1 - z^{-1}}{1 - e^{-\frac{GT_{\mathrm{s}}}{T_{\mathrm{d}}}} z^{-1}} \tag{8}$$

If the sudden change of plate position is required, the feedback compensation using control error gives good results [5]. On the contrary, if slow rotation of throttle is necessary, the better performance is obtained using feedforward friction compensator based on differentiation of reference.

$$u_{\mathrm{FC}} = u_{\mathrm{F}}\ \mathrm{sign}(d), \quad d = \begin{cases} d - \varepsilon, \text{if } d > \varepsilon \\ d + \varepsilon, \text{if } d < \varepsilon \\ 0, \text{ else} \end{cases} \tag{9}$$

6 Conclusion

In this paper, the controller with nonlinear spring and friction compensator has been introduced. Fig. 5 shows the performance if smooth and stairs reference is used. All compensators are based on dynamic plant model with off-line estimated parameters.

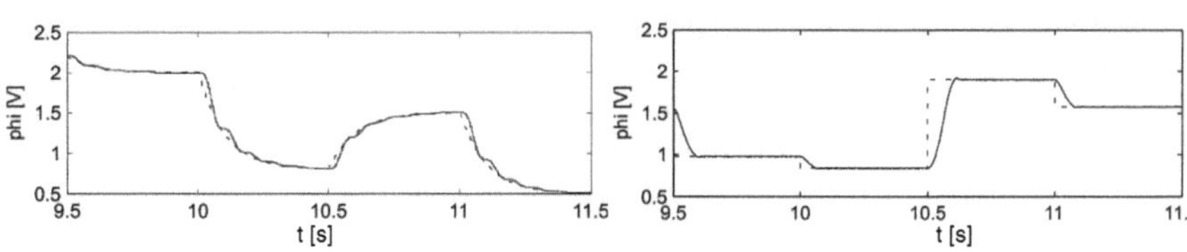

Fig. 5. Response of the system with smooth (left) and stairs (right) reference signal [3]

Experimental results have been obtained using modular dSPACE hardware with processor board DS 1005 PPC, board DS 2003 with 16 bit A/D converters and board DS 2103 with 14 bit D/A converters.

Acknowledgments. The presented work has been supported by research project MSM 0021630518 "Simulation modelling of mechatronic systems".

References

[1] Beghi, A., Nardo, L., Stevanato, M.: Observer-based discrete-time sliding mode throttle control for drive-by-wire operation of a racing motorcycle engine. IEEE Transactions on Control Systems Technology 14, 767–775 (2006)
[2] Grepl, R., Lee, B.: Modelling, identification and control of electronic throttle using dSpace tools, Technical Computing Prague (2008)
[3] Grepl, R., Lee, B.: Modelling, identification and Nonlinear Control of Electronic Throttle. In: The KSAE Daegu Gyungbuk Regional Conference, Daegu, Korea, pp. 45–53. Korean Society of Automotive Engineers (2008)
[4] Hadilebbal, M., Chafouk, H., Hoblos, G., Lefebvre, D.: Modeling and identification of non-linear systems by a multimodel approach: application to a throttle valve. International Journal of Information and Systems Sciences 3(1), 79–99 (2007)
[5] Isermann, R., Lachmann, K., Matko, D.: Adaptive Control Systems. Prentice-Hall, Englewood Cliffs (1991)
[6] Krejsa, J., Houfek, L., Věchet, S.: The Influence of GA Initial Boundaries on the Identification of Nonlinear Damping Characteristics of Shock Absorber. In: Recent Advances in Mechatronics, pp. 411–415. Springer, Heidelberg (2007)
[7] Pavković, D., Deur, J., Jansz, M., Perić, N.: Adaptive Control of Automotive Electronic Throttle. Control Engineering Practice 14, 121–136 (2006)
[8] Pivonka, P., Schmidt, M.: Comparative Analysis of Discrete Derivative Implementations in PID Controllers. Systems Theory and Applications,WSEAS 2, 33–37 (2007) ISBN: 978-960-8457-90-4
[9] Stence, R.W.: Digital By-Wire Replaces Mechanical Systems in Cars, in Electronic Braking, Traction, and Stability Controls. In: Society of Automotive Engineers, Inc., USA, pp. 29–36 (2006)
[10] Wierciak, J.: Improvement of performance of precision drive systems by means of additional feedback loop employed. In: Recent Advances in Mechatronics, pp. 495–499. Springer, Heidelberg (2007)
[11] Yang, C.: Model-Based Analysis and Tuning of Electronic Throttle Controllers. In: SAE World Congress, Detroit, Michigan, (March 8-11, 2004)
[12] Zhang, P., Yin, C., Zhang, J.: Sliding Mode Control with Sensor Fault Tolerant for Electronic Throttle. In: International Conference on Automation Science and Engineering, Shanghai, China (2006)

The Solution of 3D Indoor Simulation of Mobile Robots Using ODE

V. Ondroušek

Brno University of Technology, Faculty of Mechanical Engineering,
Institute of Automation and Computer Science,
Technická 2896/2, Brno, Czech Republic
ondrousek@fme.vutbr.cz

Abstract. This contribution deals with the development of the complex solution of various mobile robots simulations in 3D indoor spaces. Proposed solution enables to create miscellaneous complex inner spaces with furniture of office visual appearance. Furthermore, it allows including different mobile robots and saving the state of simulation using standard XODE format. It is also possible to include and test different algorithms, e.g. walking gait generation algorithms or methods for autonomous robots localization, path planning and many others. The designed project is based on open source libraries, like Open Dynamic Engine or OpenGL.

1 Introduction

Our mechatronic team is focused on autonomous mobile robots development, wheeled robots as well as legged robots. There are many problems, which have to be solved in development of hierarchical control architecture, which our team uses for mobile robots design. The lowest layer is represented by hardware, i.e. construction of a robot with driving gears. The higher (second) layer consists of low-level control units for driving gears and motors. The third layer ensures basic movements of the robot, in the case of legged robot the walking gait generation. The highest layer usually involves suitable algorithm of global localization. For more details about such architecture see [1].

The main goal of this issue is to describe the designed solution, which can be used to verify abilities of various algorithms of the two highest layers. Thus, in our case, it is used mainly for testing the walking gait generation algorithms based on state space search methods, see [2], and also for global localization algorithms [3]. However, the submitted solution can be also used for path planning algorithms, investigation of dynamic influences in mobile robot control, testing algorithms for unknown spaces mapping, see [4], and many others.

2 Used Approach

One of the main aims was to find suitable tools that meet our requirements, e.g. high accuracy, incorporation of existing programming code of walking gait generation algorithms, 3D graphical representation of simulation, involving static and dynamic effects from acting forces, etc. We can choose from many various ready-made solutions for robot modeling and walking gait simulations. There are at least two different approaches to creating model of mobile robots. First approach uses commercial modeling software, e.g. Adams or Inventor. These simulation packages represent very powerful and sophisticated tools, which can be used to very precisely modeling. But even these very expensive programs have limitations. The main disadvantage in our case is inability of integration of our own programming code, used for walking gait generation or robot localization. Another approach is based on special 3D engines, e.g. Irlicht or ODE. These engines are able to represent dynamic world with colliding rigid bodies, joints, surface or body forces and torques. However, they do not usually provide graphical interface for 3D representation of simulating world. Most of them are provided for free using dll libraries and detailed API description.

We have decided to build our own solution based on combination of Open Dynamics engine (ODE) and OpenGL. The engine is used to create virtual world with colliding bodies and acting forces. The OpenGL library is used for 3D graphical representation of simulation. The Python is used as the main programming language for implementing whole solution. The point at issue is the design of architecture that enables incorporating these tools into one project and also enables simple integration of program code for robot manipulation. Furthermore, another problem is choosing an appropriate format of data, which could be used for storing, holding and loading the state of simulation at any time, for details see next chapter.

3 Implementation

The designed solution represents stand alone application, which was designed for the purpose of different mobile robots acting in different inner spaces. This application software is based on Open Dynamics Engine. Using the ODE in a programming code of python language is enabled through PyODE. The PyODE project is a set of open-source libraries, which are used to call ODE functions from Python code and also includes a parser of XODE (extensible markup language for open dynamics engine) format. Thus, the Xml ODE format can be used to describe simulated world with all its bodies, geometries, joints, acting forces and many others. The designed application uses XODE format to save, store and load states of simulation.

Simplified scheme of the inputs and outputs of whole application is shown on Fig. 1. The proposed application enables to load any inner spaces with the model of mobile robot from XODE format file. The application contains preprepared blank subroutines for including the code to manipulate the robots, e.g. inner space mapping. The application enables to run simulation with numerical output, graphical output is optional. It is possible to save the state of simulation at defined time

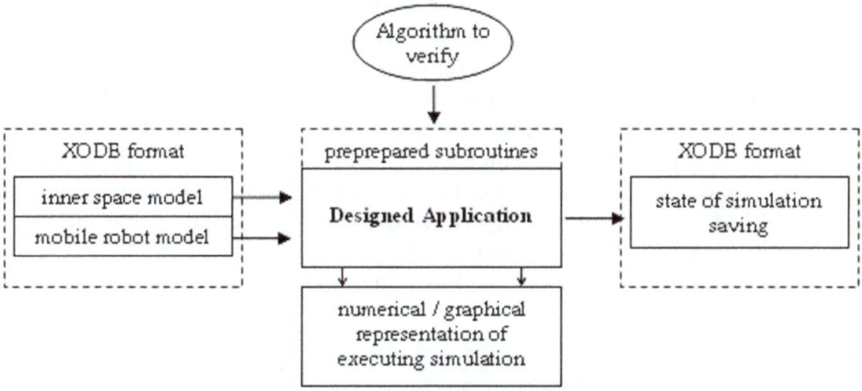

Fig. 1. Simplified scheme of proposed solution

steps into XODE format file. The application has no graphical user interface, thus all mentioned settings has to be accomplished using handwriting of the python program code.

The generator has been developed for the purpose of creating miscellaneous complex inner spaces with furniture of office visual appearance, as you can see on Fig. 2. This generator enables to create various inner spaces in an easy way using predefined set of parameters and save these spaces in XODE format. It is possible to specify built-up area and number of rooms, with or without furniture, number of pieces of furniture in any room, and many others. Rooms can be of square shape or polyhedral shape. The overall arrangement of rooms can be alternatively generated using Lindenmayer system, see [5] for more information. Such created indoor spaces can be used as the inputs of the application mentioned above.

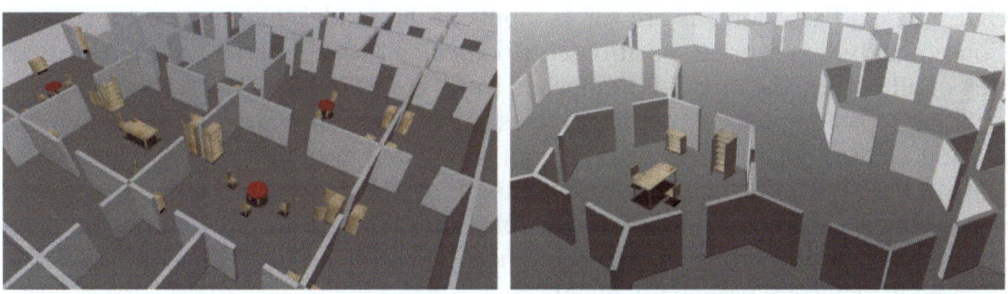

Fig. 2. Inner spaces samples generated automatically by the generator, various rooms shapes

4 Obtained Results

Several tests with the real robot and the designed models were performed to judge the accuracy of proposed simulation and suitability of used software instruments. At first, several various inner spaces was created using the generator mentioned above, see Fig. 2. Thereafter the selected models of the robots were incorporated into these indoor environments and stored as XODE files. Such files were used as the inputs of the proposed application. The walking gait task, see Fig. 3a, and the balancing task, see Fig. 3b, were used for testing whole solution.

The robot of interest is the walking robot Qasimodo, developed on Brno Technical University, Institute of Automation and Computer Science. This four legged robot has two degrees of freedom on each leg. Movements of each leg are realized using 2 servomotors HS 300. The Qasimodo weights about 1000 grams (190g for each leg). Working space of each leg is 190 degrees in horizontal plane and 80 degrees in vertical plane. Whole model of this robot is composed of 21 rigid bodies and 24 joints, see [6] for more details. Resulting model is shown on Fig. 3.

Fig. 3. Resulting dynamic simulation of a) walking robot task, b) balancing wheeled robot task

The walking gait used by the robot can be considered as a pseudo-dynamic cyclic forward gait composed of two cycling types of long steps. As soon as the first long step is finished, the second type of long step starts, for details see [2]. The walking gait simulation is realized by placing torques into the hinge joints of the ODE model. These torques are set as inputs of the model in requested time steps. Consequently, appropriate positions, velocities and accelerations in the next time step of all bodies, the model is composed from, are calculated by the model.

None of accomplished simulation tests froze or ended with the "simulation explosion". Some details about walking gait simulation tests follow. The visual comparison based on 3D graphical representation of simulating walking gait with the real walking gait showed approximate consistency. It means the walking gait diagram of both gaits correspond to each other. The sequence of foot steps as well as covered distance are practically equal.

The measured data showed admissible consistency between the simulated walking gait and the real one. The difference is not cumulative, because there is no dependency on the number of steps. However, existing differences between measured and calculated data can be explained using special parameters of the ODE simulation. There are two parameters, error correction parameter (ERP) and constraint force mixing (CFM), which have to be set for each simulation. The ERP value controls how much error correction is performed in each time step of simulation, typically from 0.1 to 0.8. The CFM value can reduce numerical errors in the simulation, whenever the system is near singularity, typically from 10E-9 to 1. These non-dimensional parameters do not have any physical basement and must be set experimentally, which can be difficult and time-consuming.

5 Conclusion and Discussion

This contribution introduces the designed application, which can be used for dynamic simulations of various mobile robots in 3D indoor spaces. The designed solution is represented by the stand alone application based on Open Dynamics Engine, which is able to represent dynamic world with colliding rigid bodies, joints, surface or body forces and torques. The OpenGL libraries are used for 3D graphical representation of simulation. The XODE format of data has been chosen for storing, holding and loading the state of simulation. The proposed application enables to load any inner spaces with the models of mobile robots from XODE file and contains prepared blank subroutines for incorporating code to manipulate the robots. The application enables to run simulations with numerical output, graphical output is optional. The automatic generator of inner spaces was design too. This application is able to create various inner spaces and store these models in XODE format. Such files can be used as the inputs of the main application.

Several tests with the real robot and the designed models were performed to judge the accuracy of proposed simulation and suitability of used software instruments, mainly the test of walking gait generation of four legged robot. None of accomplished simulation tests froze or ended with the "simulation explosion". The measured data showed admissible consistency between the simulated walking gait and the real one.

Thus proposed solution can be used to verify abilities of various algorithms from indoor mobile robots domain, especially for testing walking gait generation algorithms, methods for autonomous robots localization, path planning, unknown spaces mapping and many others.

Acknowledgements. Published results were acquired using the subsidization of the Ministry of Education, Youth and Sports of the Czech Republic, research plan MSM 0021630518 "Simulation modelling of mechatronic systems".

References

[1] Vlachy, D., Zezula, P., Grepl, R.: Control unit architecture for biped robot. In: Recent Advances in Mechatronics, Berlin, pp. 6-10. Springer, Heidelberg (2007)

[2] Ondroušek, V.: Using the Open Dynamics Engine for Walking Robot Simulation. In: Modelling of Mechatronic Systems IV, pp. 121–128. Brno University of Technology (2008)

[3] Věchet, S., Krejsa, J., Houška, P.: The Enhancement of PCSM Method by Motion History Analysis. In: Recent Advances in Mechatronics, Berlin, pp. 107–110. Springer, Heidelberg (2007)

[4] Krejsa, J., Věchet, S.: Dealing with Sensor Errors in Scan Matching for Simultaneous Localization and Mapping. Engineering Mechanics 15, 337–344 (2008)
[5] Schreiber, P.: Automatic Environment Generator Design for Mobile Robot bachelor's thesis, Brno University of Technology, Faculty of Mechanical Engineering, Brno (2008)
[6] Seriš, R.: Using ODE for Design of Dynamic Model of a Four-legged Robot. bachelor's thesis, Brno University of Technology, Faculty of Mechanical Engineering, Brno (2008)

Sensors Data Fusion via Bayesian Network

S. Věchet and J. Krejsa

Brno University of Technology, Faculty of Mechanical Engineering,
Technická 2896/2, Brno, 616 69, Czech Republic
vechet.s@fme.vutbr.cz

Abstract. Presented paper deals with the fusion of information obtained from different kinds of sensors placed on autonomous mobile robot. The method is based on Bayesian network. Implementation details and verification simulation experiment that fuses three different means to determine robot orientation are given in the paper. The method is easily extendable for higher number of sensors of different kind together with higher dimension output data to be fused.

1 Introduction

Data fusion is one of essential issues in mobile robotics [1]. It deals with the fusion of different kinds of information measured by various sensors. For example laser range finder, ultrasonic, infrared, compass, GPS sensors are mounted on common robot. The sensors are used for different purposes as they have variant sensing principles.

Data fusion can generally be divided into the three main groups which can be classified according to sensors configuration as follows:

- competitive – different kinds of sensors are used to measure the same environment attribute; possible information redundancy,
- complementary – each sensor reads different attribute of the same environment,
- cooperative – in that case one sensor depends on the other; they have to work together

So far we have been using complementary data fusion in all our robots. For such a purpose we have been developing a method of data fusion based on Bayes theorem, which is the basic method for conditional probability calculation.

Presented paper describes the basic method for fusion of data acquired via odometry, compass and steering angle. Naturally, it is easy to widespread this basic set of used sensors with any others. Therefore, this method can be used as a main tool for measured data fusion.

2 Bayes Theorem

Bayes theorem is a basic method to deal with conditional probability, more precisely it relates the conditional probability of events A and B. It is well known how to derive it from basic conditional probabilities equations (for example in [2]), so here just the final state of the Bayes theorem equation is presented:

$$P(A \mid B) = \frac{P(B \mid A)P(A)}{P(B)} \tag{1}$$

where:

- $P(A)$ is the prior probability of A ; in sense what we know about A on beginning,
- $P(A \mid B)$ is called posterior probability; the conditional probability of A given B ,
- $P(B \mid A)$ is the conditional probability of B given A ,
- $P(B)$ is the prior probability of B ; it acts as normalizing constant

As it is shown, the main principle is based on $P(A \mid B)$ calculation, if we know so-called "inverse" probability $P(B \mid A)$.

Bayes theorem can be easily used in more complicated relationships with more then two events. One of possible representations of such relationships is Bayesian network [3]. The primitive relation of two events described above could be considered the simplest network (see figure 1 - left). Basically, the Bayesian networks are primarily used for more complex relationship description (see figure 1, right). On that figure the relations between some of the sensors discussed above are shown. Those relations are naturally created as it is intuitive for human. It is easy to derive how the final orientation of the robot is influenced by the sensors.

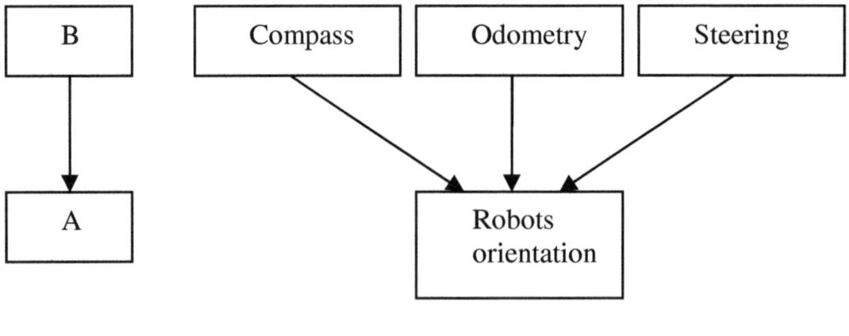

Fig. 1. The simplest Bayesian network (left) and the practical example (right)

3 The Bayesian Network Simulator

We have prepared the simulator of Bayesian network to be able to work efficiently with various Bayesian networks. This simulator is called Sybok and it is completely written in Python. The structure of Sybok is shown on figure 2.

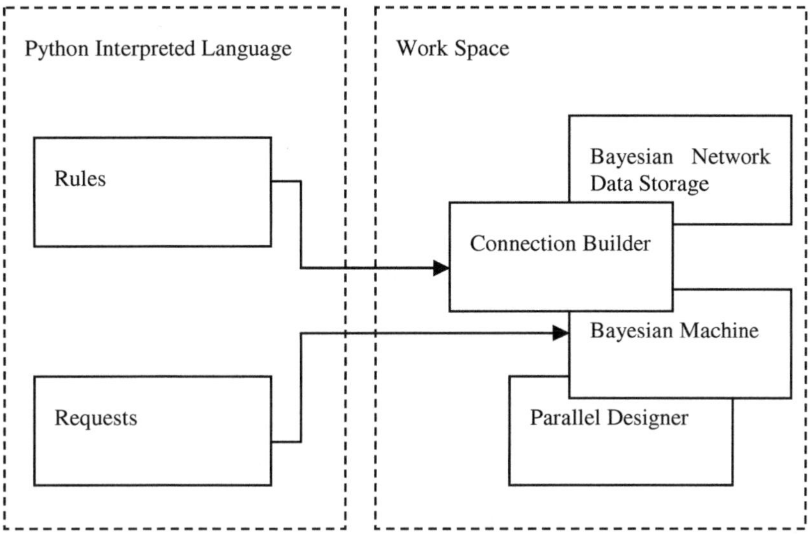

Fig. 2. Sybok internal structure

There are two main parts: Python Interpreted Language (PIL) and the Work Space (WS). The PIL acts as a program interface to the WS which is the implementation of Bayesian Network. There are six important blocks:

- Rules - user definition of network structure
- Requests – user requests to the network
- Connection Builder – interface between user inputs definition and the real data structure
- Bayesian Network Data Storage – it holds the data
- Bayesian Machine – acts as a main computation framework, it computes all necessary probabilities requested by the user via automatic inference mechanism.
- Parallel Designer – Bayesian network can be implemented as a parallel algorithm so this block organizes the parallel operations.

Experimental results obtained from Sybok are shown in next chapter as we have used it as a main tool for data fusion of robot orientation measurements.

4 Data Fusion via Sybok

To detect that our method to data fusion works properly, a simple simulation experiment with autonomous robot was prepared. The robot was equipped with three

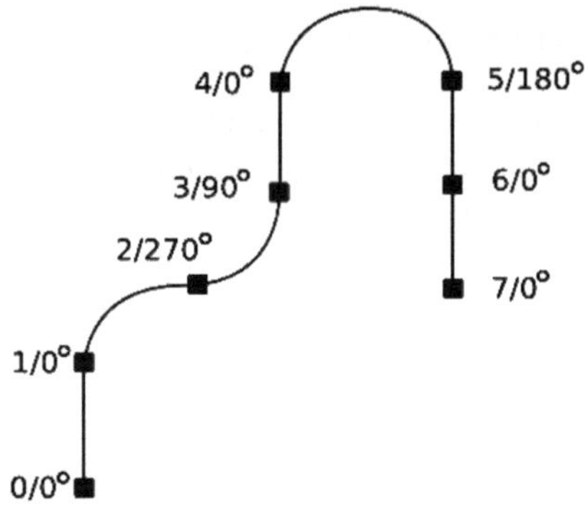

Fig. 3. Path followed by autonomous robot; Position/Robot orientation relative to robot's previous orientation

means to determine its relative orientation on followed path. First one was the compass which was used to measure absolute orientation of the robot. The second one was an odometry reading and the change in orientation was calculated from the difference in traveled distances of each single wheel. The third mechanism was the orientation of the robot calculated from steering angle.

The path traveled by the robot is shown on figure 3. There are seven points in which the orientation was measured. On that figure one can see the position number/true robot orientation in degrees relative to the robot previous location. The orientation was measured by all three methods and we use Sybok to data fusion.

The results are shown on figure 4. On the bottom graph the orientation measured by compass, odometry and steering in each position on the path is shown. The upper graph shows the probabilities that the robot is oriented 0, 90, 180 or 270 degrees. One can see that the highest probability in each point corresponds with the true orientation of the robot (see figure 3). Those probabilities were calculated by the Bayesian network with structure shown on figure 1.

The best example illustrating how the data fusion works can be seen in point 5 on figure 4. The robots real orientation is 180 degrees and the compass measured that orientation properly. On the other hand the orientations calculated from odometry and steering were wrong (odometry 270 degrees, steering 90 degrees), but the probability that the robots orientation is 180 degree is still highest. This is caused by different probabilities of correct orientation measuring. Note, that compass measures with the same probability in all directions independently $P(Compass = 0) = 0.95$. However, odometry and steering have high probabilities of success for angle 0, for example $P(Odometry = 0) = 0.95$, less for angle 90 or 270 $P(Odometry = 90) = 0.5$ and even lower probability for angle 180 $P(Odometry = 180) = 0.3$. These probabilities were measured in various practical experiments and they causes that if compass measured angle 180 degrees and odometry and steering 90 or 270 degrees, than these two values have less influence to the calculated orientation.

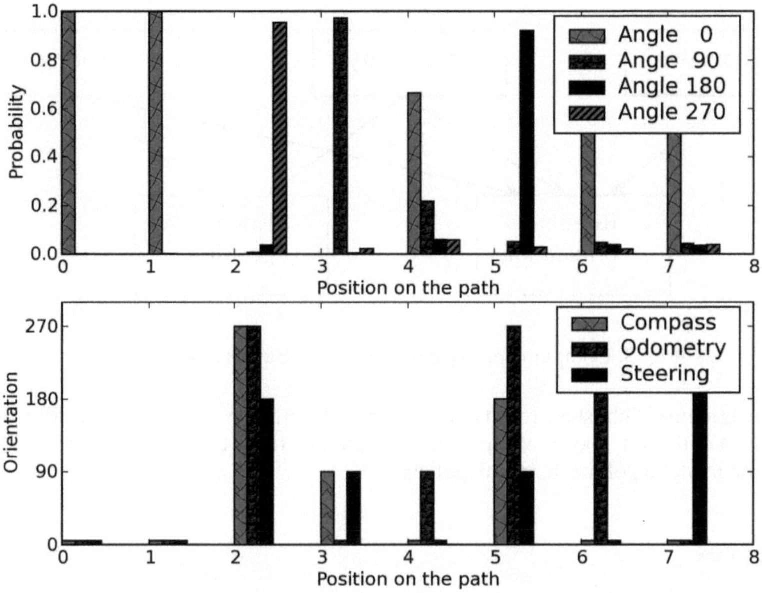

Fig. 4. Robot orientation identification via Sybok

Presented example shows how to use Bayesian network in basic task. The method can be easily extended to be used in more challenging problems including continuous variables [4]. In such a case discretization of corresponding variable is necessary.

5 Conclusions

We have presented the primary simulation experiments with data fusion via Bayesian network. Three sensors (compass, odometry and steering) were used to measure the orientation and the Bayesian network was successfully applied to sensors data fusion, determining the robot orientation relative to its previous value. The key issue in the method itself is proper determination of condition probabilities of mutually related events. The simplest way to set the probabilities is to directly utilize known imperfections in sensor measurement. More complex approach takes advantage of various learning systems and/or evaluation of robot's overall behavior.

Presented method can be used also for complementary data fusion, where the odometry or steering are used for both orientation and position estimation, see figure 5.

Future work will be focused on verification experiments with real data. For that purpose we have built the autonomous wheeled robot equipped with a number of sensors of various kinds. Bayesian network simulator Sybok is currently tested on this sensor set with encouraging preliminary results which will be announced shortly.

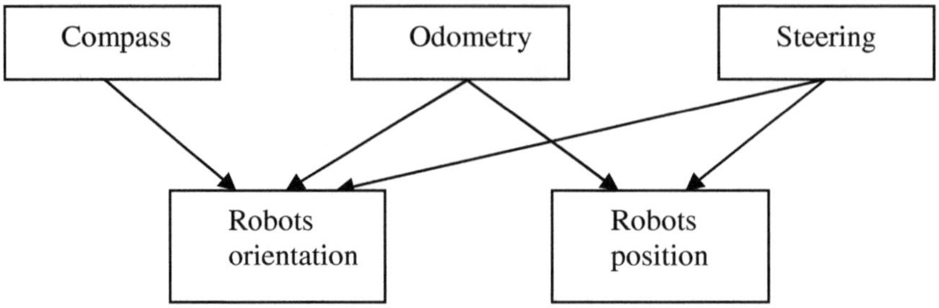

Fig. 5. Complementrary data fusion via Bayesian network

Acknowledgments. Published results were acquired with the support of the Ministry of Education, Youth and Sports of the Czech Republic, research plan MSM 0021630518 "Simulation modeling of mechatronic systems".

References

[1] Singhal, A., Brown, C.: Dynamic Bayes Net Approach to Multimodal Sensor Fusion. In: Proceedings of the SPIE - The International Society for Optical Engineering, pp. 2–10 (1997)
[2] Jensen, F.V.: Bayesian networks and Decision Graphs. In: Statistics for Engineering and Information Science. Springer, Heidelberg (2001)
[3] Murphy, K.P.: Dynamic Bayesian Networks: Representation, Inference and Learning, PhD thesis, UC Berkley, Computer Science Division (2002)
[4] Jensen, F.: An Introduction to Bayesian Networks. UCL Press (1996)

Study Model of the Snake Like Robot

M. Kelemen and T. Kelemenová

Technical University of Košice, Faculty of Mechanical Engineering,
Institute of Special Technical Science, Department of Applied Mechanics and Mechatronics and Department of Biomedical Engineering, Automation and Measurement,
Letná 9, 042 00 Košice, Slovak Republic
michal.kelemen@tuke.sk

Abstract. Paper deals with design of study model snake like robot. Robot is designed as articulated locomotion device. It consists of eight articulated joints with one degree of freedom. It is designed as didactic model for education in Mechatronic study program.

1 Introduction

Many ground locomotion devices often use endless rotating elements such as wheels or tracks. Endless rotating elements are the fascinating elements in machines. These elements have one big disadvantage. They are not suitable for extremely rough terrain. There is no analogy for wheels and tracks in the nature. In this time, there are a lot of applications with biologically inspired locomotion. Place for biologically inspired locomotion is mainly in cases, where we cannot use wheels or tracks. Living organisms are able to adapt to surround conditions, because of its physiological needs. Consequently, they are able to change own shape and locomotion type.

Snakes have interesting locomotion and they have fascinated a lot of people. The most famous research of snake like locomotion comes from Hirose & Yoneda Lab. Their results are summarized in book [1].

They have observed four basic types of snake locomotion. Using of these locomotion types depends on environment condition and purpose of locomotion. These locomotion types can be divided into these four modes [1]:

1. serpentine locomotion
2. rectilinear locomotion
3. concertina locomotion
4. sidewinding locomotion

Serpentine locomotion is the locomotion to be seen typically in almost kinds of snake, and is a gliding mode whose characteristic is that each part of body makes similar tracks. From ancient times this has been the mode which has propelled snakes like flowing water between rocks, for instance, and has surprised humans, and of the four modes this can be thought of as the most efficient.

Rectilinear locomotion is the gliding method performed with a special configuration by large snake such as boas and vipers when approaching their prey or when gliding over a smooth surface.

Concertina locomotion is gliding method used by snakes confined to a straight path over a narrow straight line, and by snakes placed on floor surfaces, for example, which are extremely slippery. In particular, the gliding configuration on such a floor surface uses the phenomenon that theoretical terms the coefficient of the static friction is greater than the coefficient of dynamic friction. For this reason, propulsion is possible even in a very slippery environment, using this gliding mode. However, the efficiency of such propulsion is extremely low.

Sidewinding locomotion is the gliding method used by snake such as the rattlesnake which live in the desert, and which lift part of their body while gliding and propel themselves like a tumbling spiral coil. In this mode of locomotion, there is no sliding movement between the body and surface glided over, its dynamic characteristic being that the body usually contacts the ground from above. Because of this characteristic, sliding friction resistance is small, and in locomotions in environments which are not firm, such as sandy ground, the locomotive efficiency is high [1].

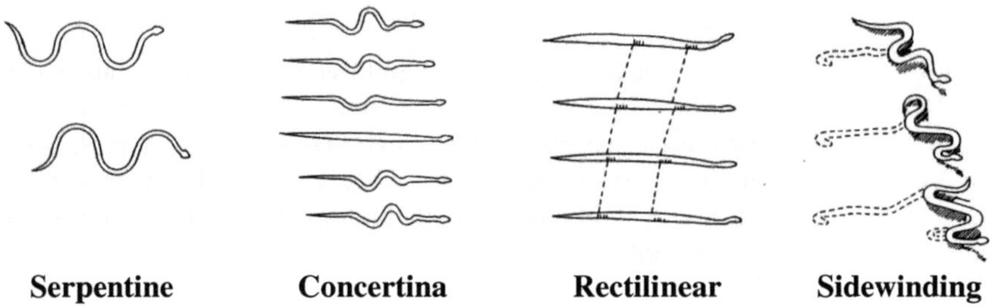

Fig. 1. Snake locomotion [1]

2 Kinematic Arrangement of the Articulated Robot

Robot consists of eight articles joined with seven plane joints (fig. 2). The base of the kinematic principle lies on alternating of vertical and horizontal plane joints. First article (head) and last article (tail) are designed with vertical plane joint, because of their possibility to cross any obstacles (fig. 3). So, every kinematic pair is designed with one degree of freedom.

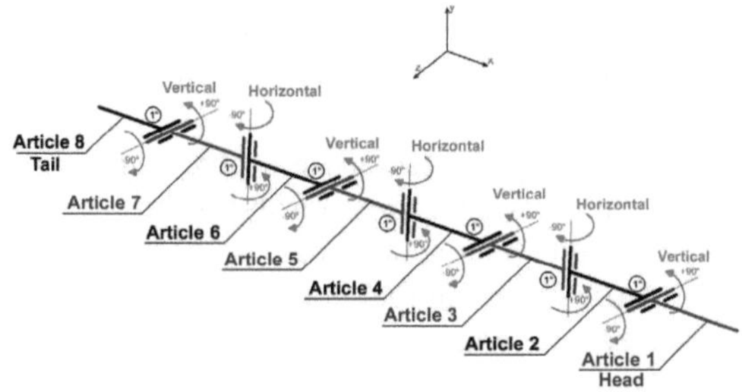

Fig. 2. Kinematic arrangement of the robot [2]

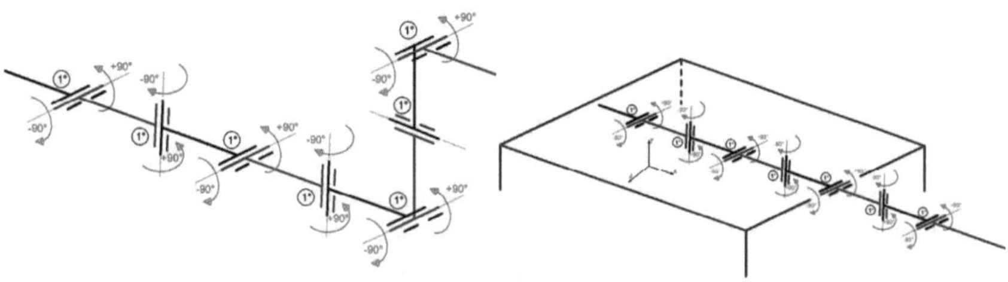

Fig. 3. Robot locomotion [2]

Preliminary model has been created from the paper. This paper model has been used for study of locomotion possibilities (fig. 4).

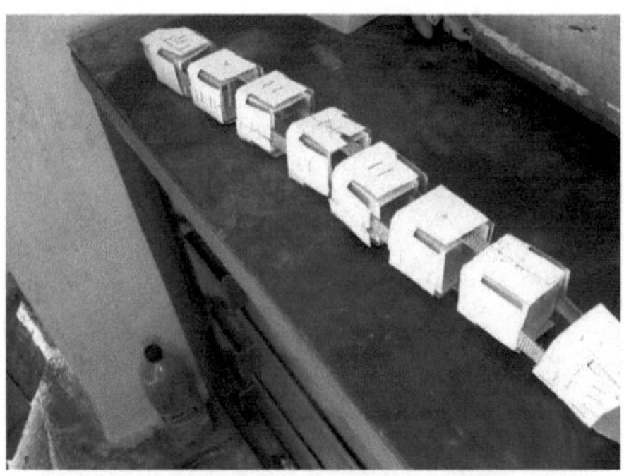

Fig. 4. Paper model of the robot[2]

3 Design of the Robot

Every article has one degree of freedom in regard to neighbour robot article. These plane joints are driven with actuators. As actuator is chosen position servomechanism from Hitec HS 645MG [3]. Desired angular position of servomechanism output shaft is entered as pulse width in range 1,0 až 2,0 ms. Every servomechanism has DC motor with gearbox, control unit and sensing potentiometer for sensing of angular position of output shaft. The information from the potentiometer is used for internal control unit.

General problem for design of locomotion algorithm is timing of sequence. Time of every joint rotation is dependant on its loading and position. So, microcomputer needs information about actual angular position of every joint. Ideal solution is to use information from internal potentiometer as feedback for our microcomputer. Consequently, we don't need additional sensor for sensing of joint angular positions. This idea reduces overall weight of robot and simplifies design.

For obtaining of this information it is necessary to connect additional cable to potentiometer slider inside the servomechanisms. Our experiment has showed that voltage on potentiometer slider depends on angular position of output shaft. This voltage also depends on supply voltage of servomechanism. So, it is necessary to use voltage stabiliser for power supplying of servomechanisms. This modification brings signals about angular position of every robot joint.

First experiments showed another problem. If one of servomechanisms has been excited, then every another servomechanisms also moved. So, servomechanisms have been disturbed each other. It was necessary to use shielded wires for control signals. All signal wires have been changed.

Our microcontroller Basic Atom Pro 28M has been used for controlling of servomechanisms based on feedback from internal servomechanisms sensors [2].

4 Arrangement of the Robot

The article of the robot consists of 13 parts. Symmetry and precision of every part has been very important requirement. It has been needed for obtaining of stabile locomotion.

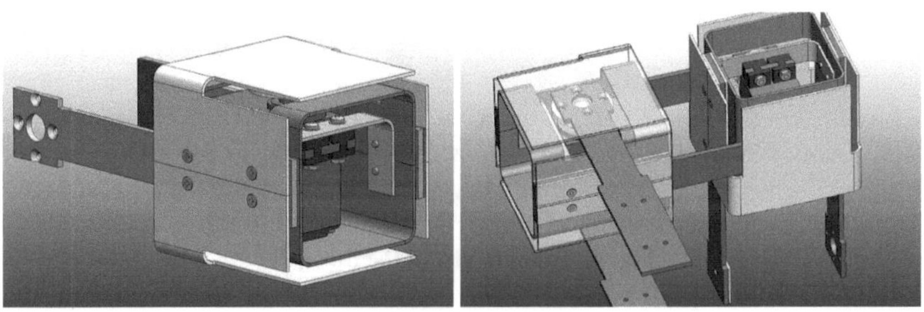

Fig. 5. Arrangement of robot article [2]

Arrangement of head and tail is different because there are stored printed circuit board with microcontroller (in head) and power supply accumulators (in tail) (fig. 6). Placement of the driven arms is shown on fig. 7. Driven arms in vertical joints are alternate because of higher stability of robot locomotion.

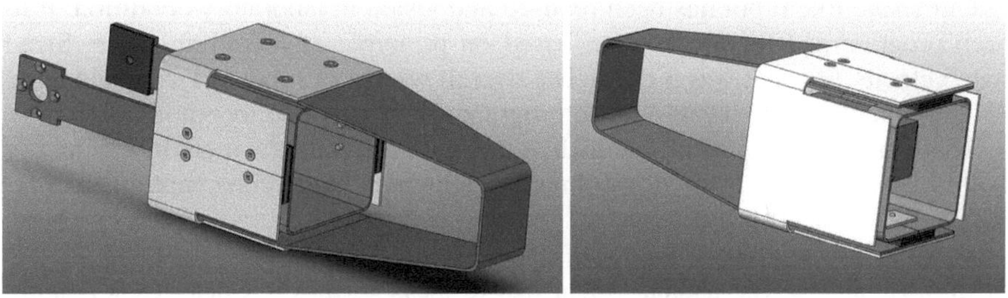

Fig. 6. Head and tail of the robot [2].

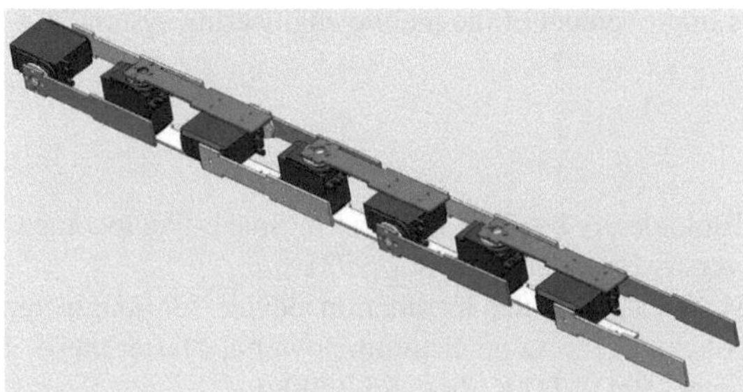

Fig. 7. Arrangement of driven arms[2].

3D model has been created for verification of functionality and possible collisions between parts in locomotion. Realisation of the robot brought several problems. It has been necessary to solve technology process, precision of bending, placement of wires etc. (fig. 8).

Fig. 8. 3D model of the robot (left) and realised robot (right) [2]

5 Conclusion

Locomotion is basic of the live of living organisms. They need locomotion for food finding, for avoiding from enemy or another dangerous. Ways of locomotion are very often as inspiration for design of various locomotion devices.

Our snake like robot has been realised and tested in laboratory condition. It has been programmed for locomotion based on principle of travelling wave. Steady state velocity of robot was 3 cm/s. The overall weight of robot is 1,2 kg and length 750 mm. In next time we will try to program another type of locomotions.

There are also a lot of problems like navigation in unknown environment, controlling, optimisation from viewpoint of weight and energy consumption etc. [4, 5, 6, 7, 8, 9, 10, and 11].

Acknowledgments. The authors would like to thank to Slovak Grant Agency -project VEGA 1/0454/09" Research on mechatronic systems imitating snake locomotion in confined and variable area" and VEGA 1/0201/08 "Research of structures and behaviour of the mechatronics mobile engineering system on the organ level and construction item level in order to properties improvement of the mobile engineering system".

References

[1] Hirose, S.: Biologically Inspired Robots. In: Snake-like locomotors and manipulators. Oxford University Press, New York (1993)

[2] Mačanga, M.: Design of the locomotion device for rought terrain. Návrh modelu zariadenia pohybujúceho sa po členitom povrchu. Master thesis. Supervisor: doc. Ing. Michal Kelemen, PhD., TU Košice, SjF (2009)

[3] Hitec HS 645MG, `http://www.hitecrcd.com/product_file/file/28/hsžčť.pdf`

[4] Kardoš, J.: Variable Structure Motion Control Systems. In: Frankovič, B. (ed.): Trends in Control Theory and Applications, Veda, Bratislava, pp. 132–158 (1999)

[5] Kardoš, J.: Reaching law modification of the time sub-optimal variable structure control. International Journal of Mechanics and Control 6(2), 39–49 (2005)

[6] Kardoš, J., Harvey, D., Howard, D.: Equivalent t-suboptimal Control of a 3-segment Leg. Journal of Electrical Engineering 52(11-12), 348–351 (2001)

[7] Vitko, A., Šavel, M., Kameniar, D., Jurišica, L.: Information processing in reactive navigation and fault detection of walking robot. In: Climbing and walking robots, pp. 794–800. Springer, Germany (2006)

[8] Markusek, J., Vitko, A., Jurišica, L.: Smart, Soft-Computing Robot navigation. In: Int. Workshop RAAD 2001, Vienna, Austria, May 16-18, 2001, pp. 435–442 (2001)

[9] Sharkawy, A.B., Vitko, A.: A State Observer For Flexible Joint Robots - Experimental Verification. In: Proc. of Int. Symp ISMCR, Prague, Czech Republic, pp. 167–173 (1998)

[10] Konečný, Z., Krys, V.: Creation of the rotational joint for the strength calculation purposes. Acta Mechanica Slovaca: Strojnícká fakulta technickej university v Košiciach, Košice, 2-A/2006 ROBTEP, ročník 10, str. 243 -247 (2006) ISSN 1335-2393

[11] Konečný,Z.-Skařupa, J.: Variational geometry in design of robots and manipulators. Sborník vědeckých prací Vysoké školy báňské – Technické univerzity Ostrava, roč. XLVII, řada strojní, č.1/2001,Ostrava (2001) ISSN 1210-0471

Relative Error Indices for Comparison of Neural Models of Different Robots

J. Możaryn and J.E. Kurek

Warsaw University of Technology, Institute of Automatic Control and Robotics,
02-525 Warszawa, ul. św. Andrzeja Boboli 8, Poland
J.Mozaryn@mchtr.pw.edu.pl

Abstract. In the article there are proposed relative error indices that can be used to evaluate neural models of different robots. The model of robot is identified using artificial neural networks with structure of the mathematical robot model in a form of the Lagrange-Euler equation. As an example the proposed indices are calculated for neural models of three different robots. Proposed indices significantly simplify an analysis and comparison of models of robots with different degree of freedom.

1 Introduction

The mathematical model of the robot has complicated, highly nonlinear multi-input multi-output structure described for instance by the Lagrange-Euler equation [1]. It requires a knowledge of the exact values of the robot physical parameters that are hard to obtain.

For the identification of robot mathematical model neural networks can be used because they can approximate the nonlinear multidimensional functions [6]. The neural network can resemble the model of the robot [4],[5] and identify nonlinear functions in the Lagrange-Euler equation without knowledge of robot physical parameters. During the identification only signals of robot joints positions and control are used.

In section 3 there is presented neural network structure for identification of mathematical model of the robot. The main problem during the analysis of results is big number of estimated signals and obtained data. Another difficulty occurs during the comparison of models of robots with different number of joints.

In section 4 there are presented indices for neural model evaluation based on estimated positions in robot joints. In section 5 there are analyzed values of proposed indices calculated for neural models of three different robots with different training parameters. Finally concluding remarks are given.

2 Discrete Time Robot Model

The discrete time model of the robot with n degree of freedom based on the Lagrange-Euler equation can be presented as follows [1]

$$\tau(k) = P(q,k) + M(q,k)\Gamma(q,k) \tag{1}$$

where

$$P(q,k) = V(q,k) + G(q,k), \quad \Gamma(q,k) = T_p^{-2}[q(k+1) - 2q(k) + q(k-1)]$$

and $\tau = [\tau_i] \in R^n$ is vector of control signals, $q = [q_i] \in R^n$ vector of generalized joint coordinates, $M(q,k) = [m_{ij}(k)] = M[q(k)] \in R^{n \times n}$ robot inertia matrix, $V(q,k) = [v_i(k)] = V[q(k), q(k-1)] \in R^n$ vector of *Coriolis* and centrifugal forces, $G(q,k) = [g_i(k)] = G[q(k)] \in R^n$ vector of the gravity loading, k discrete time and T_p sampling period ($t = kT_p$).

Equation (1) can be rewritten into following form

$$\Gamma(q,k) = \overline{M}(q,k)\tau(k) + \overline{P}(q,k) \tag{2}$$

where $\overline{M}(q,k) = M^{-1}(q,k)$ and $\overline{P}(q,k) = -M^{-1}(q,k)[P(q,k)]$

3 Neural Model of Robot

For the identification, the model (2) can be presented as a set of n equations

$$\gamma_i(k) = \overline{p}_i(q,k) + \sum_{j=1}^{n} \overline{m}_{ij}(q,k)\tau_i(k) \quad i = 1,\ldots,n \tag{3}$$

For identification of $\overline{p}_i(q,k)$ and $\overline{m}_{ij}(q,k)$ one can design neural networks, Fig. 1. Inputs to the neural network are positions $q(k)$ and control signals $\tau(k)$, and the output approximate the signal $\gamma_i(k)$. Therefore identified model will be denominated as *the model of robot position*.

For model (2) joints positions can be calculated as follows

$$q(k+1) \cong q_{NN}(k+1) = 2q(k) - q(k-1) + T_p^2[\overline{P}_{NN}(q,k) + \overline{M}_{NN}(q,k)\tau(k)] \tag{4}$$

where $\overline{M}_{NN}(q,k) = [\overline{m}_{NNij}(k)]$, $\overline{P}_{NN}(q,k) = [\overline{p}_{NNi}(k)]$ are matrices estimated by neural networks as it is shown in Fig.1 .

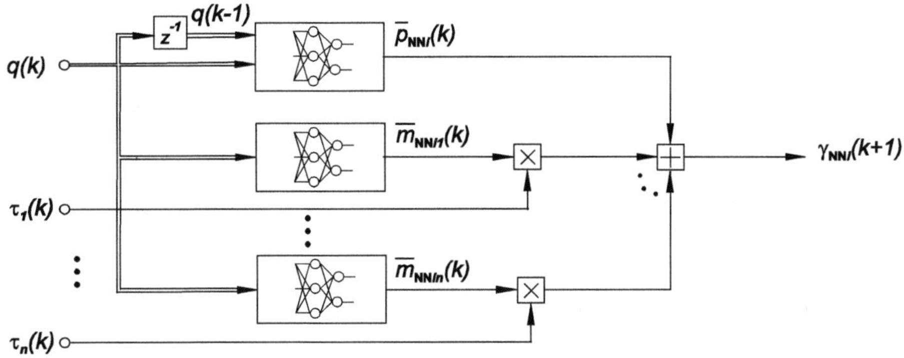

Fig. 1. Structure of the neural network for the identification of the model (3), z^{-1} denotes delay

4 Relative Position Estimation Indices

Let us denote error of the position estimation along the reference trajectory as follows

$$e_{qi}(k) = q_{NNi}(k) - q_{ri}(k) \tag{5}$$

where q_{ri} is reference position for joint i and q_{NNi} is position estimated by neural networks in joint i.

Then we propose the following indices to define the relative accuracy of joints positions estimation:

1. Estimation of the expected value of maximal relative average error of the position approximation

$$m_{savq} = \frac{1}{L}\sum_{j=1}^{L} \max_{i=[1,..,n]} \left| \frac{\frac{\sum_{k=1}^{N} e^{j}{}_{qi}(k)}{N}}{\frac{\sum_{k=1}^{N} q_i(k)}{N}} \right| = \frac{1}{L}\sum_{j=1}^{L} \max_{i=[1,..,n]} \left| \frac{\sum_{k=1}^{N} e^{j}{}_{qi}(k)}{\sum_{k=1}^{N} q_i(k)} \right|, \tag{6}$$

2. Estimation of the expected value of maximal relative average absolute error of the position approximation

$$m_{savaq} = \frac{1}{L}\sum_{j=1}^{L} \max_{i=[1,..,n]} \frac{\frac{\sum_{k=1}^{N} |e^{j}{}_{qi}(k)|}{N}}{\frac{\sum_{k=1}^{N} |q_i(k)|}{N}} = \frac{1}{L}\sum_{j=1}^{L} \max_{i=[1,..,n]} \frac{\sum_{k=1}^{N} |e^{j}{}_{qi}(k)|}{\sum_{k=1}^{N} |q_i(k)|} \tag{7}$$

where L is number of neural models with the same structure and e_{qi}^{j} is error (5) of the estimation of q_i calculated for model j.

Quality index (6) allows to estimate the relative steady error for neural models, whereas index (7) allows to estimate the relative absolute error for neural models, it is a measure of dispersion of relative error around the average value. Proposed indices can be used to compare neural models of robots with different number of joints.

5 Computer Simulations

In the example indices (6), (7) are used to analyze neural models of three different robots: the first with 2 degree of freedom which physical parameters can be found in [2], the second with 3 degree of freedom (first three links of robot PUMA 560) and the third robot with 6 degree of freedom (PUMA 560) which physical parameters can be found in [3].

For calculation of training and testing data reference trajectories for every joint were set as the sum of three different cosine functions according to the following formula

$$q_{ri}(k) = \frac{1}{2}\left|q_{\max i} - q_{\min i}\right| + q_{\min i} + \sum_{j=1}^{3} a_{ij}\cos(\omega_{ij} k T_p + \varphi_{ij}) \tag{8}$$

where $i = 1,\ldots,n$ is joint number, $q_{\min i}$ and $q_{\max i}$ are lower and upper boundary of the position in the joint i, a_{ij} amplitude, ω_{ij} an angular velocity and φ_{ij} a phase.

Training and testing trajectories were different for every robot. There were 10 000 data samples for training and 10 000 data samples for testing of neural models of robot with 2 degree of freedom, and 20 000 data samples for training and 20 000 data samples for testing of neural models of robots with 3 and 6 degree of freedom.

Every element (nonlinear function) of the robot mathematical model i.e. $\overline{p}_i(q,k)$ and $\overline{m}_{ij}(q,k)$ was identified by two layer neural sub-network with one nonlinear hidden layer and one linear output layer with 1 neuron. There were investigated neural networks with 2, 3, 5 and 10 neurons in hidden nonlinear layers. We have generated sets of 11 $(L = 11)$ neural models identified with neural networks. Neural networks were trained using conjugate gradients method [6]. In Table 1 there are gathered values of indices m_{savq} and m_{savaq} calculated for sets of neural networks with different training parameters identifying models of considered robots.

From the obtained results it follows that if neural models of robots were trained using conjugate gradients method the best accuracy of position estimation can be obtained for neural models with two neurons in hidden nonlinear layers of sub-networks estimating nonlinear functions. It can be seen that the accuracy of position

Table 1. Values of indices m_{savq}, m_{savaq} ($\times 10^{-8}$) calculated for neural models of robots

	Training trajectory				Testing trajectory			
Neurons	2	3	5	10	2	3	5	10
Robot with 2 degree of freedom, (100 training iterations)								
m_{savq}	0.4	0.3	0.2	0.2	0.2	0.1	0.1	0.05
m_{savaq}	0.2	0.2	0.2	0.2	0.2	0.1	0.1	0.1
Robot with 3 degree of freedom, (100 training iterations)								
m_{savq}	3.6	3.7	8.0	10.0	0.3	0.4	0.7	1.5
m_{savaq}	0.1	0.1	0.2	0.2	0.1	0.2	0.2	0.4
Robot with 6 degree of freedom, (700 training iterations)								
m_{savq}	1.5	3.0	2.9	2.3	0.7	0.8	0.2	1.8
m_{savaq}	0.1	0.1	0.1	0.1	0.1	0.1	0.2	0.3

estimation measured using index m_{savq} was best for neural model of the robot with 2 degree of freedom and decreased when the number of joints increased. For all considered neural models dispersion of relative error of position estimation around the average value measured using index m_{savaq} was small.

The position signals from testing set were approximated with better accuracy than position signals from the training set. This difference can be a result of minimized quality index of the output signal (γ_i) during neural model training. Output signal (γ_i) of neural network was different than estimated position (q_i).

6 Concluding Remarks

In the paper the relative error indices are presented. There are compared models of three different robots identified using neural networks with different training parameters. Proposed indices simplify analysis of identified models and allow to compare neural models of different robots.

The presented criterions can be used for general comparison of learning techniques or network structures used for design of neural networks for different tasks, e.g. identification of neural models for 2 and 6 degree of freedom robots. This can help to find better solution, e.g. learning methods, for a class of problem like neural network identification of robot.

References

[1] Fu, K.S., Gonzalez Lee, R.C.: Robotics: Control, Sensing, Vision and Intelligence. McGraw - Hill Book Company, New York (1987)

[2] Tang, K.M.W., Tourassis, V.D.: Mathematical Deficiencies of Numericaly Simplified Dynamic Robot Models. IEEE Transactions on Automatic Control 34(10), 1109–1111 (1989)

[3] Corke, P.I., Armstrong-Hélouvry, B.: A Search for Consensus Among Model Parameters Reported for the PUMA 560 robot. In: Proc. IEEE International Conference on Robotics and Automation, San Diego, vol. 1, pp. 1608–1613 (1994)
[4] Lewis, F.L.: Neural Network Control of Robot Manipulators and Nonlinear Systems. Taylor & Francis, London (1999)
[5] Możaryn, J., Kurek, J.E.: Comparison of Neural Network Robot Models with Not Inverted and Inverted Inertia Matrix. In: Duch, W., Kacprzyk, J., Oja, E., Zadrożny, S. (eds.) ICANN 2005. LNCS, vol. 3697, pp. 417–422. Springer, Heidelberg (2005)
[6] Osowski, S.: Neural Networks for the Information Retrieval (in polish). In: OWPW (2006)

HexaSphere with Cable Actuation

M. Valášek and M. Karásek

Czech Technical University in Prague, Faculty of Mechanical Engineering, Department of Mechanics, Biomechanics and Mechatronics, Karlovo namesti 13, Praha, Czech Republic
{Michael.Valasek,Matej.Karasek}@fs.cvut.cz

Abstract. The paper deals with details of development of cable actuated version of spherical mechanism HexaSphere, a redundantly actuated mechanism with parallel kinematic structure. The cables can only transfer tension forces, stress load of the cables would make the mechanism collapse. That can be achieved by anti-backlash control principle. Anti-backlash area of mechanism HexaSphere is mapped first. Secondly actuation forces required for given trajectory and robot's dynamics are computed and minimised by dimension optimisation.

1 Introduction

Spherical mechanisms have many fields of application. They are used for swivel heads of machine tools for 5 axis machining, for assemblies of telescopes or positioning of antennas. Spherical mechanism HexaSphere (Fig. 1.) has been developed as an equivalent to traditional Cardan hinge. It is a representative of redundantly actuated mechanisms with parallel kinematical structures and has very promising properties - large workspace (±100°) with no singularities [1], high accuracy, stiffness and dynamics [2].

Very high movability of the mechanism requires use of spherical joints with large extent of motions. Present spherical joints have limited movability to 90° at least in one axis. Therefore new spherical joints with innovative design were developed [3]. They are based on traditional Cardan joint with an extra rotation axis and require mechatronical controlled break of inner joint axis or at least measurement of its movement. These joints fulfil the requirement of high movability but showed up to be very complex, hence cannot be miniaturized and their stiffness is problematic as well.

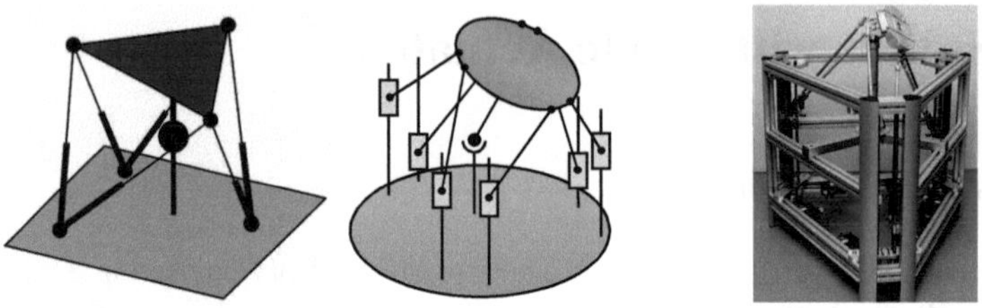

Fig. 1. Spherical mechanism HexaSphere - possible concepts and a functional model

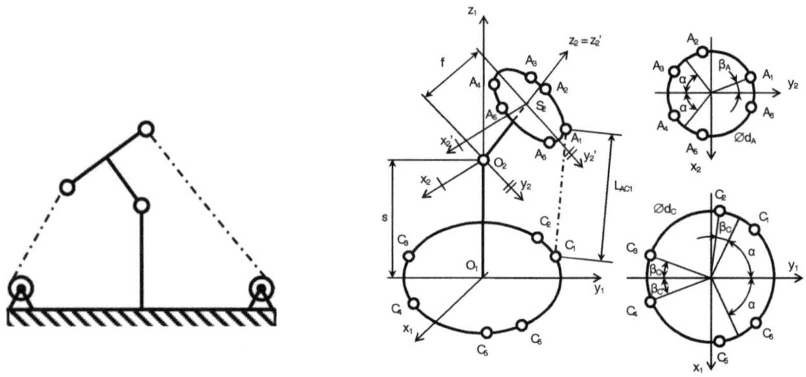

Fig. 2. 2D concept of cable actuated HexaSphere and kinematic scheme (with only 1 cable)

Thus a concept of cable actuated HexaSphere was proposed (Fig. 2.). Using cables instead of struts has many advantages. Cables are lighter so less mass is being moved, likewise they are capable of torsion and therefore their attachment to the platform can be realized without problematic spherical joints. Besides this solution allows miniaturization of the whole mechanism, e.g. for surgical applications.

However for cable actuation unique control algorithm is required. The cables cannot transfer stress, they need to be under tension all the time otherwise the platform would collapse. Hence the anti-backlash control strategy has been applied.

2 Anti-backlash Control Strategy

Anti-backlash control strategy was proposed in [4] and is applicable only on machines with redundant actuation. Its original purpose is to exclude drives backlashes. That can be obtained by fulfilling a simple condition - the orientation of drive's force effects must remain constant for the whole working cycle. A sign vector representing orientation of actuator forces in vector $\mathbf{T_a}$ has been defined as

$$\mathbf{s} = \operatorname{sign}(\mathbf{T_a}) \ . \tag{1}$$

assuming that all drives are in action, i.e. the elements of vector s are either +1 or -1. Than the condition of backlash exclusion can be written as

$$\mathbf{s} \stackrel{!}{=} const \ . \tag{2}$$

For mechanisms with no actuator redundancy the sign vector is explicitly given in each position by the interaction and dynamics thus it cannot be changed. On the other hand for robots with redundant drives the actuator forces vector $\mathbf{T_a}$ is a solution of m linear equations with n unknowns, where m < n (m is number of degrees of freedom, n is number of actuators). Hence for parallel machines with redundant actuators the actuator forces vector $\mathbf{T_a}$ and so the sign vector $\mathbf{s}$ can be influenced to some extent. Algorithm for finding an optimal sign vector $\mathbf{s}$ was proposed in [4], however for the cable actuation problem the sign vector is given implicitly. To load all the cables by tension, orientation of all drive forces must be negative thus

$$\mathbf{s} \stackrel{!}{=} \begin{bmatrix} -1 & -1 & \cdots & -1 \end{bmatrix}^{\mathrm{T}} \ . \tag{3}$$

Using diagonal sign matrix

$$\mathbf{D}(\mathbf{s}) = \begin{pmatrix} s_1 & & \mathbf{0} \\ & \ddots & \\ \mathbf{0} & & s_n \end{pmatrix} \tag{4}$$

vector $\mathbf{T_{as}}$ with only positive elements has been introduced

$$\mathbf{T_{as}} = \mathbf{D}(\mathbf{s})\mathbf{T_a} \ . \tag{5}$$

Concerning (3) the right side of the equation can be reduced to

$$\mathbf{T_{as}} = -\mathbf{T_a} \ . \tag{6}$$

Even with specified sign vector $\mathbf{s}$ the vector $\mathbf{T_{as}}$ does not have a unique solution. However it can be obtained by using more complex method, e.g. optimization of operating characteristic in terms of drive forces.

3 Anti-backlash Area

To map spaces where the drives backlashes are excluded apart from interaction and dynamics an algorithm has been created. The condition of force effects equilibrium can be written as

$$\mathbf{J}^{+\mathrm{T}}\mathbf{T_a} = \mathbf{F_x} \tag{7}$$

where $\mathbf{F_x}$ is a vector of resultant reduced load in the robots workspace coordinates $\mathbf{X}$. For $\mathbf{\Phi}$ representing vector of actuators coordinates pseudo inverse Jacobi matrix $\mathbf{J}^+$ is defined as follows

$$\mathbf{J}^{+} = \frac{d\mathbf{\Phi}}{d\mathbf{X}} \tag{8}$$

and specifically for mechanism HexaSphere in modified form with time differentiation of numerator and denominator

$$\mathbf{J}^{+} = \frac{d\dot{\mathbf{L}}_{\mathbf{AC}}}{d\boldsymbol{\omega}_{\mathbf{12}}} \tag{9}$$

where $\boldsymbol{\omega}_{\mathbf{12}}$ is vector of platform angle velocities and $\mathbf{L_{AC}}$ is vector of lengths of cables between points A on the platform and C on the frame. For expressing $\mathbf{L_{AC}}$ transformation matrix method [5] has been used. The spherical motion has been described by Euler angles ψ, ϑ and φ with a condition that the angle φ is an opposite of angle ψ, which assures the rolling of platform and prevents the cables from interfering.

Using equation (6) the problem of finding anti-backlash area can be transformed to a solubility problem of system

$$-\mathbf{J}^{+\mathrm{T}}\mathbf{T}_{\mathbf{a}s} = \mathbf{F}_{\mathbf{x}} \quad \text{where} \quad \mathbf{T}_{\mathbf{a}s} > \mathbf{0} \tag{10}$$

where Fx is an arbitrary vector of resultant reduced load and Tas is the unknown solution, if it exists. A form where Fx would not appear would be more appropriate for testing the solubility of the system. Hence we substitute the arbitrary vector Fx with a zero vector

$$-\mathbf{J}^{+\mathrm{T}}\mathbf{T}_{\mathbf{a}s} = \mathbf{0} \quad \text{where} \quad \mathbf{T}_{\mathbf{a}s} > \mathbf{0} \tag{11}$$

It was proven in [4] that if a solution of this system exists, the system (10) is also soluble. Generally a soluble system (11) has infinite number of solutions. However we can easily transfer the problem to a problem of quadratic programming by adding a criterion of finding minimum of quadratic function of vector $\mathbf{T_{as}}$.

A disadvantage of this method is in need of discretization of the workspace, which brings the risk of undiscovered inhomogeneities. The final form of quadratic programming problem has been modified by specifying vector of minimal acceptable force effects absolute values $\mathbf{T_{a\,min}}$ that guarantees certain amount of reserve for real control with feedback component and inaccurate model of the mechanism

$$\begin{aligned} &\min_{\mathbf{T_{as}}}\left(\mathbf{T}_{\mathbf{a}s}{}^{\mathrm{T}}\mathbf{T}_{\mathbf{a}s}\right) \quad \text{for} \\ &-\mathbf{J}^{+\mathrm{T}}\mathbf{T}_{\mathbf{a}s} = \mathbf{0} \quad \text{where} \quad \mathbf{T}_{\mathbf{a}s} > \mathbf{T}_{\mathbf{a}\,\mathrm{min}} \end{aligned} \tag{12}$$

4 Anti-backlash Control for a Given Trajectory

To obtain time behaviour of required actuation forces for a given operating cycle, i.e. for specified trajectory, mechanism dynamics and time behaviour of external load, it is necessary to calculate vectors of resultant reduced load $\mathbf{F_x}$ in every point of discretized trajectory. For unloaded mechanism HexaSphere and given trajectory where only dynamics is taken into account (e.g. a telescope assembly) the vector of resultant reduced load can be expressed from Euler dynamic equation as

$$\mathbf{F_x} = -\mathbf{I}\dot{\boldsymbol{\omega}}_{12} - \boldsymbol{\omega}_{12} \times (\mathbf{I}\boldsymbol{\omega}_{12}) \tag{13}$$

where **I** is the inertia matrix of the platform formulated to the centre of spherical motion. The dynamic effect of moving cables has been omitted for its low significance. For maximal absolute values of actuation forces stored in vector $\mathbf{T_{a\,max}}$ the vector $\mathbf{T_{a\,s}}$ can be in every step obtained as a solution of quadratic programming problem

$$\begin{aligned} &\min_{\mathbf{T_{as}}}\left(\mathbf{T_{as}}^{\mathrm{T}}\mathbf{T_{as}}\right) \quad \text{for} \\ &-\mathbf{J}^{+\mathrm{T}}\mathbf{T_{as}} = \mathbf{0} \quad \text{where} \quad \mathbf{T_{a\,min}} < \mathbf{T_{as}} \le \mathbf{T_{a\,max}} \end{aligned} \tag{14}$$

Anti-backlash area for mechanism HexaSphere with dimensions deduced from functional model and time behaviour of actuator forces absolute values for a testing trajectory is in Fig. 3. It is obvious that the cable driven HexaSphere with chosen dimensions is not able to cover the whole required workspace (ϑ up to 100°).

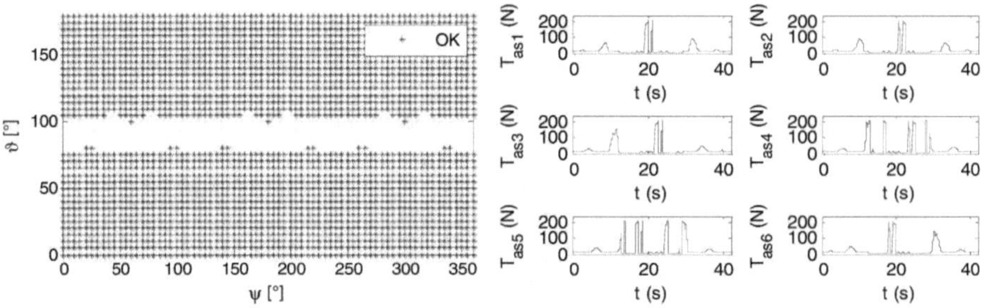

Fig. 3. Antibacklash area and actuator forces for testing trajectory before optimization

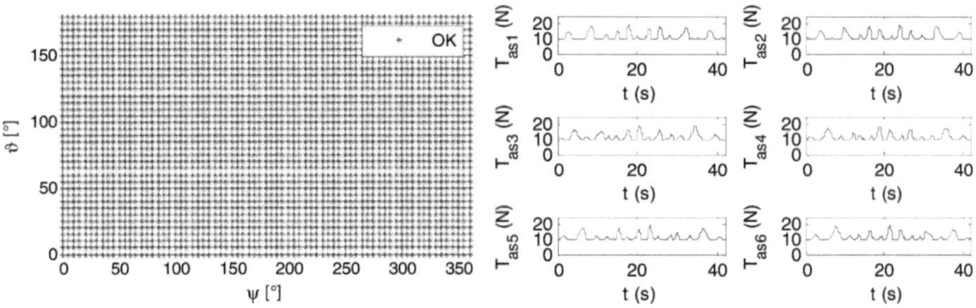

Fig. 4. Antibacklash area and actuator forces for testing trajectory after optimization

Thus three most important dimensions were chosen as parameters and optimised for maximisation of anti-backlash area and minimisation of maximal actuator force absolute value. The results after optimisation (Fig. 4.) show that theoretically the entire spherical space can be reached and the maximal actuation forces have been significantly lowered.

5 Conclusion

The paper demonstrates that the anti-backlash control principle can be successfully implemented for redundantly actuated parallel cable driven mechanism HexaSphere. Its dimensions were optimized with respect to maximizing the anti-backlash area and minimizing actuation forces required for motion along testing trajectory. It has been proven that a cable driven spherical mechanism HexaSphere can theoretically reach the entire spherical space. Even when concerning design limitations of the central spherical joint it can be used for tilting angles up to 100° as has been shown on the testing trajectory.

References

[1] Valasek, M., Karasek, M.: Kinematical Analysis of HexaSphere. In: Engineering Mechanics 2009 Conference Proceedings, pp. 1371–1378. ITAM Academy of Sciences of Czech republic (2009)

[2] Valasek, M.: Redundant Actuation and Redundant Measurement: The Mechatronic Principles for Future Machine Tools. In: Proc. of International Congress on Mechatronics MECH2K4, FME CTU, Prague, pp. 131–144 (2004)

[3] Sulamanidze, D.: Spherical Joints with Increased Mobility, PhD Thesis, FME CTU in Prague (2007)

[4] Florian, M.: Active Stiffness and Anti-backlash Control of Redundant Parallel Robots, PhD Thesis, FME CTU in Prague (2006)

[5] Stejskal, V., Valasek, M.: Kinematics and Dynamics of Machinery. Marcel Dekker, New York (1996)

Optimization of Vibration Power Generator Parameters Using Self-Organizing Migrating Algorithm

Z. Hadaš[1], Č. Ondrůšek[2], and J. Kurfürst[3]

[1] Brno University of Technology, Faculty of Mechanical Engineering, Institute of Solid Mechanics, Mechatronics and Biomechanics, Technicka 2896/2, Brno, Czech Republic
hadas@fme.vutbr.cz
[2] Brno University of Technology, Faculty of Electrical Engineering and Communication, Department of Power Electrical and Electronic Engineering, Technicka 8, Brno, Czech Republic
ondrusek@feec.vutbr.cz
[3] Brno University of Technology, Faculty of Electrical Engineering and Communication, Údolní 53, Brno, Czech Republic
xkurfu02@stud.feec.vutbr.cz

Abstract. This paper deals with a self-organizing migrating algorithm (SOMA) for an optimization of vibration power generator parameters. The vibration power generator is an energy harvesting device, which is capable of harvest electrical energy from an ambient mechanical vibration. The generator consists of a precise mechanical part, electro-mechanical converter and electronics. It creates a complex mechatronic system, where parameters of individual parts are mutually affected. For effective harvesting of energy all parameters have to be tuned up optimally to nature of an excited vibration and required output power. A generator model can be used for optimization study of maximal output power and minimization of generator volume. Main problem is complexity of this system and number of parameters in mutual feed back of this mechatronic system. Thus the SOMA is applied to the optimization problem of the vibration power generator.

1 Introduction

This paper deals with an optimization of vibration power generator parameters. The vibration power generator is energy harvesting device, which generates electrical energy from an ambient mechanical vibration. This autonomous source of energy, which is placed in an environment excited by ambient vibrations, can be used for feeding of wireless sensors and remote application without physical connection to outside world. This device is based on a resonance operation of an oscillating mass and an electromagnetic principle of an electro-mechanical conversion provides sufficient generating of electrical energy for wireless applications. The output power depends on a nature of the ambient vibration and a design of the generator.

The generator consists of a precise mechanical part, electro-mechanical converter and electronics with powered device. This assembly creates a complex mechatronic system with feed back between each part, where optimal setting-up of parameters for each individual part is mutually affected. Parameters have to be tuned up optimally to a level of excited vibration and required output power for effective harvesting energy.

Nowadays a broad class of existing algorithms can be used for optimization studies. Due to a complexity of this mechatronic system and a number of generator parameters in mutual feed back is suitable to use some sophisticated optimizing methods. Genetic algorithm methods [1] are used for the optimization of such electromechanical [2] and electromagnetic systems [3, 4 and 5]. Therefore the self-organizing migrating algorithm (SOMA), which is type of genetic algorithm used for optimization of these problems [6], is applied to optimization of the complex vibration power generator.

2 Vibration Power Generator

The first developed vibration power generator [7], performance data and picture are shown in Table 1, was tuned up in accordance with mechatronics approach of system development and the generator design passed through several development cycles [8].

Table 1. Performance data of vibration power generator with electronics [7]

Parameter (Condition)	Value	Unit
Weight	125	g
Volume 50x40x40 mm	80	cm³
Coil	2000	turns
Inner resistance of coil	1600	Ω
Operating frequency	17	Hz
Voltage DC 0.3 G	13.5	V rms
Output Power 0.3 G	20	mW

3 Model of Vibration Power Generator

Fundamental parts of the vibration power generator, shown in Fig. 1, are the resonance mechanism (seismic mass m suspended on spring k with damping b) with the suitable electro-magnetic converter (oscillating magnetic circuit against fixed coil L with inner resistance R_C) and the electronics (resistance of electrical load R_L). The simplified diagram of generator is shown in Fig. 1 a) schematic diagram of a simplified linear system is shown in Fig. 1 b).

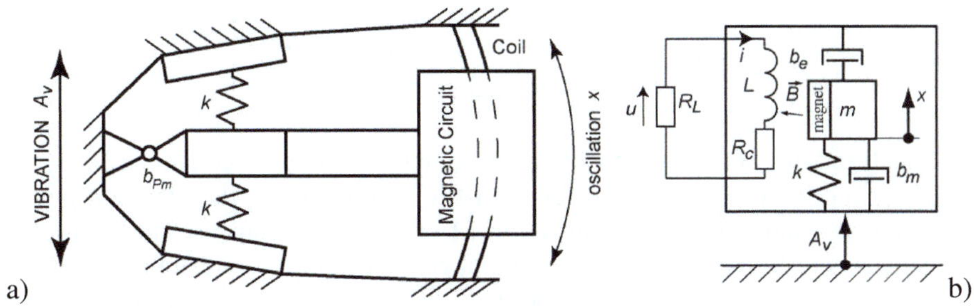

Fig. 1. a) Simplified diagram of generator; b) Schematic diagram of model

The differential equation (1) that describes the oscillating movement x of a mass m against mechanism housing with a fixed coil is derived from dynamic forces on the mass. The quality factor of the oscillating system with operating frequency Ω is described by mechanical damping ratio b_{Pm}. The system is excited by acceleration A_V of the ambient vibration.

$$\ddot{x}+2\Omega\left(b_{Pm}+b_{Pe}\right)\dot{x}+\Omega^2 x = A_V \sin\left(\omega t\right) \tag{1}$$

The oscillating of a magnetic circuit against the fixed coil induces voltage due to Faraday's law. Current flows through the coil and the electric circuit if the electrical load is connected (2). This current through the coil provides feed back to the mechanical system as an electro-magnetic damping ratio b_{Pe} in equation (1).

$$N \cdot B \cdot l_{act} \cdot \dot{x} = R_C \cdot i + R_L \cdot i \tag{2}$$

The induced voltage is proportional on velocity of the oscillation movement, active length of coil l_{act} and number of coil turns N and magnetic flux density B through coil. The coil is placed in an air space of the magnetic circuit, Fig. 2. A magnetic field inside the coil space was analyzed and the relation (3) describes the magnetic flux through the coil space in Fig. 2. The coefficients in relation (3) correspond with specific FeNdB permanent magnets and their demagnetizing curve.

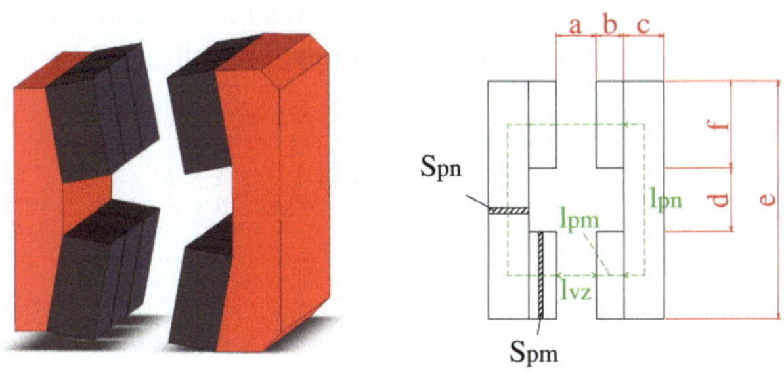

Fig. 2. Magnetic circuit of generator and optimized dimension parameters

$$B = \frac{1,3074}{1 + 1,3687 \cdot 10^{-6} \left(\frac{1}{2\mu_0} \cdot \frac{a}{b} + \frac{1}{2\mu_0 \mu_{pn}} \cdot \frac{f(d+f+c)}{bc} \right)} \tag{3}$$

The analyzed magnetic flux density in equation (3) is input for equation (2).

This simplified model is used for simulation modelling of the vibration power generator [9] and optimizing analysis in MATLAB environment. The additional equations, concerning relation between real geometry, Fig. 1, and simplified model, Fig. 2, and relation between geometry of the magnetic circuit and weight of the oscillating mass or the active length of the coil, have to be included in optimizing analysis. This complex electromechanical model with total 22 system parameters is optimized for maximal output power with minimal volume of the device. The electronic part is simplified as a resistance load for this study. The optimized generator can operate with frequency of vibration 17 Hz and level of vibration around 0.1 g. The required voltage on connected load 3.5 kΩ is minimally 4 volts RMS.

4 Self-Organizing Migrating Algorithm – SOMA

Our optimization study uses algorithm, which is labeled a "memetic" algorithm where no new generations are created. This algorithm was inspired by the behavior patterns of groups of wild animals in the wild and it has been termed “the Self-Organizing Migrating Algorithm” – or SOMA [10]. Soma is stochastic search algorithm for global optimization. This algorithm is powerful tool for solving of engineering problems when analytical solving of a given problem is very difficult or unrealistic.

Main aim of this algorithm is extremes searching of a cost function. A definition of a suitable cost function (fitness function) is very important for successful optimization. The algorithm can use several strategies and our optimization uses basic strategy AllToOne. This strategy means that all individuals move toward to leader. The leader remains at its current position during migration loop.

The control parameters of SOMA are perturbation, length of path, step, MinDiv (largest allowed difference between the best and the worst individual position) and size of population. The quality of optimized results depends on the selection of these parameters. Values of optimized parameters are limited by user in boundary conditions.

The SOMA consists of following steps:

- Parameters definition.
- Creation of population.
- Migrating loop.
- Test for stopping condition.
- Stop, which recall the best solution during searching.

5 SOMA for Optimization of Vibration Power Generator

On the base of a theoretical description of SOMA [10] and free source codes on website [11], an optimization tool for the vibration power generator construction was created. A graphic interface of the SOMA optimization tool was built under thesis [12]. This optimization tool is performed by the following three issues.

- Optimization of generator with known connected load resistance.
- Optimization of generator with maximal volume.
- Optimization of generator to minimal volume and maximal power.

For each optimization issue is designed different cost function for successful searching of a global extreme. The presented results correspond with the third optimization issues - generator with minimal volume and maximal output power. This cost function minimize overall volume of the generator with maximal output power and required output voltage (4 volts and higher). All cost factors have approximately equivalent effect to the cost function. A calculated value of the cost function depends on a choice of individual function coefficients.

The generator parameters were optimized for operating conditions (frequency 17 Hz, level of vibration 0.1 g and b_{Pm} 0.025). The graphic output with several representative parameters from this optimization study is shown in Fig. 3.

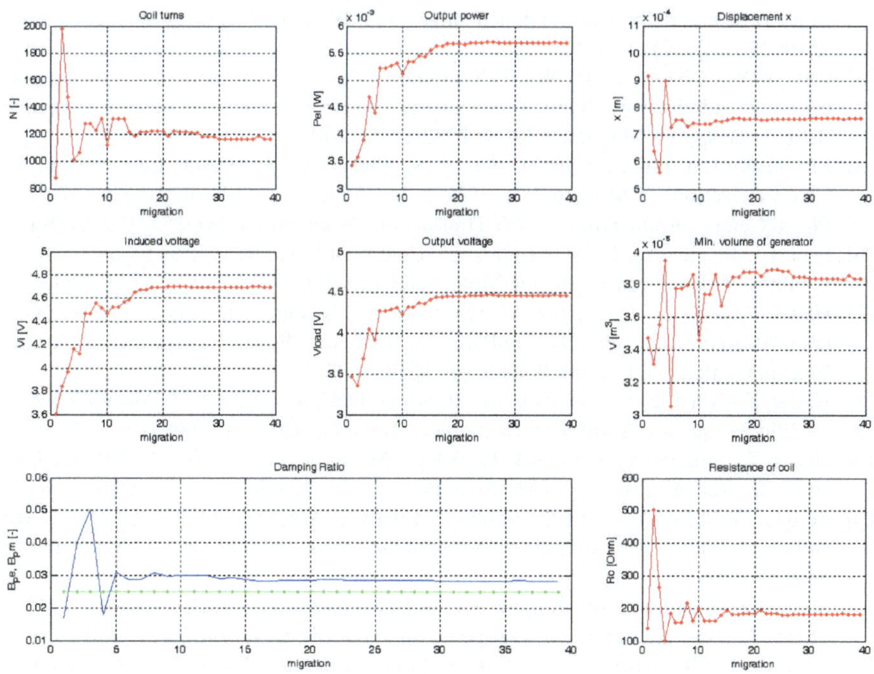

Fig. 3. Graphic output of SOMA optimization tool

6 Conclusions

The SOMA appears to be suitable optimization tool for such complex mechatronic systems. This study shown that optimized generator, which is placed in environment with mentioned conditions, can harvest output power 5.7 mW inside overall volume 38 ccm. The optimal geometry of the simplified resonance mechanism, optimal dimension parameters of the coil and the magnetic circuit were calculated.

The used simulation model of the mechanical part for this optimization can be improved with respect of the quality factor and consequently this optimization study can be used for a primary proposal of real generators design for new operating conditions and customer (wireless application) requirements.

Acknowledgements. Published results were acquired using the subsidization of the Ministry of Education, Youth and Sports of the Czech Republic, research plan MSM 0021630518 "Simulation modelling of mechatronics systems".

References

[1] Skaar, S.E., Nilssen, R.: Genetic optimization of electric machines, a state of the art study. In: NORPIE 2004, NTNU, Trondheim, Norway (2004)

[2] Alkhatib, R., Jazar, G.N., Golnaraghi, M.F.: Optimal design of passive linear suspension using genetic algorithm. Journal of Sound and Vibration 275, 665–691 (2004)

[3] Chao, P.C.-P., Wu, S.C.: Optimal design of magnetic zooming mechanism used in cameras of mobile phones via genetic algorithm. IEEE Transactions Magnetics 43(6), 2579–2581 (2007)

[4] Chiu, C.W., Chao, P.C.-P., Wu, D.Y.: Optimal design of magnetically actuated optical image stabilizer mechanism for cameras in mobile phones via genetic algorithm. IEEE Transactions Magnetics 43(6), 2582–2584 (2007)

[5] Liu, C.-S., Lin, P.D., Lin, P.-H., Ke, S.-S., Chang, Y.-H., Horng, J.-B.: Design and Characterization of Miniature Auto-Focusing Voice Coil Motor Actuator for Cell Phone Camera Applications. IEEE Transactions Magnetics 45(1), 155–159 (2009)

[6] Senkerik, R., Zelinka, I.: Optimization and Control of Batch Reactor by Evolutionary Algorithms. In: European Simulation Multiconference 2005, Riga, Latvia, pp. 59–65 (2005)

[7] Hadas, Z., Zouhar, J., Singule, V., Ondrusek, C.: Design of Energy Harvesting Generator Base on Rapid Prototyping Parts. In: IEEE 13th Power Electronics and Motion Control Conference Poznan, Poland, pp. 1688–1692 (2008)

[8] Hadas, Z., Singule, V., Ondrusek, C.: Optimal Design of Vibration Power Generator for Low Frequency. Solid State Phenomena 149-147, 426–431 (2009)

[9] Hadas, Z., Singule, V., Ondrusek, C., Kluge, M.: Simulation of Vibration Power Generator. In: Recent Advances in Mechatronics, pp. 350–354. Springer, Heidelberg (2007)

[10] Zelinka, I.: SOMA - Self-Organizing Migrating Algorithm. In: Babu, B.V., Onwubolu, G. (eds.) New optimization techniques in engineering, ch. 7, Springer, Heidelberg (2004)

[11] Soma, I.Z.: Faculty of Technology. Tomas Bata University, Czech Republic, http://www.ft.utb.cz/people/zelinka/soma/ [cit. 25.5.2009]

[12] Kurfürst, J.: Optimalization of Vibration Micro-Generator. Thesis, Brno University of Technology, Faculty of Electrical Engineering and Communication Supervisor Prof. C. Ondrusek, 67 (2009)

Recent Trends in Application of Piezoelectric Materials to Vibration Control

P. Mokrý, M. Kodejška, and J. Václavík

Institute of Mechatronics and Computer Engineering, Technical University of Liberec, Studentská 2, CZ-46117 Liberec, Czech Republic
pavel.mokry@tul.cz

Abstract. Experimental investigation and modeling of the application of piezoelectric actuators to semi-active piezoelectric shunt vibration control devices are presented. The considered devices utilize the principle that the transmission of vibrations through an interface between two solid objects is controlled by the ratio of their mechanical impedances. Since the mechanical impedance of a material is proportional to its stiffness, an extremely soft element placed between the source of vibrations and the object being isolated from vibrations works as an interface with a high transmission loss of vibrations. It is also known that by connecting a negative capacitor to the piezoelectric actuator it is possible to create an element whose elastic stiffness can be arbitrarily tuned. Therefore, by combining these two principles, very simple and efficient mechatronic systems for the vibration suppression, which consist of a piezoelectric actuator connected to a negative capacitor, have been realized.

1 Introduction

Piezoelectric materials have become a very promising group of materials, which can be used in complex mechatronic systems. Their advantages stem from a unique possibility to construct simple and low-cost electroacoustic transducers with superb functional properties such as high sensitivity, high precision and fast response [1]-[3]. These are essential feature that can be beneficially used in vibration control devices.

In this Article we demonstrate the use of a piezoelectric actuator in a vibration control device that is based on a recently developed technique called active elasticity control [4]. This technique is based on the fact that by connecting the piezoelectric actuator to the electric circuit, which behaves as a negative capacitor, it is possible to actively control the effective elastic properties of the piezoelectric actuator to a large extent and in a broad frequency range. This offers a technique for suppressing the vibration transmission through a bulk piezoelectric actuator. The reason for this is due to the principle that the vibration transmission through the interface between two solid objects is controlled by the ratio of their mechanical impedances. Because the mechanical impedance of the piezoelectric actuator is proportional to its spring constant, extremely soft piezoelectric actuator works as an interface with a high acoustic transmission loss of vibrations.

Early applications of this system have been reported by Okubo et al. [5] and Kodama et al. [6]. Later, Imoto et al. [7] and Tahara et al. [8] demonstrated the great potential of this method on a system for suppressing vibrations by 20 dB in the broad frequency range from 1 to 100 kHz. The presented Article is focused on a comprehensive modeling of the vibration isolation system. In the next Section, the principles of the active elasticity control are presented.

2 Active Elasticity Control

The basic idea by Date [3] for controlling the elasticity of piezoelectric materials is based on the superposition of direct and converse piezoelectric effects with the Hooke's law, which is achieved by connecting to an active shunt circuit that behaves as negative capacitor. Its principle is indicated in Fig. 1.

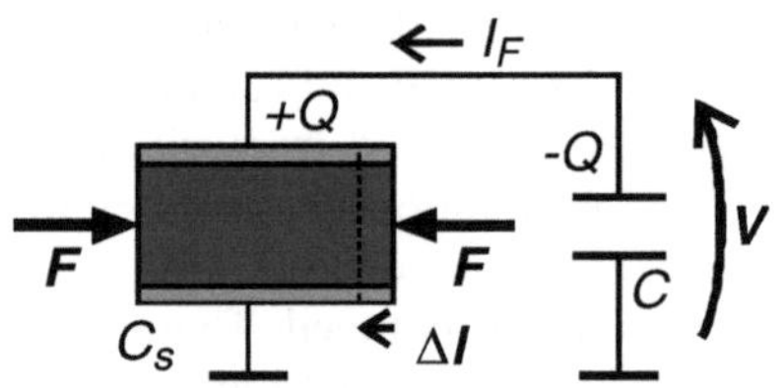

Fig. 1. Principle of the Active Elasticity Control: Force F applied to the piezoelectric actuator produces a charge Q due to the direct piezoelectric effect, which is introduced to the shunt capacitor C through current I_F. Voltage V is applied back to the piezoelectric actuator, whose deformation Δl is controlled by the sum of the Hooke's law and the converse piezoelectric effect.

To obtain the effective spring constant K of the piezoelectric actuator actively shunted by the capacitor, one should start with the piezoelectric equations of state for the charge Q and the change of the piezoelectric actuator length Δl:

$$\begin{aligned} Q &= d\,F + C_S\,V, \\ \Delta l &= (1/K_S)\,F + d\,V, \end{aligned} \tag{1}$$

which should be appended by the by the formula for the voltage V applied back to the piezoelectric actuator from the shunt capacitor of capacitance C:

$$V = -Q/C, \tag{2}$$

where symbols d, C_S, and K_S stand for the piezoelectric coefficient, the capacitance, and the spring constant of a mechanically free piezoelectric actuator, respectively.

Combining Eqs. (1) and (2), one can obtain the formula for the effective spring constant of the piezoelectric element connected to the external capacitor [4]:

$$K = \frac{F}{\Delta l} = K_S \left(\frac{1 + C / C_S}{1 - k^2 + C / C_S} \right), \tag{3}$$

where k is the electromechanical coupling factor of the piezoelectric element ($0<k<1$). It can be concluded from the above formula that great changes in values of the effective spring constant K of the piezoelectric actuator can be achieved only when the capacitance of the external circuit is negative. It should be also noted that our analysis is restricted to harmonic time dependence of all state quantities. In this case, it is reasonable co consider that both capacitances C_S and C are complex numbers, as well as the effective spring constant, i.e. $K = K' + iK''$.

The next Section demonstrates the application of the active elasticity control technique to semi-active piezoelectric shunt vibration control device.

3 Semi-Active Piezoelectric Shunt Vibration Control Device

Figure 2 shows the scheme of the semi-active vibration control device, which consists of the piezoelectric actuator of the effective spring constant K and the capacitance C_S. The piezoelectric actuator separates the object of a mass M from the source of vibrations.

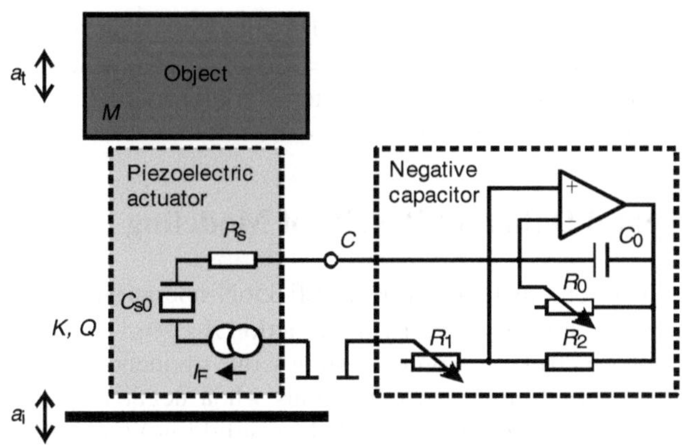

Fig. 2. Mechanical and electrical scheme of the semi-active piezoelectric shunt vibration control device, which consists of the piezoelectric actuator connected to the negative capacitor.

Transmissibility TR of vibrations transmitted through the piezoelectric actuator is defined as a ratio of the transmitted acceleration a_t over the incident acceleration a_i at the reference source point, i.e. TR = $|a_t/a_i|$. In this work we consider the transmission of vibrations of angular frequency ω through the viscoelastic element of a spring constant K and the mechanical quality factor Q on an object of a mass M. By solving the equations of motion in the steady state, one can immediately arrive at the formula for the transmissibility of vibrations:

$$TR = \sqrt{\frac{(K'+QK'')^2 + (QK')^2}{(K'+QK'')^2 + Q^2\left(K'-M\omega^2\right)^2}}. \quad (4)$$

It is seen that by reducing both the real and imaginary parts of the effective spring constant K, it is possible to easily reduce the transmissibility of vibrations.

Straightforward calculations of the effective capacitance of the negative capacitor shunt circuit with an operational amplifier shown in Fig. 2 yield the formula:

$$C(\omega, R_0, R_1) \equiv \frac{R_1 + R_2 - A_u R_2}{R_1 + R_2 + A_u R_1}\left(C_0 - \frac{i}{\omega R_0}\right) \approx -\frac{R_2}{R_1}\left(C_0 - \frac{i}{\omega R_0}\right), \quad (5)$$

where A_u is the output voltage gain of the operational amplifier and the approximate formula on the right hand side stands for the case of the ideal operational amplifier, i.e. $A_u \rightarrow \infty$. Considering the frequency dependence of the piezoelectric actuator capacitance

$$C_S(\omega, C_{S0}, R_S) \equiv \frac{C_{S0}}{1 + i\omega C_{S0} R_S}, \quad (6)$$

by proper adjustment of resistances R_0 and R_1, it is possible to achieve at given frequency ω_0 the condition $C_S(\omega_0, C_{S0}, R_S) = -C(\omega_0, R_0, R_1)$, which yields the situation when the effective spring constant K tends to zero (see Eq. (3)) resulting in the suppression of the transmissibility of vibrations TR.

4 Experimental Data and Results of Modeling

The negative capacitor was realized using LF 356N operational amplifier. Its output voltage gain was approximated by the function $A_u(\omega) = A_0/(1 + i\omega/(2\pi f_0))$, where $A_0 = 105$ dB and $f_0 = 100$ Hz. Figure 3 shows the frequency dependence of the transmissibility of vibrations in two situations when negative capacitor was disconnected from (dashed line) and connected to (solid line) the piezoelectric actuator. In the latter case the negative capacitor was adjusted to achieve the minimal values of the transmissibility of vibration at the frequency 800 Hz.

At the first step in our analysis, we performed fitting the experimental results in the former case to our model. We extracted values for the mass $M = 1.69$ kg, spring constant $K = 1.75 \times 10^7$ F/m, and mechanical quality factor $Q = 11$. The fitted value of the mass well agrees with its value of 1.77 kg obtained by direct measurement. At the next step, we measured the capacitance of the piezoelectric actuator using ESCORT ELS-3133A LRC-meter at the frequency 1 kHz, which gives the values $C_s = 6.94$ μF and $R_S = 0.87$ Ω. Capacitance of the reference capacitor was $C_0 = 4.7\ \mu F$. Then, we performed fitting the experimental data in the case when the negative capacitor shunt was connected to the piezoelectric actuator, which yields the values of $R_0 = 1\ 317\ \Omega$, $R_1 = 40\ 393\ \Omega$, $R_2 = 59\ 607\ \Omega$, $R_S = 0.92\ \Omega$, and value of the effective electromechanical coupling factor $k = 0.06$. It is seen that the fitted value of R_S corresponds to the directly measured value.

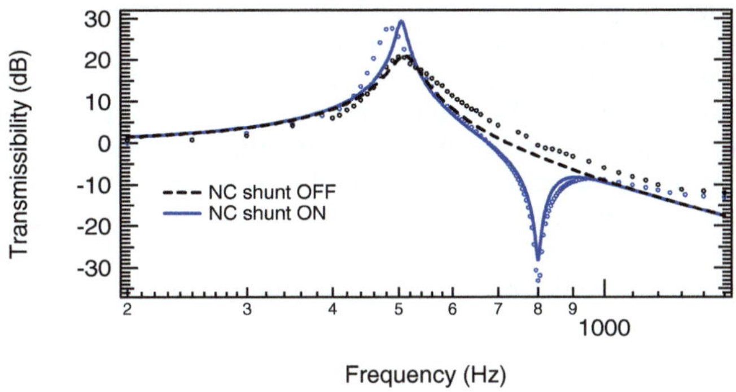

Fig. 3. Frequency dependence of the transmissibility of vibrations transmitted through the piezoelectric actuator. Open circles represent experimental data. Dashed and soled lines represent the theoretical prediction of our model calculated from Eqs. (3)-(6) where the negative capacitance shunt circuit was ON and OFF, respectively.

Using the fitted values, we calculated from Eqs. (3), (5), and (6) the frequency dependencies of the real and imaginary parts of the effective spring constant of the piezoelectric actuator, which is shown in Fig. 4. It is seen that at the frequency 800 Hz at which the negative capacitor was adjusted to achieve the minimal transmissibility of vibrations, the value of K' reaches minimum and K'' equals zero. It is also seen that at mechanical resonance frequency 500 Hz the value of K''

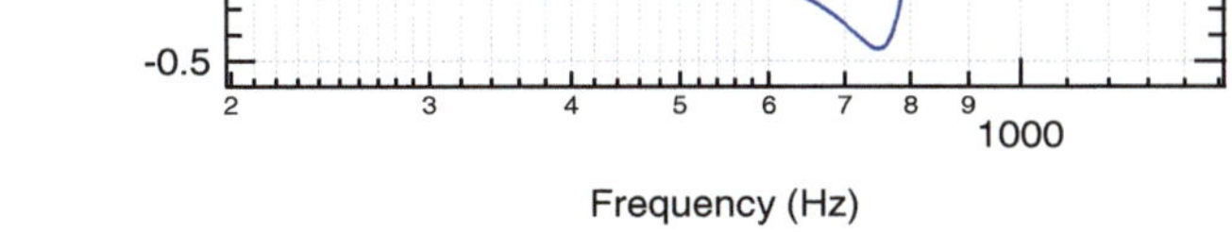

Fig. 4. Frequency dependence of the real and imaginary parts of the effective spring constant K of the piezoelectric actuator, which is connected to the negative capacitor shunt circuit, normalized to the value of the intrinsic spring constant of the piezoelectric actuator K_S calculated from Eqs. (3), (5), and (6).

is negative, which yields to the increase of the transmissibility of vibrations at 500 Hz compared to the case of the disconnected negative capacitor.

5 Conclusions

We have demonstrated a simple but very efficient way to realize a vibration control system using a single piezoelectric actuator connected to the negative capacitor. We have performed frequency measurement of the transmissibility of vibrations and the theoretical modeling of the semi-active vibration control device. Using our system, we have achieved a 35 dB reduction of the transmissibility of vibrations at the frequency 800 Hz. We have demonstrated an excellent agreement of the prediction of our theoretical model with the experiment, which we believe is important for deep understanding the function of the vibration control devices.

Acknowledgments. This work has been supported by the Czech Science Foundation Project No. 101/08/1279.

References

[1] Uchino, K.: Piezoelectric Actuators and Ultrasonic Motors. Kluwer Academic Publishers, Norwell (1996)
[2] Wang, T.T., Herbert, J.M., Glass, A.M.: The Applications of Ferroelectric Polymers. Blackie and Son Ltd. (1988)
[3] Setter, N.: Piezoelectric Materials in Devices. EPFL, Switzerland (2002)
[4] Date, M., Kutani, M., Sakai, S.: Electrically controlled elasticity utilizing piezoelectric coupling. J. Appl. Phys. 87, 863–868 (2000)
[5] Okubo, T., Kodama, H., Kimura, K., Yamamoto, K., Fukada, E., Date, M.: Sound-isolation and vibration-isolating efficiency piezoelectric materials connected to negative capacitance circuits. In: Proc. 17th Int. Congr. Acoustics, pp. 301–306 (2001)
[6] Kodama, H., Okubo, T., Date, M., Fukada, E.: Sound reflection and absorption by piezoelectric polymer films. In: Proc. Material Research Society Symp., vol. 698, pp. 43–52 (2002)
[7] Imoto, K., Nishiura, M., Yamamoto, K., Date, M., Fukada, E., Tajitsu, Y.: Elasticity control of piezoelectric lead zirconate titanate (pzt) materials using negative-capacitance circuits. Jpn. J. Appl. Phys., 44(9B), 7019–7023 (2005)
[8] Tahara, K., Ueda, H., Takarada, J., Imoto, K., Yamamoto, K., Date, M., Fukada, E., Tajitsu, Y.: Basic study of application for elasticity control of piezoelectric lead zirconate titanate materials using negative-capacitance circuits to sound shielding technology. Jpn. J. Appl. Phys., 45(9B), 7422–7425 (2006)

Piezo-Module-Compounds in Metal Forming: Experimental and Numerical Studies

R. Neugebauer[1,3], R. Kreißig[2], L. Lachmann[3], M. Nestler[1], S. Hensel[1], and M. Flössel[4]

[1] Fraunhofer-Institute for Machine Tools and Forming Technology IWU, Chemnitz, Germany
[2] Chemnitz University of Technology, Faculty of Mechanical Engineering, Institute for Mechanics and Thermodynamics, Professorship of Solid Mechanics, Chemnitz, Germany
[3] Chemnitz University of Technology, Faculty of Mechanical Engineering, Institute for Machine Tools and Production Processes, Professorship of Machine Tool Design and Forming Technology, Chemnitz, Germany
[4] Dresden University of Technology, Institute of Material Science, Chair of Inorganic Non-Metallic Materials, Dresden, Germany

Abstract. The integration of piezoceramics in metal-structures offers the possibility to achieve different functionalities like active control of dynamic behaviour, health monitoring or energy harvesting in structural parts. To avoid the time-consuming additional application of piezo-modules on a formed semi-finished part, piezoceramic-modules were embedded in a double-layer-sheet. The presented method of integration allows the forming of the piezo-metal-compounds after the integration process. The use of a semi-cured adhesive offers the possibility to form the compound avoiding tensile stresses due to shear loads in the piezoceramic-module. The process limits and the level of operability of integrated piezo-ceramic-modules after forming were determined experimentally by examining different functionalities. In numerical studies using homogenization-localization theories the stresses and strains of the piezoceramic-module during different forming processes are evaluated. A homogenization of the local periodical substructure with a unit-cell-model is performed. Global loads of the piezo-module due to forming are transferred in a submodel to obtain local loads of the piezo-ceramic.

1 Introduction

The industrial use of adaptronic components in different areas of production technologies belongs to the state of the art. Therefore a lot of piezoceramic materials in various structural shapes and with separate characteristics are commercially available. Fuel injectors are typical applications in the high-quantity-segment, whereby PZT-multilayer-actuators are often integrated in this assembly group [1].

The integration of piezoceramics in sheet metals enables interesting applications like health monitoring, the control of the vibration behaviour, an integration of crash sensors [2] or energy harvesting for example in the field of automotive industry. The lack of appropriate production technologies is one reason for the rare use of piezo-sheet-metal-compounds. To apply laminar piezo-modules on sheet metals, the manual affixing with adhesives is necessary [3]. To avoid the time-consuming and sometimes impossible application of piezo-modules on formed semi finished parts, the integration of piezo-elements in sheet-metal-compounds before the forming process is focused. In this process, the crucial point is the connection of the hard and brittle ceramic material with a well formable metallic part. In the following, an integration method to overcome this problem is presented.

2 Integration of a Piezoceramic-Module in a Double-Layer-Sheet

Figure 1 shows laminar piezo-modules integrated between two metal sheets with different thicknesses.

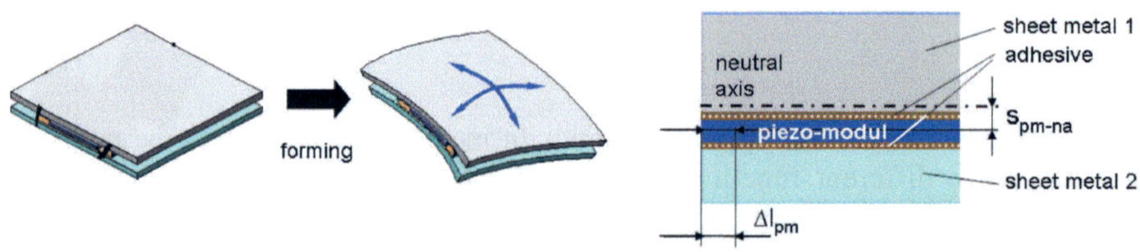

Fig. 1. Double-layer-sheet

To realise a strong bond between module and sheets, the use of adhesive is required. Epoxy adhesives are capable of bonding metals with high adhesive strength. The chosen 2-component epoxy (Sika Icosit 220/15) offers low viscosity during the forming process and high stiffness after curing.

The actuator and sensor action of the compound depends on the distance between piezo-module and the neutral axis $s_{pm\text{-}na}$ (fig. 1) which can be controlled by selecting sheet metals with different thicknesses. In actuator mode an increased distance $s_{pm\text{-}na}$ effects more actuation of the compound with a given displacement of the piezo-module Δl_{pm} (fig. 1). Analogue, in sensor mode a larger distance $s_{pm\text{-}na}$ enables sensing of smaller actuation of the compound.

In addition to the mechanical characteristics of the piezo-metal-compound the functionality depends on the following properties of the integrated piezo-module:

1. Plane build-up
2. Secure formability
3. High performance
4. Possibility for adhesive machining

The use of Macro-Fibre-Composites (MFC), which consist of rectangular piezo ceramic rods integrated in a compound with adhesive and an electroded polyimide film, turned out to be suitable.

3 Forming of Piezo-Metal-Compounds

With regard to a use in automotive industries, different forming technologies have to be evaluated. For basic researches of the formability of piezo-metal-compounds simple bending has been analysed. Therefore different specimens were manufactured, whereat the bending radii and the combination of sheet thicknesses were varied to realise different load conditions of the laminar-piezo-modules. To achieve a 3-dimensional load condition of the piezo-modules stretch-forming, deep-drawing and hydroforming operations are accomplished. Figures 2 and 3 show an assortment of specimens with bending radii of 40 mm and 10 mm as well as a stretch-formed piezo-metal-compound.

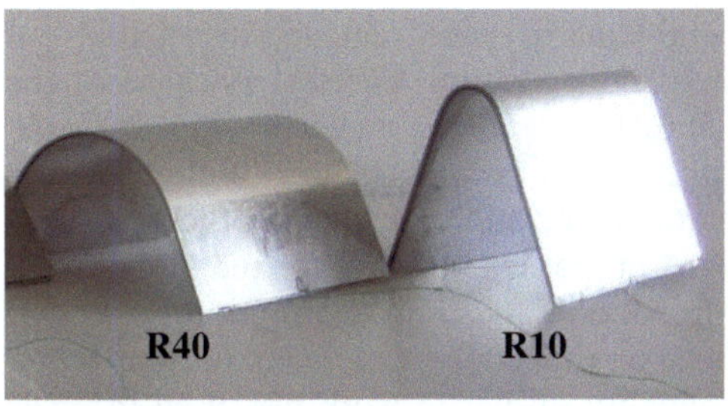

Fig. 2. Piezo-metal-compounds bended with different radii (40 mm and 10 mm)

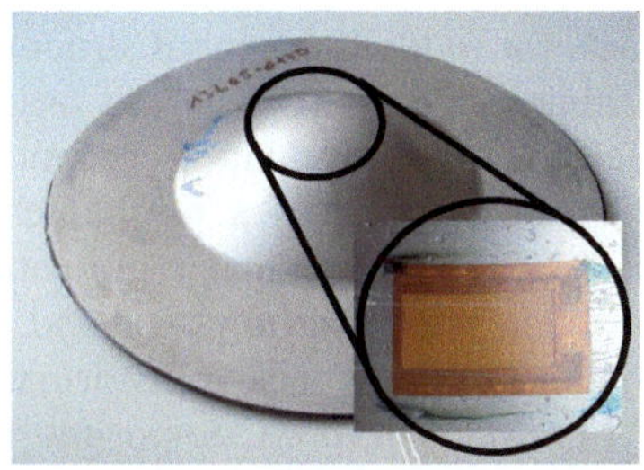

Fig. 3. Stretch-forming of piezo-metal-compounds

4 Functional Tests of Formed Piezo-Metal-Compounds

For the evaluation of functionality, different test procedures are implemented. A measurement of capacitance allows examination of the copper electrodes at PZT-fibres and periphery. A damage of electrodes would reduce the active volume of whole fibre-material and is indicated by a reduction of capacitance.

Visualisation and localisation of the damaging of the piezo-module takes place by the use of nondestructive computer tomography at Fraunhofer IKTS Dresden. This examination shows cracks in the fibres as well as a damage of cords or electrodes (fig. 4).

To get extensive information about the functionality, furthermore it is necessary to analyse the actuator and the sensor effects.

The research of the actuator effect is realised by a defined electrical load of the piezo-module with a trapezoidal voltage of about 1500 V and the measurement of the caused displacement also realised at Fraunhofer IKTS Dresden. For checking the sensor effect, the compound is deflected by a low-frequent signal of a shaker.

Certain parts have been analysed with different testing methods. It is shown, that bending radii from 75 mm, 50 mm, 40 mm and even 10 mm are realisable with a low degradation of functionality.

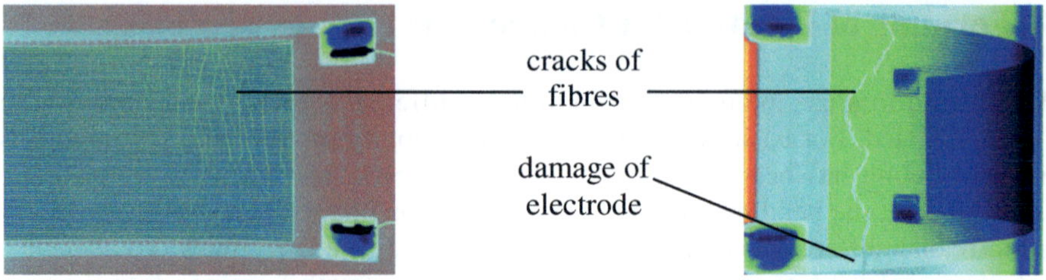

Fig. 4. Cracks in damaged MFCs illustrated by computer tomography

5 Numerical Investigation

Basis for modelling global structures with equally spaced detailed distributions of mechanical properties is the usage of homogenization theories. Because of the necessary high amount of computational resources, it is not possible to solve a detailed description of a mechanical-mathematical problem with a high-resolution model. In the numerical examination a homogenization-localization theory is used to obtain both, the homogenized material properties of MFC for the global solution and within a back-transformation the local stress/strain distributions.

For Macro-Fibre-Composites in the experimentally examined forming compounds a homogenization among the equality of elastic work was used [4]. The elastic properties of the orthotropic plate with its strain components are equated with the volumetric elastic work. The amount is determined by element-wise summarization. With load combinations ε_{ii} applied in a model of the smallest periodically repeated structural unit – the unit cell (Fig. 5) – the elastic properties Q_{ii} are determined as averaged global values. Index 11 denotes the main direction, 22 transverse direction and 66 pure shear in the 12-plane.

$$W_{elast} = \frac{1}{2}\int(\varepsilon^T Q \varepsilon) dV = \frac{V}{2}\begin{pmatrix}\varepsilon_{11} & \varepsilon_{22} & \gamma_{66}\end{pmatrix} \cdot \begin{pmatrix} Q_{11} & Q_{12} & 0 \\ Q_{12} & Q_{22} & 0 \\ 0 & 0 & Q_{66} \end{pmatrix}\begin{pmatrix}\varepsilon_{11} \\ \varepsilon_{22} \\ \gamma_{66}\end{pmatrix} = \sum_{elem} W_{elast,elem} \cdot V_{elem} \qquad (1)$$

The material characteristic of the components in the unit cell (e.g. electrodes, fibre and epoxy-embedding) are assumed to be isotropic-elastic except for the piezo-fibre-material. The fibre was modelled with orthotropic-elastic material characteristics from data sheets of manufacturer and without electro-mechanical coupling. As a result the homogenized material has also an orthotropic characteristic.

In the continuum each node on the boundary planes has the same displacement as the according node on the opposite plane. Hence unit loads and boundary conditions were applied to the unit-cell as constraint equations. The bending modes were implemented under use of anti-cyclic conditions with change of sign of the displacement over the height of the cell and zero-crossing at the cell's middle [5].

As a result of the homogenization with separate MFC-components the averaged orthotropic material properties for the utilization in the global model are obtained. Young's moduli in fibre and transverse electrode's direction were calculated as

$E_{11} = 31.01 GPa$ and $E_{22} = 20.37 GPa$, shear modulus in 12-plane and Poisson's ratio as $G_{12} = 6.74 GPa$ and $\nu_{12} = 0.301$ in very good agreement with literature [6].

The geometry of the unit-cell-model was deduced from micrographs.

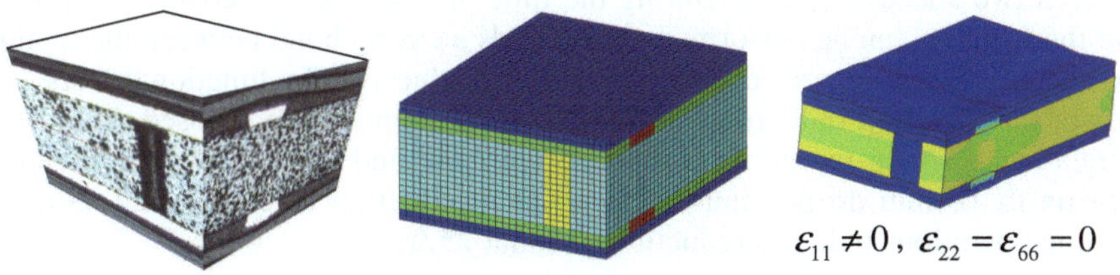

Fig. 5. Micrograph geometry MFC; model of unit cell; load-case tension 11

Initiated loads of the MFC due to the forming operation are modelled with global simulations of the forming processes: bending and stretch-forming. The two global process models were validated by the experimental results (fig. 6).

The global total strains of the MFC are subdivided in membrane and bending portions and according to the two leading perpendicular material axes. They are transferred back to the unit-cell-model. The already computed stresses / strains of the load cases within homogenization are then scaled with the global loads and combined by linear superposition. As a result the local stresses and strains of the MFC-components due to global forming operation are obtained and thus the load level can be characterized in detail (fig. 7).

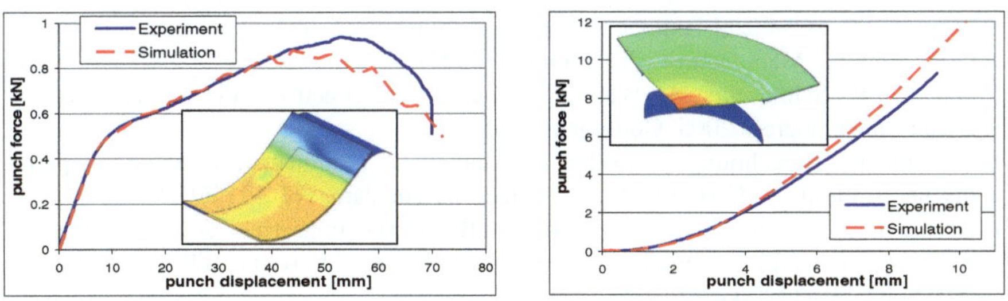

Fig. 6. Comparison experiment and simulation – left: bending; right: stretch-forming

Fig. 7. Piezo-fibre with stresses in fibre direction – left: bending; right: stretch-forming

6 Summary and Outlook

In order to resolve the challenging task of the creation of a formable piezo-metal-compound, the principle of embedding is used. It allows the integration of a MFC between two sheets of metal. During the time of a semi-cured condition of adhesive the forming can be performed. Afterwards a strong bond between the module and the blanks is generated by curing of the adhesive. In functional tests it is shown, that bending radii from 75 mm up to 10 mm can be realised. There is a correlation between the active volume of ceramic and the capacitance. Bending radii up to 10 mm do not cause a significant reduction of capacitance. Stretch-forming operations induce a reduction of about 25 %.

In a numerical study averaged material properties of the piezoceramic-module are provided in a homogenization step. Global simulations of the forming are validated with experimental results. The global forming loads of the ceramic-module are transferred to a unit-cell-model to obtain the local stress/strain distributions.

Acknowledgements. This work was supported in part by Deutsche Forschungsgemeinschaft (DFG) in the research project SFB/Transregio 39 PT-Piesa.

References

[1] Senousy, M.S., Li, F.X., Mumford, D., Gadala, M., Rajapakse, R.K.N.D.: Thermo-electro-mechanical Performance of Piezoelectric Stack Actuators for Fuel Injector Applications. Journal of Intelligent Material Systems and Structures (2009)
[2] Drossel, W.-G., Hensel, S., Kranz, B., Nestler, M., Goeschel, A.: Sheet metal forming of piezoceramic–metal-laminar structures - Simulation and experimental analysis. CIRP Annals - Manufacturing Technology (2009)
[3] Wierach, P., Schönecker, A.: Bauweisen und Anwendungen von Piezokompositen. In: Adaptronic Congress 2005, Göttingen (2005)
[4] Schmidt, R.: Berechnung elastischer Konstanten für inhomogene Bauteile mit periodischer Struktur. In: CAD-FEM User's Meeting, Potsdam, vol. 19 (2001)
[5] Kranz, B., Drossel, W.G.: Rechnerische Beanspruchungsermittlung bei adaptiven Kompositen mit Piezokeramischen Fasern, Darmstadt 15./16.03.2006, DVM-Bericht 901, Seite 43-52 (2006)
[6] Williams, R.B.: Tensile and Shear Behavior of Macro Fiber Composite Actuators. Journal of Composite Materials 38(10), 855–870 (2004)

Commutation Phenomena in DC Micromotor as Source Signal of Angular Position Transducer

M. Bodnicki[1] and H.J. Hawłas[2]

[1] Warsaw University of Technology, Faculty of Mechatronics, Institute of Micromechanics and Photonics, 8 Sw. A. Boboli Str., 02-525 Warszawa, Poland
m.bodnicki@mchtr.pw.edu.pl
[2] Warsaw University of Technology, Faculty of Mechatronics, Institute of Metrology and Bioengineering, 8 Sw. A. Boboli Str., 02-525 Warszawa, Poland
h.hawlas@mchtr.pw.edu.pl

Abstract. This paper presents transducer based on idea of detection of angular position of the rotor in electric micromachines by determination of electrical signals. Changes of the resistance cause pulsation of the current passing in the armature circuit, with angular frequency proportional to the number of commutator sectors. The LEM-type (based on Hall Effect) sensors are used as current/voltage transducers. Then new method of the elimination of component proportional to mechanical load from the signal is used. Proposed method was first testing in computer simulation and then the prototype was built and physically tested. New transducer is proposed for control application in high dynamic DC drive systems of small size mechatronic devices.

1 Introduction

A requirement of a closed-loop control, e.g. in the case of problems connected with positioning of objects, extorts integration of the micromotor with transducers of angle and/or velocity. At the same time, one can note a tendency - maybe not fully applicable at present - to eliminate the traditional transducers of the above mechanical quantities in favour of identification of position and velocity of the rotor by application of processing easy-measurable electrical quantities, which characterize operation of the micromotor.

Pursuit for miniaturization of the structure ruled out application of a standard measuring unit offered by the manufacturer, i.e. transducer of angular position with resolution of e few pulses per revolution, integrated with the micromotor and the reducer, since that would result in elongating the actuator too much. In the case of typical rotation-to-pulse angle transducers, a quasi-sinusoidal signal from the detector (photodetector, inductive sensor or Hall-effect device) is converted into series of digital signals [5]. Resolution of the system is dependent on number of the markers (pulses per revolution), while accuracy of angle measurement is dependent, first of all, on the accuracy of performing the markers, repeatability of the shape of the digitized course and selection of the triggers applied in the system for signal digitalization.

2 Pulses of the Current in DC Micromotor and Their Use for Control

During the work of micromotor the number of supplying sections is changed. In case of motor characterized by the odd number of commutator sections (*k*) two cases occur:

- when both brushes contact themselves with only one segment of commutator, and the resistance of total resistance of armature circuit R_t assumes the maximum value R_{tmax};
- when one of brushes contacts with one segment of commutator, and the second one short-circuits neighboring two segments - the resistance of total resistance of armature circuit R_t assumes the minimum value R_{tmin}.

Changes of the resistance cause pulsation of the current passing in the armature circuit (see Fig. 1). Its frequency is dependent on the number of commutator sectors. Number of pulses per one revolution of the rotor is determined by the formula:

$$n = 2k \quad (1)$$

where: n – number of pulses, k – number of commutator sectors.

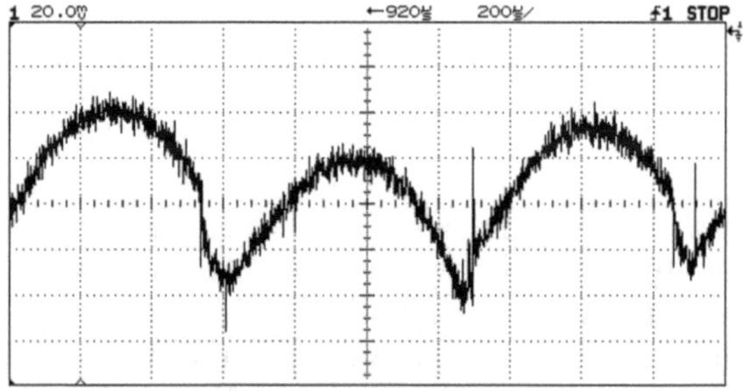

Fig. 1. Pulsation of the DC motor armature current (an example)

Idea of practical use of such phenomena for detecting of control signals is described in papers [1-4].

3 Structure of the Transducer

First, the high limit of the frequencies of the commutation phenomena could be find. Exemplary - for the DC motors with no-load speed to 10.000 rpm (estimated) and 11 sections in commutator the high limit of the pulsation is equal circa 3.7 kHz. For transducer the higher components in the current signal could be taken as noise.

The proposed algorithm involves the following steps:

A. Detecting voltage signal from LEM [6] transducer.
B. Filtering of the high frequency noises.
C. Detecting and separation of the component proportional to load of the micro-motor, determined on way of LP filtering of the signal.
D. Subtraction of the signal "after step C" from the signal "after step B"– generation of the analogue signal of the pulses.
E. Digitalization of the differential signal - generation of the TTL length-constant pulses.

Two analyzed structures for realization of maintained algorithm are presented in Fig. 2.

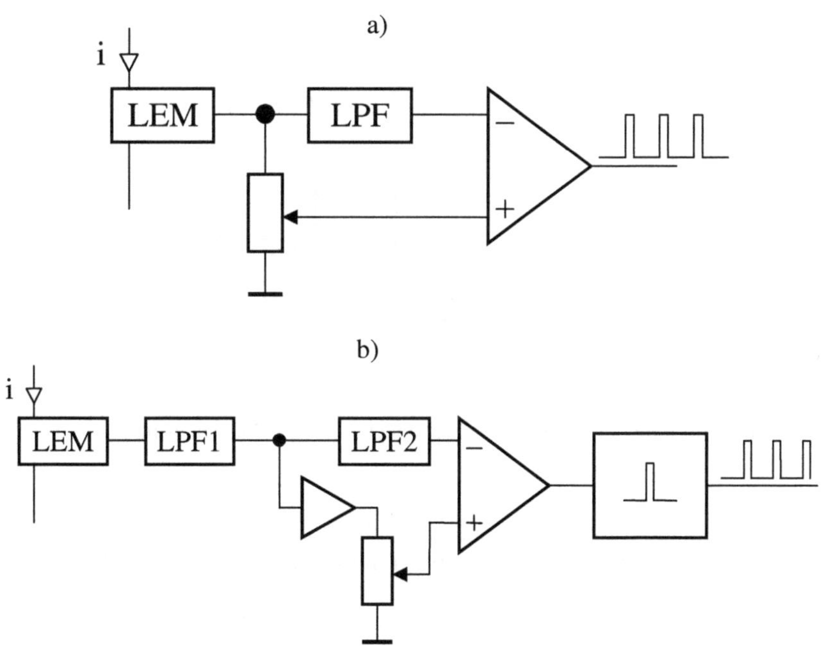

Fig. 2. Structure of the transducer: a) basic solution; b) advanced version i – current in armature; LEM – current/voltage transducer, LPF – low-pass filters

4 Testing of the Measuring Channel

First, the simulation model of the micromotor/transducer system has been elaborated in *MATLAB/SIMULINK* environment (see Fig. 3). In order to model the micromotor (*DCmotor* block), one used a classical system of equations of voltages and torques complemented with the additional equation that represented pulses of the armature resistance. Coefficients of the model for specific micromotors types were accepted on the basis of catalogue data provided by manufacturers. The load can be added in a form of reduced quantities - moment of inertia and breaking torque (constant or variable). Mathematical formulas of the micromotors model are analogue as presented in [1]. For counting of pulses was used model of the typical digital counting system, reference position signal was calculated as number of the rotor's revolutions.

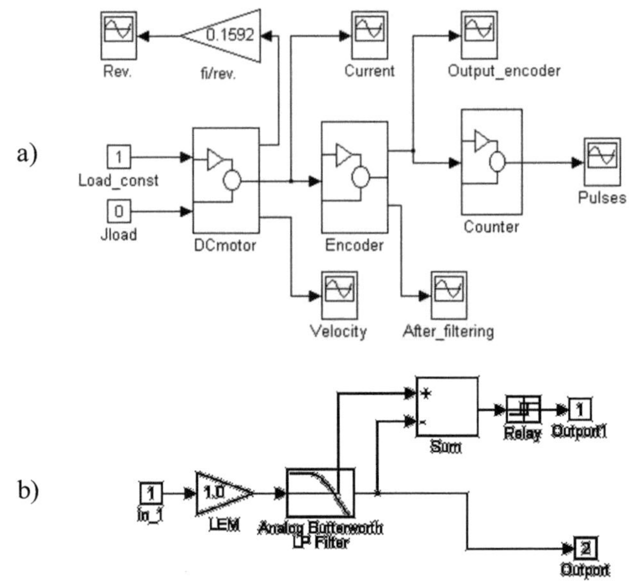

Fig. 3. Simulation model a) transducer connected to the DC micromotor and counter, b) encoder

During simulation test the parameters of the filters were fitted. Too high elimination of the pulses from in the "C step" of algorithm could change phase between both signals transformed by differential block. Characterized signals obtained in computer simulation test are shown in Fig. 4. After the stage of model studies, one designed and built a prototype of electronic circuit of the encoder. The transducer of angular position, simplified this way, was laboratory tested (test stand is presented in Fig. 5) within the range of:

- observing the shape of the output signal from the transducer while starting the micromotor, as well as under steady-state conditions of its operation;
- comparative studies of the output signal related to indication of a standard revolution-to-pulse transducer (also while staring the micromotor and under steady-state conditions of its operation).

In both cases, as in the case of the simulation studies, one focused mainly on determining a so-called "dead zone", where the transducer does not respond to changes of the angular position of the rotor during electromechanical transient states of the micromotor. Characteristic courses are shown in Fig. 6.

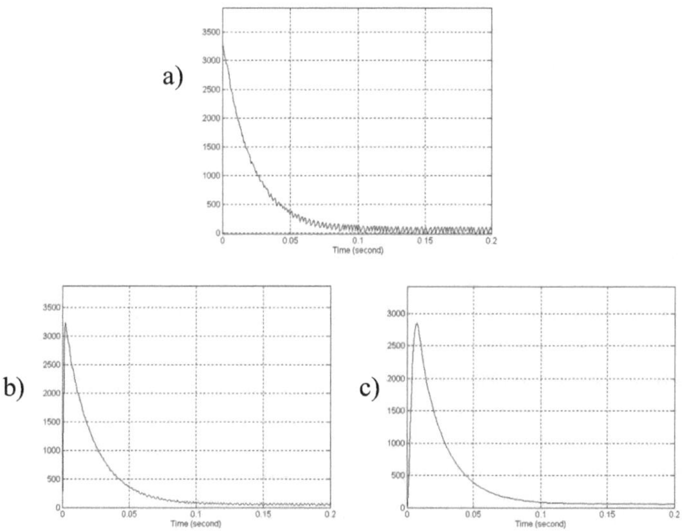

Fig. 4. Exemplary results of the simulation tests – fitting of the low pass filter parameters using for determination of component proportional to the load (step C of algorithm) a) current of the start-up process, b) not effective elimination of pulses of current, c) elimination of pulses but too much change of dynamics of the signal

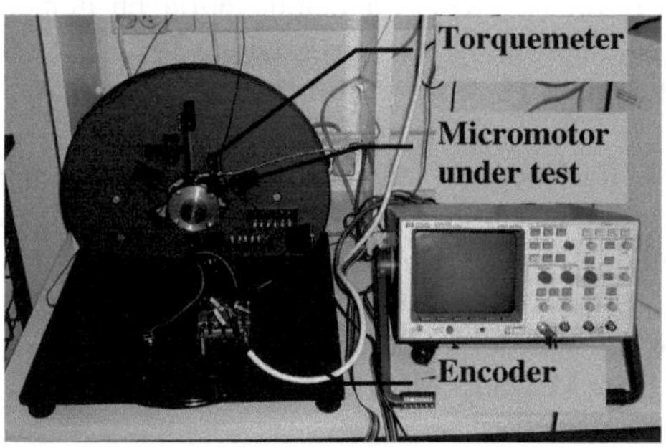

Fig. 5. General view of the test stand (micromotor is connected to the additional measuring torque channel used for model identification)

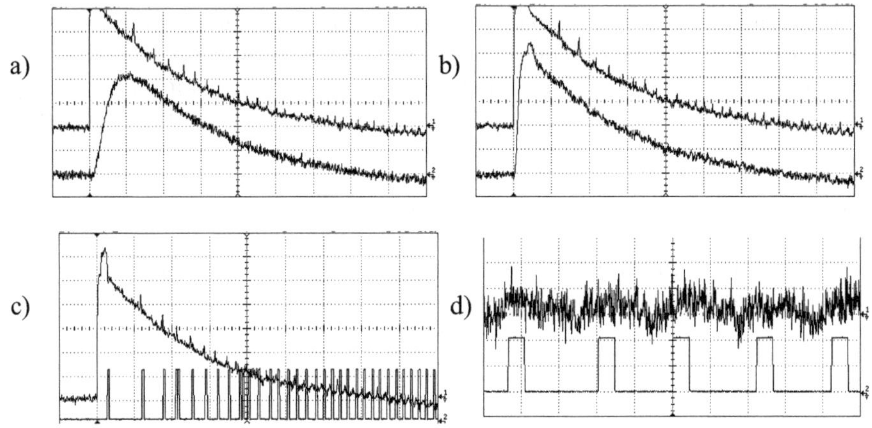

Fig. 6. Tests of prototype: a), b) changes of the filter frequency, c) operation during start-up, d) operation during steady angular speed (inaccuracy of commutator system has influence on time intervals between output pulses)

5 Summary and Conclusion

The tests of the prototype of encoder based on commutation effects fully proved that it is possible to apply the transducer for determining angular position of a micromotor rotor. During work - the gradual wear of the commutator changes also operation conditions of the system, and therefore it is necessary to perform a periodical calibration. However, in many cases the system may be an economical supplement of techniques for measuring the revolution angle of *DC* micromotors specially for mechatronic systems with requirement of small M.I. of rotating parts.

References

[1] Bodnicki, M.: Encoder Using Pulsation of the Current in DC Micromotor - Model Tests. Hydraulika a Pneumatika 2(2), 35–36 (2000)

[2] Bodnicki, M.: Additional effects of commutation phenomena in DC micromotor - identification and application for positioning. In: Proc. of the 47. Int. Wissenschaftliches Kolloquium TU Ilmenau, Ilmenau (Germany), September 23-26, 2002, pp. 143–144 (2002)

[3] Acarnley, P., Hill, R., Hooper, C.: Detection of rotor position in stepping and switched motors by monitoring of current waveforms, Sensorless Control of AC Motor Drives. IEEE Press, Los Alamitos (1996)

[4] Goryca, Z., Jaworski, M.: Sposób bezczujnikowego wyznaczania prędkości i położenia wirnika silnika prądu stałego. In: Proceedings of SENE 2001, p. 4 (2001)

[5] Products Catalogues: MAXON MOTORS, MINIMOTOR, API-PORTESCAP

[6] `http://www.lem.com`

PWM Controlled DC Drive with ADuC812 Microcontroller

M. Dub and R. Jalovecký

University of Defence, Faculty of Military Technology,
Department of Aerospace Electrical Systems, Kounicova 65, Brno, Czech Republic
michal.dub@unob.cz, rudolf.jalovecky@unob.cz

Abstract. The paper deals with design and construction of PWM controlled DC drive for commercial door motion control. The drive control element is ADuC812 microcontroller, which not only controls Valeo 0273 GML type DC motor but also process information from several sensors such as inductive proximity sensors, passive infrared motion sensors and fire sensor. Integration of used sensors into the drive control system was necessary to solve the problem of commercial door motion control according to the user requirements.

1 Introduction

The use of the PWM control of DC motor is considered to be one of the modern as well as ecological approaches to the motion control of the electrical drives. These new designed systems have been replacing out of date drives in last years. The whole commercial door system is represented by ten independently opened sliding doors and basic requirements for described particular application were given by customer.

The basic function of primary system is to open single sliding door according to the detection of human body (or any obstacle) in front of the door area and to close it after some defined time with no obstacle detection in the door area. Advanced function is featured by dependence of primary systems motion when dangerous situation occurs and the doors are opened in sequences with possible dead time control.

The other user door control is possible via manual control when the detection system is intentionally set out of order. The reliability of the system is enhanced with power supply backup. The 12 Volts DC power system with backup battery is fed from the 230 Volts AC power supply. The battery is able to open all doors in the case of external power supply failure.

2 System Design

PWM controlled DC drive can be designed in two basic ways. Fist way is the logic circuits based solution unit and can be represented by [1]. The second way is the microcontroller based solution we used. The primary system of the commercial door motion system is designed from several parts according to the Fig. 1.

Basic part is the ADuC812 microcontroller, which not only controls sliding door motion but also process information from several sensors. The PWM driver is based on BTS780GP power module [2] and the DC motor (M) armature is directly fed from the power MOSFETs of the PWM driver. The Valeo 0273 GML type DC motor (part 1 on Fig. 2) rotation motion is converted to the sliding door motion via toothed belt (part 2).

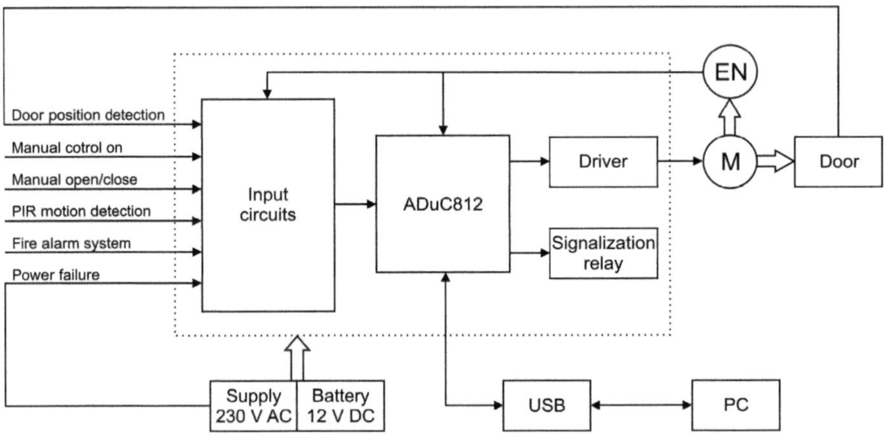

Fig. 1. Primary system block diagram

Motor rpm is measured by HEDR 542 encoder (EN) and the velocity feedback is connected to ADuC812 microcontroller where computing programme for the door motion velocity stabilization is running. The door position is indicated by signal from two GL-N12 series inductive proximity sensors (part 4 on Fig. 2) that cause Interrupt ReQuest of the running program. The obstacle in the door area is detected by four MS-112 passive infrared motion sensor modules. Fire alarm system is not integral part of the commercial door motion system. Signal from fire sensor opens the doors one after another or in time sequences.

The initial conditions such as door motion velocity and PWM signal frequency of the full bridge power transistors are entered from user PC via USB interface. It is also possible to use USB interface to load the program into control processor and to get information from it during program compilation.

Although ADuC812 microcontroller (part 1 on Fig. 3) has three internal interval timers, PWM signals generation for power module was solved by three external programmable interval timers (PIT) 8254 (part 2a) and by delay logic circuits (part 2b). Special attention has to be focused on dead time control of the complementary PWM signals for power module transistors due to possible crossover

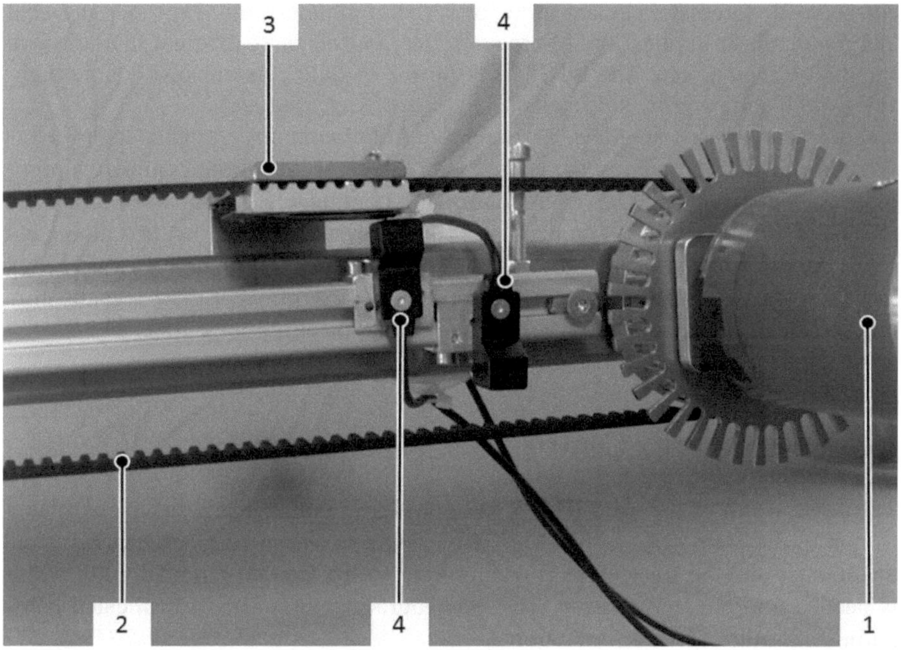

Fig. 2. Mechanical design (top view) – Valeo drive (1) with toothed belt (2), sliding door mounting (3) and inductive proximity sensors (4)

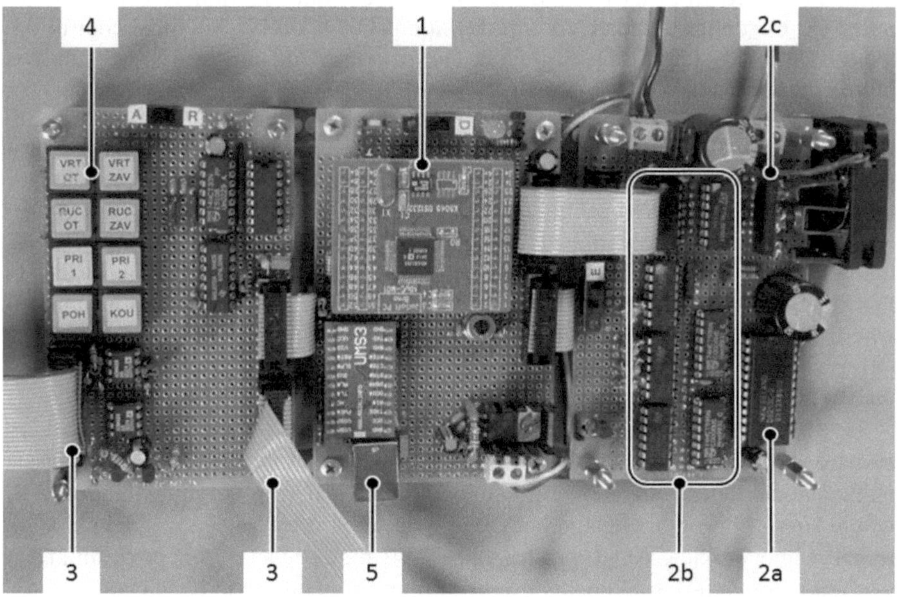

Fig. 3. Electronic design – ADuC812 microcontroller (1) with PWM driver (2), input signals (3), input signals simulator (4) and USB interface (5)

currents [2]. Internal PITs are used for serial communication, rpm measurement and run of whole application. Microcontroller controls three external PITs through standard SPI interface. The BTS780GP power module contains one double high-side switch and two low-side switches in one package (part 2c).

Each primary system ADuC812 microcontroller contains communication interface for cooperation among the control processors of the whole commercial door motion system. ADuC812 microcontrollers communicate with each other by means of simplex transmission. First primary system opens the first door and sends the information via interface to the second primary system when emergency signals from fire alarm system or power supply system occur. The signalization relay for external signalization system is switched on at the same time. The second and next primary systems act similarly.

3 Motion Velocity Control

The pulse width of the used HEDR 542 encoder output signal is inversely related to rpm of the Valeo DC motor. This pulse width is measured by ADuC812 internal counter and the final value is rpm inverse value. Certain minimal pulse width is needed for rpm stabilization. This procedure just simplifies complicated computing algorithm of the door motion.

Stabilization of the door motion velocity is shown on Fig. 4. The information of door motion (Door_Close and Door_Open) is derived from PWM pulse width for BTS780GP power module. Difference between measured (HEDR 542 signal) and wanted (entered from user PC via USB interface) door motion velocity is computed (K_Ot_delta) to create control deviation (O_Ot_delta). Maximal control deviation (PWM__step_max) is used if the computed difference (K_Ot_delta) is higher than set value (K_Ot_max).

We need to know plus or minus sign (plus_minus) to change PWM pulse width for BTS780GP power module correctly. For example, we need to slow down when the door is being opened and the measured door motion velocity is higher than wanted – the PWM width would be shorter (PWM_minus_step). We also need slow down when the door is being closed and the measured door motion velocity is higher than wanted but PWM width would be longer (PWM_plus_step) in this case. This interesting fact is caused by unipolar PWM generation unit [2] where the PWM duty cycle of the full bridge power transistors is 0.5 when DC motor shaft doesn't rotate. It is necessary to connect motor armature properly to the PWM driver or the program would not work.

Described program design can appear to be too much complicated but all instructions are written in assembly language and therefore the use of the classic mathematical operation is not possible. Use of the programming language C would simplify program design but there is problem with the control processor internal memory space. ADuC812 has only 2 kB memory space for program and it was really tough task to prepare designed program according to it.

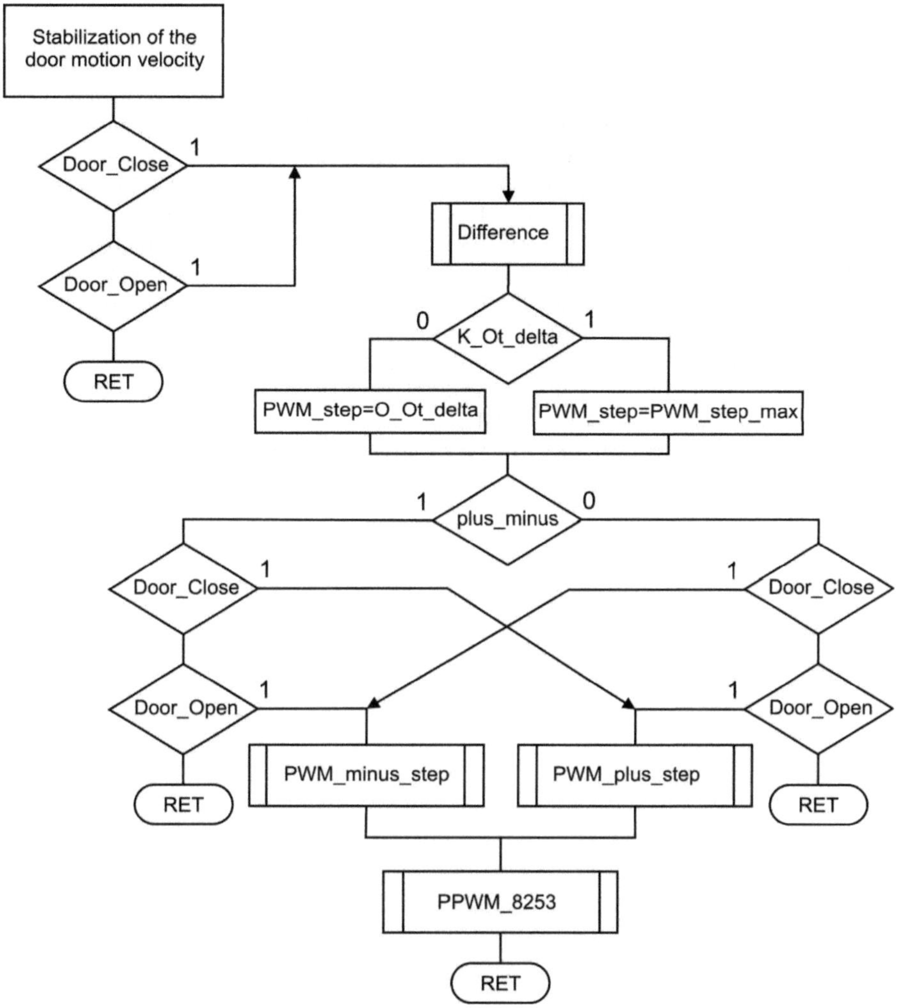

Fig. 4. Stabilization of the door motion velocity using ADuC812 microcontroller

4 Conclusion

To carry out particular tasks from theoretical knowledge can sometimes cause practical problems. The authors of the article have tried to show one possible successful solution of PWM controlled DC drive. The function of the primary systems was verified during several hours of testing. The whole commercial door system has not been yet put into service because of hall building delay.

Acknowledgments. The work presented in this paper has been supported by the Ministry of Defence of the Czech Republic (Research Plan No. MO0FVT 0000403).

References

[1] Dub, M.: PWM Controlled DC Drive Construction. In: Proceedings of reviewed papers of 1st International Scientific Conference Special Technology 2006, pp. 131–138. Alexander Dubcek University of Trencin, Bratislava (2006)

[2] Dub, M., Jalovecký, R.: Unipolar PWM Generation Unit for BTS780GP Based Driver. In: Jalovecky, R., Stefek, A. (eds.) Proceedings of the International Conference on Military Technologies 2009, pp. 978–980. University of Defence, Brno (2009)

[3] Pechlaner, A., Kern, H., Auer, F.: Throttle Control with Smart Power Bridges and Microcontrollers of the C500 and the C16x-families. System Engineering Automotive Application Note (2000-08-29), http://www.infineon.com/dgdl/SE_0899_electronic_throttle_control.pdf?folderId=db3a304412b407950112b41714ad221d&fileId=db3a304412b407950112b41714e9221e [cit. 2009-04-08].

[4] Patočka, M.: Selected Articles on Power Electronics II (in Czech), 2nd edn. University of Technology, Brno (2004)

Sensor BLDC Motor Model in Simulink Environment

V. Hubík and V. Singule

Brno University of Technology, Faculty of Mechanical Engineering,
Institute of Production Machines, Systems and Robotics,
Technická 2896/2, Brno, Czech Republic
vhubik@nbox.cz, singule@fme.vutbr.cz

Abstract. This paper deals with mathematical model of electrically commutated motor (BLDC) made up of standard components in MATLAB/Simulink© environment and its toolbox SimPowerSystems©. The model is intended to create and verify of control algorithms for sensor and sensor-less operating of BLDC motor. It is possible to insert it to any complex control systems as a functional block. Simulated motor is described by approximately 15 electrical and mechanical parameters that makes possible to model particular type of BLDC motor as faithful as possible. Inserting of parameters and post-processing is realized through well-arranged "Graphical User Interface - GUI".

1 Introduction

A fundamental element that drives electrical actuator is the DC motor. Nowadays, brushless DC motors (BLDC) are widely used, mainly because of their better characteristic and performance. BLDC motors can be controlled in sensor or sensor-less mode. The advantage of sensor-less BLDC motor control is that the sensing part can be omitted, and thus overall costs and mechanical precision during sensor assembly can be considerably reduced. The main disadvantages of sensor-less control are higher requirements for control algorithms and more complicated electronics. An analysis of the input parameters has been performed and a BLDC motor MATLAB/Simulink model has been designed. Performing simulation and test on the BLDC motor model dramatically speeds up development of a new type of electrical drive. The BLDC motor model is based on parameter analysis and matches the real motor as closely as possible. Development and perfection of the model parameters have been running almost 2 years. During this period has been created several versions of mathematical model, such as magnetic Hall sensor one. Reached records are applied in real hardware by use dSPACE environment.

2 BLDC Motor Structure

Brushless Direct Current motors [3] are one of the motor types rapidly gaining popularity. BLDCs were designed to replace the electro-mechanical commutator sub-system in a conventional Brushed DC motor due to several advantages that are:

- Higher reliability
- Lower maintenance, lower audible noise and electromagnetic emission
- More power per unit volume due to more thermal efficiency
- Higher speed range, etc.

In addition, the ration of torque delivered to the size of the motor is higher, making it useful in applications where space and weight are critical factors - especially in aerospace. To ensure the reliability of electric drives it is normally used sensor variation of the BLDC motor. From a modeling perspective it looks exactly like a DC motor, having a linear relationship between current and torque, voltage and rpm. The following chapter deals with detail description of the mathematical model of the BLDC motor with integrated Hall sensors.

3 Mathematical Model of BLDC Motor

The mathematical model of the BLDC motor [3], created in the MATLAB/Simulink environment, consists of several independent blocks, which describe its real behavior. For easier orientation it can be separated into three parts - *electrical*, *mechanical* and *sensor*.

The mathematical model is designed to implement as many as possible of the parameters supplied by the BLDC motor's manufacturer. The aim is to have a model that reliably matches a real BLDC motor. Input parameters are set in special M-file, in numbers about 20 values.

The electrical part models the internal conditions and wiring of the direct current motor. It provides fundamental electrical parameters at its outputs such as currents and voltages on individual windings. The component's design parameters and appropriate driving signals from superior blocks are required as the input parameters.

The differential equation (1) describes electrical behavior of direct current motor and is solved by numerical methods in Simulink environment. Individual phase voltages are evaluated and could be used for sensorless detection of actual rotor position.

$$u_{U,V,W} = R_{U,V,W} \cdot i_{U,V,W} + L_{U,V,W} \cdot \frac{di_{U,V,W}}{dt} + u_{bemfU,W,W} \tag{1}$$

The electrical part of the model is created from parts of the SimPower integrated environment. It can be thought of as an electrical schematic of a motor, Fig 1. Even if Simulink libraries have the better switching components, such as power MOSFET, IGBT or bipolar transistors, it was used only basic ideal switching device with inverse diode. That will dramatically decrease computation time to real value.

$$\frac{d\omega}{dt} = \frac{M_U + M_V + M_W - M_{EXT} - B \cdot \omega}{J} \tag{2}$$

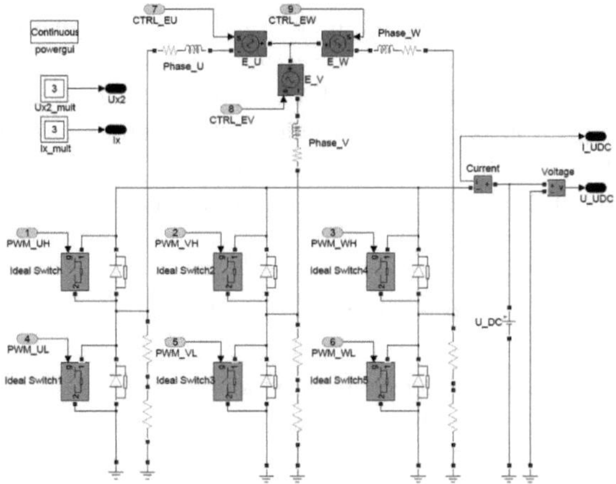

Fig. 1. Electrical part of the model with power stage

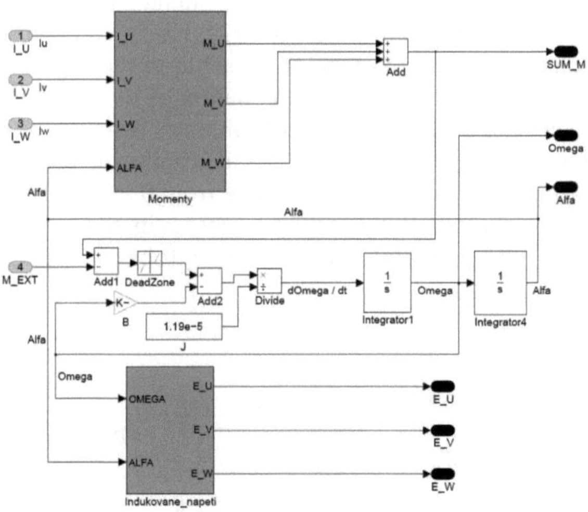

Fig. 2. Mechanical part of the model

The mechanical part Fig 2 of the model is based on equation of motion (2) and simulates the interaction between mechanical and electrical values of the system. Actual kinetic torques of individual windings is evaluated according to input currents, equation (3) from the electrical part. The model also takes into account attenuation forces arising from friction during rotation or from appropriate external braking torques on the rotor shaft:

$$M_{U,V,W} = K_m \cdot i_{U,V,W} \cdot \sin\left(\left(\alpha - F_{U,V,W}\right) \cdot POL_{num}\right) \tag{3}$$

An important output value is the amplitude of induced voltages in particular windings. This is used for detection of the correct moment for commutation during sensor-less control conditions. These back EMF voltages are evaluated in subsystem. The appropriate equation is mentioned below (4). Integration of the equation of motion is possible to enable evaluation of instantaneous rotor position and angular velocity. This part of model is created in the Simulink environment by means of basic integrated blocks, Fig 3.

$$u_{bemfU,V,W} = -K_e \cdot \omega \cdot \sin\left(\left(\alpha - F_{U,V,W}\right) \cdot POL_{num}\right) \tag{4}$$

The last block of the model is sensor part, which consists of three Hall effect sensors, Fig 3. They are placed on the stator part of BLDC motor. This kind of electrically commutated sensor motor is the most common. Use of any special sensors such as IRC, resolver or other Hall effect one is rare due to overall cost. It is possible to ensure resolution of 60 degree between rotor and stator with precise position of all Hall sensors, so the next commutation state can be determine. Input value for the sensor block is the actual rotor angel (alfa) evaluated in the previous blocks by double integration of the equation of motion (2). Every Hall sensor has determinated angel of activity by two constant, also visible in the Fig (4). Difference between on and off state is 180 degrees. Combination of three output codes in 3 bits digital form indicates the actual position and simulates behavior of real commercial sensor BLDC motor. The block with 'Alfa_recalc caption' is used to recomputation of the evaluated angel (alfa) to the specific range from 0 to 360 degrees.

Simple condition is used. If the actual angel is in the range determined by Hall_ON and Hall_OFF constant, output value will be in logic 1, else in logic 0 state.

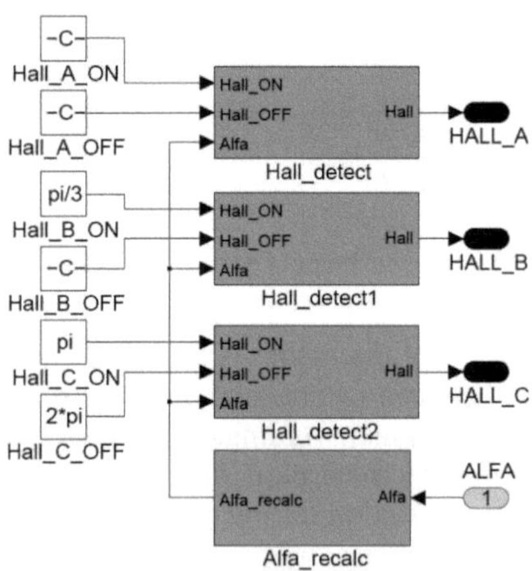

Fig. 3. Sensor part of the model with 3 Hall sensors.

4 Graphical User Interface – GUI

The graphical user interface - GUI is possible to execute by double-click on the appropriate block in the model, such as BLDC motor block or data store block. All GUI windows are created in MATLAB GUI editor and are used for setting the basic parameters of BLDC motor, Fig 4, parameters of the simulation and final data presentation. This will reduce overall development time and make simulation more effective and comfortable.

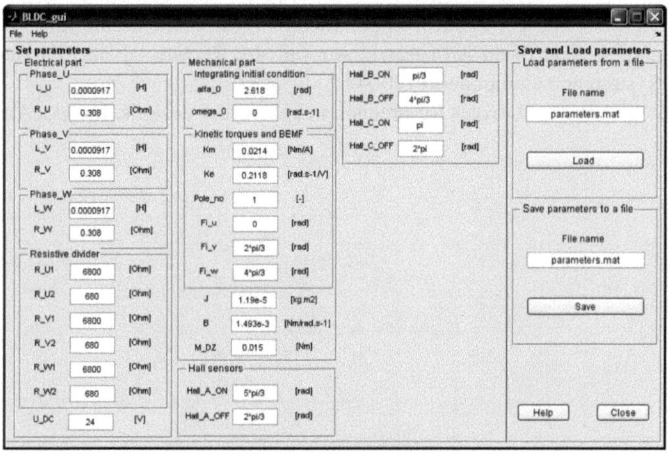

Fig. 4. GUI for parameters setting

It is possible to choose from the preselect courses or from the rollout menu any relation between two physical values. Button Multi-plot for the overall picture is also implemented in this window. In the Fig 4 is depicted GUI interface for the motor parameters setting. Whole window is separated into several parts - electrical, mechanical and sensor. The right side of the GUI is dedicated to load the parameters from file or to save.

5 Conclusion

The project involves the development of control algorithms and electronics to improve run up performance and reliability of the BLDC motor in actuating devices for safety critical applications.

Because of this high demand, simulation and modeling tool are widely used to accelerate the development of control algorithms. The simulation results are tested on real hardware in dSPACE environment. This type of development is called Model based design and is the integral part of every development department.

Main advantage of this mathematical model is its complex, integrity and possibility to use it in any higher system without need to understand the principle of

BLDC motor. From the highest point of view, whole motor model is only the other Simulink component in complex Simulink model.

Acknowledgments. This work was also supported by project No. MSM 0021630518 "Simulation modeling of mechatronic systems" solved on the Faculty of Mechanical Engineering, Institute of Production Machines, Systems and Robotics, Brno Technical University.

References

[1] Leonhart, W.: Control of Electrical Drives, 3rd edn. Springer, Heidelberg (2001)
[2] Greissner, C., Carl, U.: Control of an Electro-Hydrostatic Actuation System for the Nose Landing Gear of an All Electric Aircraft. In: Recent Advance in Aerospace Actuation System and Components (2004)
[3] Compter, J.: Electrical drives for precision engineering designs. In: specAmotor (2007)

Automatic Control, Design and Results of Distance Power Electric Laboratories

D. Maga, J. Sitár, and P. Bauer

AD University of Trencin, Faculty of Mechatronics,
Studentska 2, 911 50 Trencin, Slovakia
maga@yhman.tnuni.sk
Delft University of Technology, Mekelweg 4, 2628 CD Delft, The Netherlands
P.Bauer@its.tudelft.nl

Abstract. The paper gives a detailed view on three laboratories of electromechanical actuators operated within the frames of 18 modules based on *Edipe* project. The project has been realized with cooperation of 13 partners located in 11 different European countries as a Leonardo da Vinci II pilot project. The project is available at www.PEMCWebLab.com. PEMCWebLab offers a set of remotely controlled real and virtual experiments from fields of electrical engineering mainly from power electronics, electrical drives and motion control. The brief overview on project can also be found in *The Parliament Magazine*.

1 Introduction

According to the Leonardo da Vinci pilot project, a set of distance laboratories based on power electronics and motion control has been realized. This has been done within the frames of cooperation between 11 universities from 9 different European countries. The project (known as EDIPE -Electrical DIstance Practical Education), has been finished in december 2008 with independent testing of cooperating industrial partners.

The basic ideas of the project are to allow the users to connect to the partners laboratories through the internet and realize a set of measurement wherever in Europe they are. The authors put together a wide set of power electronics oriented laboratories connected by developed access technology (booking system). In addition, a universal LabView oriented approach has been applied to obtain a unique and unified connection of measurement apparatus to web and to end users. The measurement process itself has several stages: data collection, data analyze, data presentation and data storage. These functions are usually implemented in data acquisition systems – DAQ measurement cards.. Designed virtual laboratory system are based on *LabView*, *ControlWeb* and *DasyLab* technologies.

2 Overview on the Modules

The laboratories are covering four different fields, depending on authors interests, completed with different number of concrete laboratory modules:

- Fundamentals of electrical engineering,
- Power electronics,
- Electrical machines,
- Electromechanical and motion control systems.

Two modules built within the field of *Fundamentals of Electrical Engineering* contain remote experiments on rectifier topologies, resonant circuits and semiconductors. Six modules in *Power Electronics* are based on measurements on power converters, their characteristics, operation and behavior, and control techniques. Four modules in *Electrical Machinery* are dedicated to measurements of basic characteristics of synchronous machine, DC machine or asynchronous machine. Six modules in *Electromechanical and Motion Control Systems* are based on mobile robots behavior and control, experiments on high dynamic operation, design of control procedures.

The user is able to book a required set of measurements. After passing through the labs, the user will not only have the measured results, but also will understand the theory of measured machinery and became familiar to basic physical principles needed to explain the measured technique and technology.

3 Laboratories of Mechatronical Actuators

Three of above mentioned modules (laboratories) have been realized by Faculty of mechatronics, AD University in Trenčín, Slovakia within the frames of Leonardo da Vinci project. One of the objectives of this Leonardo da Vinci project is to facilitate the use of the distance laboratories in the field of electrical machinery and power electronic. Simulations and remote-controlled experiments, together with accompanying of the educational material, support self learning. In this way the users can be either students, or unemployed engineers and/or all those who want to refresh or update their knowledge can do so on their own and freely select the time optimal for them. The experiment can be realized on any web-browser platform. For correct functionality it is necessary to accept and install the plug-in for distance measurements from *LabView*. *Cookies* necessary for user number limitations must be accepted.

3.1 Synchronous Generator Module

The users of this module are able to verify the principles of 3 phase AC power source generators. The experiment will demonstrate the influence of excitation and rotor speed on quality and quantity of obtained voltage (current). The user will also understand the possibilities and (dis-)advantages of synchronous machinery and became familiar to the principles of synchronous electrical machinery. Two tasks are prepared for the user – the measurement of load characteristic $U=f(I)$ of synchronous generator loaded with active resistance and constant excitation current I_b=*const*. and the measurement of the regulation (working) characteristics $I_b = f(I_Z)$ of synchronous generator loaded with active resistance and constant supply voltage U=*const*, where I_Z is the required output current.

Passive load is connected to the synchronous generator. The properties of the generator are strongly dependent on the type of load and can be observed when the machine is loaded by ideal resistance, capacitance or inductance (*R, L, C*).

The synchronous generator speed must be set to nominal by controlled servomotor using the controls from Fig. 1. The generator is loaded by three-phases six-pulse bridge rectifier with resistance load. The output voltage on generator U is measured, together with output current I_Z. The excitation winding is supplied by DC current source with current I_b. As the speed of the generator has been set, the output voltage is within the range from 0 to 400 V (typical value is equal to nominal output voltage). This can be done by control of rectified AC source. The first value of output voltage U has to be measured with $I_z \cong 0$ (R_z set to maximum). The value of output current I_z is change by change of R_z value.

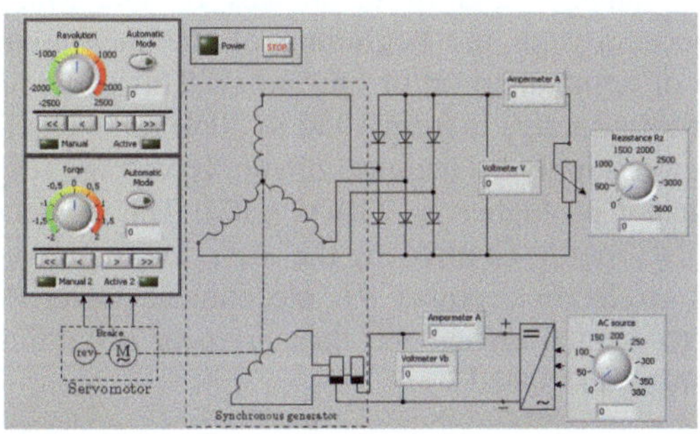

Fig. 1. Synchronous generator – measurement scheme and web-layout

The load characteristic will be obtained by measuring the output parameters, while the excitation current is at constant value. Changing the resistivity of the load also the output current is changed. The obtained results could be written into the table. These can be exported to any conventional table processor with typical and well-known services of the computer operating system.

The regulation (working) characteristics of synchronous generator is the dependence of excitation current I_b on output current I_z, while the output voltage of the generator is constant. The measurement can be done by change of output load and change of excitation voltage. The required parameters could again be written into the table and postprocessed afterwards.

3.2 Asynchronous Motor Module

The users of this module are able to verify the basic principles of induction (asynchronous) machinery, demonstrate the influence of applied mechanical load on motors speed, as well as the response of motor to direct start procedure. They will understand the possibilities, advantages and disadvantages of asynchronous machinery as well as the stable and unstable balance of torques. The task is to measure the torque characteristics of the asynchronous motor $M=f(n)$.

After reaching the required value of voltage supply, the first point (zero load condition) can be added to results table. The next step is in loading of the asynchronous motor by connected servomotor. After each change of load the measured data can be recorded into the table. While it is not possible to measure the complete torque characteristics when the motor is supplied by rated voltage, it is necessary to reduce the value of supply voltage. To complete the measurement, the user must become familiar to dependence of torque on supply voltage and different operation modes of connected load (asynchronous servo-drive) – torque control and/or speed control.

3.3 DC Motor Module

The user of this module will be able to verify the basic principles of DC machinery, describe the principles of DC motor with shunt excitation, demonstrate the influence of applied mechanical load on values of current supply and understand the possibilities, advantages and disadvantages of DC Machinery with shunt excitation. It is necessary, at the beginning of the measurement, to reach the approximate value of nominal speed of DC motor. This can be done by control of servo drive. The required range is from 1950 to 2050 rpm. The DC motor can now be connected to power supply by increase of supply voltage (Fig. 2). The value of supply voltage can not be higher as 220 V (DC). The asynchronous servo now works as a load of DC motor. Controlling the servo the required load torque can be set up. The values of electrical power *P1*, mechanical power *P2* and efficiency η are automatically calculated during the measurement according to the values of voltage, current, torque and speed.

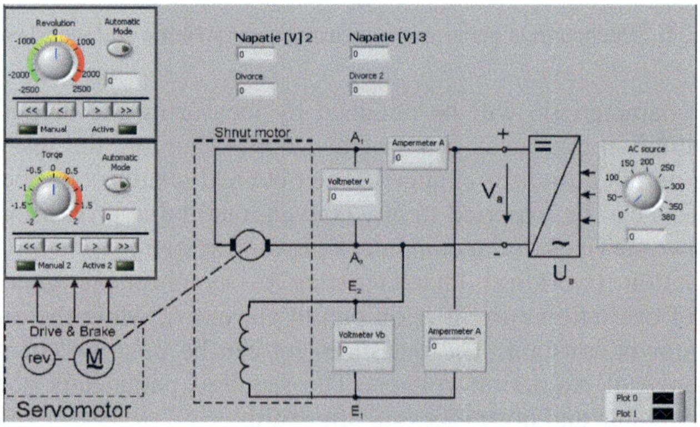

Fig. 2. DC motor – measurement scheme and web-layout

4 Pilot Testing

The pilot testing has been realized by project subcontractors (industrial partners). Each module has been tested by two different users (usually from different country). The testers after passing through the module evaluated the quality of offered services [8] with 5 scale expression (5-excelent; 1-worts) applied to 12 categories. Their opinion can be seen in Tab. 1.

Table 1. Pilot testing results

Category	Average expression
Clarity of Manual	4,31
Completeness of the manual	4,34
References or recommended literature	4,04
Clarity of the booking process	4,41
Lay-out of the booking process	4,31
User friendliness of the booking process	4,21
Layout, user friendliness, compatibility of the laboratory	4,33
Fulfilling educational goals	4,57
Measurement assignment clear	4,33
Clarity of the instructions during the measurement	4,17
Evaluation of the measurement	4,29
Robustness, operating links	4,17

5 Conclusions

The power electric laboratories of *EDIPE* project partners have been presented in the paper. A brief overview on 18 modules has been given, followed by detailed description of 3 modules realized at laboratories of Faculty of Mechatronics in Trenčín. As the project partners are from 9 different European countries, also the pilot testers of the project are Europe-wide. Their opinion, also presented in the paper, is an important part of the project realization and evaluation.

Acknowledgement. This work has been performed within the project "Elearning Distance Interactive Practical Education (EDIPE)". The project was supported by the European Community within framework of Leonardo da Vinci II programme (project No CZ/06/B/F/PP-168022). The opinions expressed by the authors do not necessarily reflect the position of the European Community, nor does it involve any responsibility on its part.

References

[1] EDIPE [online]. [ref. 2009-05-20], `http://www.PEMCWebLAb.com`
[2] Bauer, P., Dudak, J., Maga, D., Hajek, V.: Distance Practical Education in Power Electronics. International Journal of Engineering Education 23(6), 1210–1218 (2007)
[3] Uran, S., Hercog, D., Jezernik, K.: Remote Lab Experiment RC Oscillator for Learning of Control. International Journal of Online Engineering 2(4) (2006)
[4] Bauer, P., Dudak, J., Maga, D.: Distance practical education with DelftWebLab. In: EPE-PEMC 2006, pp. 1528–1535. University of Maribor, Maribor (2006)

[5] Sobczuk, D.: E-Learning Presentation Of Pulse Width Modulation Methods for Two And Three-Level Converters. In: International conference EDPE, Stara Lesna, Slovakia (September 2007)
[6] Sziebig, G., Korondi, P., Suto, Z., Stumpf, P., Jardan, R.K., Nagy, I.: Integrated E-learning Projects in the European Union. In: 34th Annual Conference of the IEEE Industrial Electronics Society (IECON 2008), Orlando, Florida, U.S.A. (November 10-13, 2008) ISBN: 978-1-4244-1766-7
[7] Sobczuk, D., Kamiski, B.: Distant Learning of Pulse Width Modulation Techniques for Voltage Source Converters. In: Proceedings of 13th EPE-PEMC. Ptetis Publishers, Poznań (2008) IEEE Catalog Number: CFP0834A-CDR
[8] Chovanec, A., Balog, J., Kianicova, M.: Technology and Management of Maintenance' (in Slovak). In: Trenčín: TnUAD, 2002, p. 182 (2002) ISBN 80-88914-65-5
[9] Huzlik, R., Kuchyňková, H., Kučera, P.: Virtual Laboratory of Power Electrical in BUT. In: International conference on Low Voltage Electrical Machines, pp. 1–148. VUT Brno, VUT (2008)

Identification of Parametric Models for Commissioning Servo Drives

S. Hofmann, A. Hellmich, and H. Schlegel

Chemnitz University of Technology, Faculty of Mechanical Engineering,
Institute for Machine Tools and Production Processes, Mechatronics,
Reichenhainer Str. 70, Chemnitz, Germany
{stefan.hofmann,arvid.hellmich}@mb.tu-chemnitz.de
holger.schlegel@mb.tu-chemnitz.de

Abstract. Modern production machines contain up to 100 position controlled servo drives. The setting-up operation of these servo drives is a time-consuming and cost-intensive process. The paper presents a new method, which supports the user by justifying the position controller automatically. Based on the auto relay feedback experiment by Åström and Hägglund, the introduced technique is applied to identify a parametric model of the position controlled system by means of gradual pole compensation. Therefor, a mechanism has been developed, which automatically adjusts the model parameters according to the time behaviour of the controlled system. The presented algorithm has successfully been implemented in the motion control system SIMOTION. In addition, it has been evaluated concerning the achievable identification accuracy. Furthermore, examples for the identification procedure and the parametrization of the position controller are given.

1 Introduction

In respect of the commissioning and optimization of drives today's operators are supported by numerous applications [5] or tuning algorithms [4]. Most of them are based on iterative algorithms or are only utilized efficiently by skilled users. Additionally, not all existing industrial applications can be covered with the known tools. As a result, an unsatisfactory tuning as well as a time consuming manual optimization is required. If this is estimated to a complete machine with a large number of position controlled servo drives, the generated costs and necessary time rise enormously.

The relay feedback experiment [7] is a well known identification approach, which and is generally applied to identify a process $G_s(s)$ on its limit of stability. For this purpose, the controller $G_r(s)$ is replaced by a relay controller (Figure 2).

The method provides a nonparametric model with the characteristics ultimate gain (k_u) and ultimate frequency (ω_u). Based on these two values, several tuning rules for PID controllers have been developed.

With regard to the presented application a parametric model of the plant needs to be identified to provide a basis for the controller design. Furthermore, the plant parameters (time constant and dead time) are located within the range of the controller cycle time. As a result, the accuracy of the relay feedback experiment is affected negatively.

2 Formulation of Problem

In order to obtain a convenient design of a proportional controller (parameter: k_v) with velocity precontrol (Figure 1) (PV position controller) in the motion control system SIMOTION, the controller has to be adjusted very precisely to the subordinate transfer response. The pre-control is mainly used to reduce the following error and to allow a separate design for command response and disturbance response.

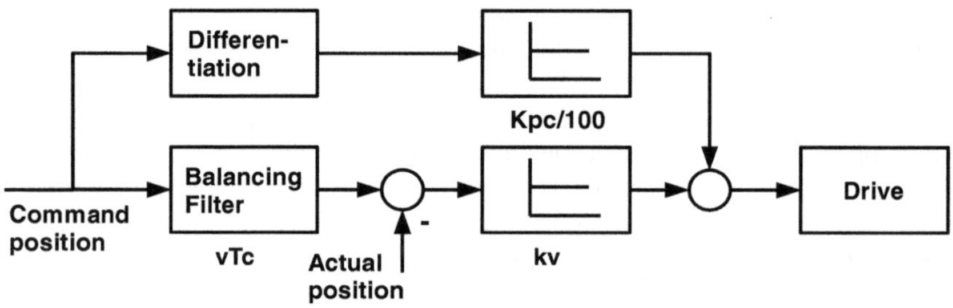

Fig. 1. Structure of the PV position controller

As Figure 1 shows, a balancing filter (Finite-duration Impulse Response; parameter: **v**elocity **T**ime **c**onstant) is applied in order to separate the two branches. Basically, to represent the controlled system of the servo drive, one time constant is sufficient.

3 Solution Statement

In the presented approach, the relay feedback experiment is exclusively used as an excitation for the plant (Figure 2). The controlled variable x(t) appears as input for a compensator $G_c(s)$. The closed loop remains unaffected. The aim is to adjust the compensator to the plant parameter by using the method of gradually compensation of the dominant pole [8]. This approach combines the advantages of the relay feedback experiment (using the critical point on the process frequency response) with those of the parametric model identification.

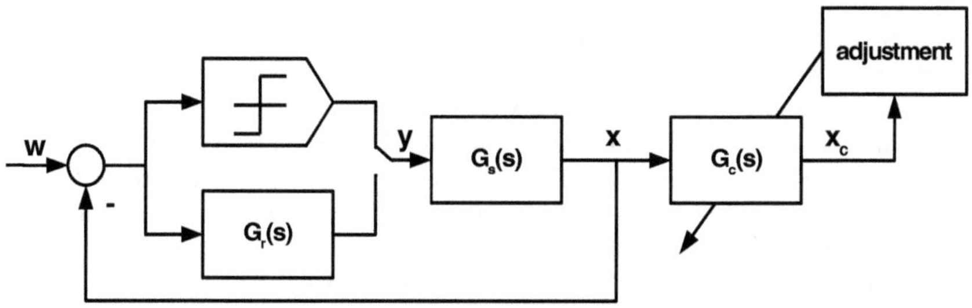

Fig. 2. Scheme of Identification

For a first order transfer function with dead time (FOPDT)

$$G_s(s) = \frac{k}{T \cdot s + 1} \cdot e^{-T_d \cdot s} \tag{1}$$

a compensator as shown in [1], [2]

$$G_c(s) = \frac{T^* \cdot s + 1}{s} \tag{2}$$

is chosen. Applying this, the forward path of the control loop becomes:

$$G_{yx_c}(s) = \frac{X_c(s)}{Y(s)} = G_s(s) \cdot G_c(s) = k \cdot e^{-T_d \cdot s} \cdot \left[\frac{T^* \cdot s + 1}{s \cdot (T \cdot s + 1)} \right]. \tag{3}$$

Finally, this equation is expanded in terms of two components:

$$X_c(s) = k \cdot e^{-T_d \cdot s} \cdot \left[\frac{1}{s} \right] \cdot Y(s) + k \cdot e^{-T_d \cdot s} \cdot \left[\frac{T^* - T}{(T \cdot s + 1)} \right] \cdot Y(s). \tag{4}$$

In the case of T*=T the term reduces to an integral plus dead time (IPDT) system:

$$X_c(s) = k \cdot e^{-T_d \cdot s} \cdot \left[\frac{1}{s} \right] \cdot Y(s) \tag{5}$$

As mentioned above, Y(s) is a relay output signal. With this input sequence, equation (5) describes the step response of an integrator, an ideal ramp function x(t). To adjust T*, expression (4) needs to be evaluated to satisfy the equation as follows:

$$\lim_{T^* \to T}\left[k \cdot e^{-T_d \cdot s} \cdot \left[\frac{T^* - T}{(T \cdot s + 1)}\right] \cdot \frac{1}{s}\right] = 0 \tag{6}$$

The identification process can be reduced to one time constant, because all remaining process parameters (static gain of the closed velocity loop k=1, dead time caused by communication T_d) are known.

Determining the compensator time constant T*, the selected adjustment criterion is of primary importance to satisfy equation (6). It is necessary to differentiate between three cases (Figure 4) by using diverse criteria. Intuitive approaches, like determining the increase of the time behaviour of $x_c(t)$ or interpreting an error function are not suitable for electric drives, because of the expected small time constants of the controlled system (0-5 ms) and the dead time caused by communication (approx. 2 position controller cycles).

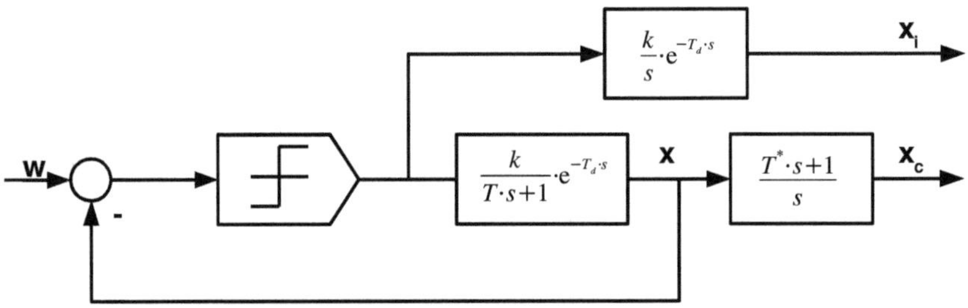

Fig. 3. Proposed criterion for compensator adjustment

In the presented method, the oscillation magnitude of $x_c(t)$ (Figure 5) is compared with the expected magnitude of the ramp function $x_i(t)$ from equation (5) and T* is adjusted according to the magnitude ratio (Equation 7).

$$T^*_{n+1} = T^*_n \cdot \frac{\hat{x}_i}{\hat{x}_c} \tag{7}$$

It needs to be pointed out, that only one single measured value of the oscillation magnitude is compared to one of the expected magnitude. For the evaluation, only the extreme values of $x_c(t)$ have to be determined, which entails a high toleration of measurement uncertainty. Due to these reasons, this criterion was proven to be very fast and highly efficient, because theoretically only one iteration step leads to an adequate convergence.

4 Illustrative Examples

The application of the presented approach for an electric servo drive (SIMODRIVE 611U) is illustrated in Figure 5:

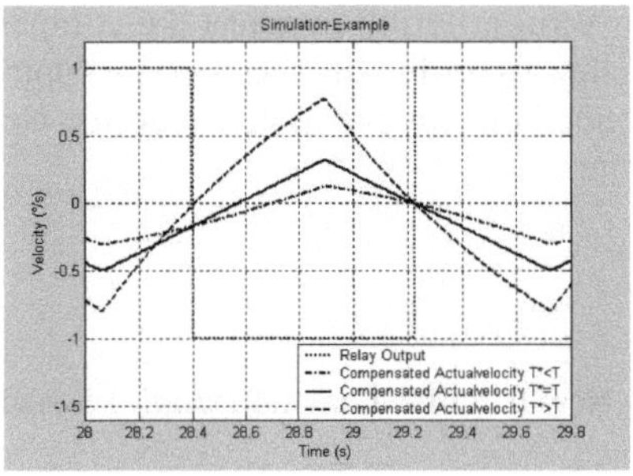

Fig. 4. Waveforms for different T*/T ratio

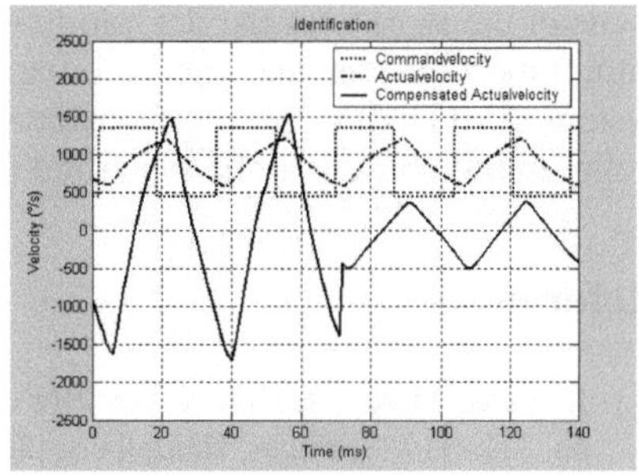

Fig. 5. Adjustment of T*

Based on the results of the identification algorithm, the parameters of a proportional controller with velocity pre-control in SIMOTION can be calculated. This can be realized by using a pole placement method for first order systems with dead time [6] as a tuning rule.

$$k_v \leq \frac{1}{2 \cdot (T^* + T_d)} \tag{8}$$

$$vTc = T^* + T_d \tag{9}$$

This procedure provides the following results with the chosen drive. The quality of the controller is usually rated based on predetermined position ramp functions as reference input [4], [5]. The graph on the right in Figure 6 shows the optimized system. The graph on the left is an example for a balancing filter, which is unsuitably adjusted to the plant.

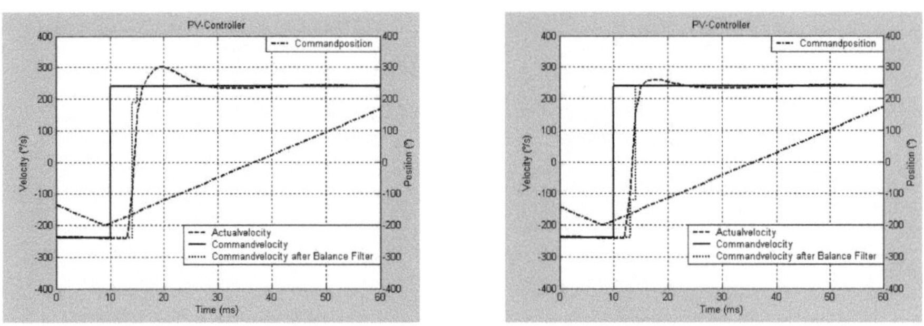

Fig. 6. Positon control based on identification result

5 Conclusion

The paper presents an automatic identification and controller tuning scheme. The result of the identification is a time constant of the closed velocity loop. The main focus of the work was the derivation of a powerful criterion to adjust the compensator time constant and the automation of the algorithm in the motion control system SIMOTION. The performance of the method was proven through the controller tuning, based on the identified parameters.

References

[1] Johnson, M.A., Moradi, M.H.: PID Control New Identification and Design Methods, pp. 311–319. Springer, Heidelberg (2005)
[2] Thyagarajan, T., Yu, C.-C., Huang, H.-P.: Assessment of controller performance: a relay feedback approach. Chemical Engineering Science 58, 497–512 (2003)
[3] Lutz, H., Wendt, W.: Taschenbuch der Regelungstechnik. Verlag Harri Deutsch (2007)
[4] Siemens, A.G.: Optimierung von Antrieben V2.0 (2004)
[5] Siemens, A.G.: SIMOTION Drive tuning with SINAMICS, SIMODRIVE or MASTER-DRIVES, applications & tools (2006)
[6] Groß, H., Hamann, J., Wiegärtner, G.: Elektrische Vorschubantriebe in der Automatisierungstechnik. Publicis Corporate Publishing (2006)
[7] Åström, K.J., Hägglund, T.: Advanced PID Control; ISA (2006)
[8] Reinisch, K.: Analyse und Synthese kontinuierlicher Steuerungs- und Regelungssysteme. Verlag Technik (1996)

Electrical Drives for Special Types of Pumps: A Review

J. Lapčík and R. Huzlík

Brno University of Technology, Faculty of Electrical and Comunication Technology,
Department of Power Electronic and Electrical Engineering, Technická 8, 616 00 Brno
{lapcik,huzlik}@feec.vutbr.cz

Abstract. Active magnetic bearing and scarceness passive magnetic bearing are being developed for many applications due to their attractive features. Axial flux, permanent magnet, slot-less machine is preferably to radial one in mechatronic systems using magnetic active bearings. An overview of these machines used in pumps is presented in this paper. Pump's structure, advantages and features of the magnetic bearing and disc type motors are clarified. Several interesting pump's solutions are also covered from a variety of perspective.

1 Introduction

Permanent magnet synchronous motors, where the magnets have a linear demagnetization characteristic give a new degree of freedom to motor designers and has indicated the general potential of these materials. In particular, for miniature precision motors, the magnet cost is reduced relative to manufacturing cost. The current high cost/volume of magnet materials becomes tolerable in terms of the benefits to be obtained. The feature of rare earth magnetic materials, which is attractive in many applications, is that it is impossible to affect the magnet characteristic by any practical current in the motor windings. If the basic equation for the force or torque developed by a motor is considered, it is only the air-gap flux density rather than the coercive force which has direct effect on performance. Even in larger sizes, the slot-less design has constructional advantages both in winding insertion and magnetic circuit tolerances.

The ability of rare-earth magnets to generate significant flux-density levels in open magnetic circuits prompted the investigation of the torque available with a completely ironless stator. The stator iron gives no inherent benefit as regards the absolute production of torque. The construction of motors without stator iron is not to be generally recommended, however. The stator iron can be thought of as fulfilling a dual purpose. First, the iron completes the magnetic circuit and secondly it provides a magnetic screen. Without the latter, operation near to magnetic bodies is of feasible and, furthermore, significant drag torques occur on the rotor owing to eddy currents induced in conducting bodies situated at considerable distances from the rotor.

2 Realization

Motors used in our applications are Axial Flux Permanent Magnet Disc Machines. Permanent Magnet (PM) machines are increasingly becoming dominant machines with the cost competitiveness of high energy permanent magnets. As for the axial flux PM machines they were used in our application because they have a number of distinct advantages over radial flux machines when used in mechatronic system using magnetic active bearings. They have planar and easily adjustable air-gaps. The noise and vibrations levels are less than in the conventional machines. Axial flux surface magnet PM machines including slot-less construction was found to be advantageous for our design. The review and theory as well as General purpose sizing equations for these types of disc motors are mentioned in [1]. Axial flux machines can be constructed in many ways. Two of our designs are presented in the Fig.1 and 2.

Within the frame of cooperation between Victor Kaplan Department of Fluid Engineering and our Department, we tried to find a solution for design and realization of the motors, magnetic bearings and drives for special pumps.

We started with the drive for Total Artificial Heart (TAH), part of which was the slot-less axial flux synchronous motor. The latest type was double turbine design. The new double turbine blood pump solution consist of a fix part of pump stator with two inlets port of the pump and two outlets for the left and the right ventricle. In the middle part of the stator is slot/less disc winding of the permanent magnet axial flux synchronous motor, which is without iron core. The rotor disc of the motor are integrated into the turbine rotor and this system is equipped with two active permanent magnet discs, which magnetic flux cross through the stator winding. The driving torque on the impeller is applied trough the electromagnetic interaction between currents in the winding fixed to the housing and embedded magnets in the impellers of the turbine. The layout of this prototype is shown in Fig. 1(A) and (B).

For motor control was need small, efficient, reliable power electronics with technology for a new and truly universal brushless DC motor concept, the technology of which must promise the universality essential for commercial development at a cost and with a reliability previously unattainable. The system must solve two key problems. First, the transducer used in previous concept to sense rotor position had to be eliminated, and second, a single electronics package had to be designed that was universal enough to provide constant speed or programmed speed changes regardless of changes in load or voltage. In this concept a single electronics package with an appropriate power module can control three-phase machine of our design. Digital electronics senses motor back EMF to provide commutation as well as control motor speed without using the transducer. The frequency converter used starts and accelerates motor along a straight line, DC, speed-torque curve like any DC commutated permanent magnet motor. The machine can be theoretically of any size, can have any number of poles, and can have three or two phases winding. The armature and field need only to produce a generated voltage when operated as a generator. The power electronics varies in size with the size of the motor. Frequency converter TMM 40e and control circuits of the magnetic bearings are supplied from the stabilized source of the DC voltage.

Input voltage of the frequency converter is 15V, input voltage in the case of system with magnetic bearings was 24V. Nominal current of the motor is 3A and current of magnetic bearing 1A.

Frequency converter employs advances solid state sensing and logic technology to control the switching of DC power to an AC synchronous machine. The result is a motor with unprecedented flexibility. With only minor control system changes, a single unit can be operated as a DC or a speed programmed electromechanical control unit.

A) Electric motor for TAH pump

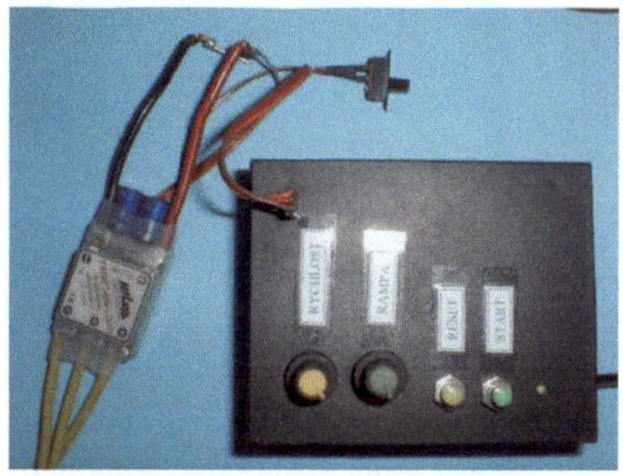

B) Drive for Total Artificial Heart

Fig. 1.

Within the development were realized several different variants and in one of these was used active magnetic bearing. Experiences derived from the TAH development and realization were used in the industrial pump design, mainly in the mechatronic system using magnetic active radial and axial bearing, drive using slot-less, disc motor and electronic for motor and bearing control. In comparison with traditional (sliding or roller) bearings the magnetic bearings have remarkable qualities: they do not require any maintenance, they have a long operating life, there are no looses through mechanic friction and they do not heat, they are noiseless and there is no danger of contamination with lubrication oil near the bearing. Magnetic bearings can be used in extreme conditions with a wide range of operating temperature and in chemically aggressive environment. Active magnetic bearings require complicated control hardware, such as digital signal processor amplifiers, digital-to-analog converters and software. Passive magnetic bearings are simple, they do not have power consumption, they take up only little space and their price is low. Their disadvantage is a lower loading capacity and dynamic stiffness. The permanent magnet by itself is not able to keep the ferromagnetic body at free and stable levitation, as was shown by Earnshow as early as 1842. To make the passive magnetic bearing stable, an additional mechanic bearing or a pair of magnetic bearing must be used. All of these mentioned possibilities were used in our projects, but for testing the systems were simplified and some of the magnetic bearings were replaced by mechanical ones in both medical and industrial mechatronic systems. These mechatronic systems are unique one for which the patent were taken out.

Fig. 2. Functional sample of the Industrial Pump Drive and Magnetic bearings.

3 Conclusion

Development in design of electrical machines using rare earth permanent magnets and production of these machines requires quite new access to the machine design.

Magnetic field solution, sometimes even in the static and plane type is the first step to the scientific solution of the electric machines design. Design was verified by field calculation and check in laboratory measurements and results of the field in the air-gap were used as a base for the torque, reactance and parasite harmonic fields of the machines [5]. Field calculation was made also for the currents (magneto motive force) produced by rotor reaction.

Next no less important contribution is possibility of the reactance calculation both in direct and quadrate axis, which can be used direct in the power equation of the synchronous machine. The problem is that this reactance depends on the current value especially in the permanent's magnets machines, where the direct axis reactance is usually smaller then quadrate axis reactance. Solution in these problems can be solved by the new methods of the design in the electrical machines using field representation.

Laboratory tests which are in progress in Department of Power Electrical and Electronics confirm all the presumptions and goals of our research.

Acknowledgments. This paper was supported by Ministry of Education, Youth and Sports of the Czech Republic research grant MSM 0021630518 "Simulation modeling of mechatronic systems" and GACR 102/09/1875 "Analysis and Modelling of Low Voltage Electric Machines Parameters".

References

[1] Aydin, M., Huang, S., Lipo, T.A.: Design and electromagnetic field analysis of non-slotted and slotted TORUS type axial flux surface mounted machines. In: IEEE International Conference on Electrical machines and Drives, Boston, pp. 645–651 (2001)

[2] Lapčík, J., Láníček, T.: Drives For Artificial Heart Pumps Systems. In: Epve 2005. Brno: Vysoké Učení Technické V Brně, Fakulta Elektrotechniky A Komunikačních Technologií, pp. S141–S145 (2005) ISBN: 80-214-3052-4

[3] Josef, L., Tomáš, L.: Design Of The Permanent Magnet Slotless Disc Motor For Tah Application. In: Low Voltage Electrical Machines Joint Czech-Polish-Slovak Co-Operation: Under Project GAČR No 102/06/1320, pp. S182–S187. Brno University of Technology, Faculty Of Electrical Engineering And Communication, Brno (2006)

[4] Huzlík, R., Pazdera, I., Lapčík, J., Fialová, S.: Sealless Industrial Pump With Magnetic Bearing And Axial Flux Motor. In: International Conference On Low Voltage Electrical Machines. VUT (2008) ISBN: 978-80-214-3795-1

[5] Lapčík, J., Huzlík, R.: Sealless Industrial Pump With Permanent Magnet Slotless Disc Motor And Magnetic Bearings. In: Prace Naukowe Instytutu Maszin, Napedów I Pomiarów Elektrycznych Politechniki Wroclawskiej, Roč. 1, Č. 62, pp. S209–S860 (2008) ISSN: 1733-0718

[6] Pazdera, I., Huzlík, R., Lapčík, J.: Magnetic Bearing For Industrial Pump Application. In: XIV Conference Computer Applications In Electricall Engineering, Poznan, pp. S47–S48 (2009) ISBN: 978-83-89333-19-3

References

Cable Length and Increased Bus Voltage Influence on Motor Insulation System

M. Nesvadba[1], J. Duroň[1], and V. Singule[2]

[1] Danaher Motion, s.r.o., Evropská 864, Brno 664 42, Czech Republic
martin.nesvadba@danahermotion.com
[2] Brno University of technology, Faculty of Mechanical Engineering, Technická 2 Brno, Czech Republic
singule@fme.vutbr.cz

Abstract. The target of this paper is to document influence of increased bus voltage of power converter on motor insulation system. This is needed for evaluation of motors for 800V DC bus based servo-amplifiers. Increase of DC bus voltage is logical consequence of increase power need for servo-drive systems.

1 Introduction

Life-time of the insulation materials evaluation is usually related to Thermal endurance. This Thermal endurance of film insulated magnet wire, stator varnishes, encapsulates and other insulating materials as components and as a system follow the combined standards of IEC60085, ASTM D2307 and UL1446.

These standards for the most part link the known Insulation Classes with 20,000 hrs as a base life from which to extrapolate from (for example per IEC Standard 60085, "Thermal Evaluation and Classification of Electrical Insulation.", the temperature rating for a class of material is defined as the temperature at which the average life is 20,000 hours). Therefore use of the extrapolations should be considered for relative thermal comparisons, not exact life performance, since many other factors impact exact winding life.

2 Technical Information

Especially stresses caused by voltage spikes (caused by line reactance also known as Traveling Wave, high frequency, high dv/dt spikes) can dramatically degrease the Insulation life-time. These voltage peaks may exceed 1700 volts for a 480-volt system and these voltage spikes are changing with the supply voltage (together with drive BUS voltage) increasing – see Fig.1.and 2. These voltage spikes may also cause an insulation break down.

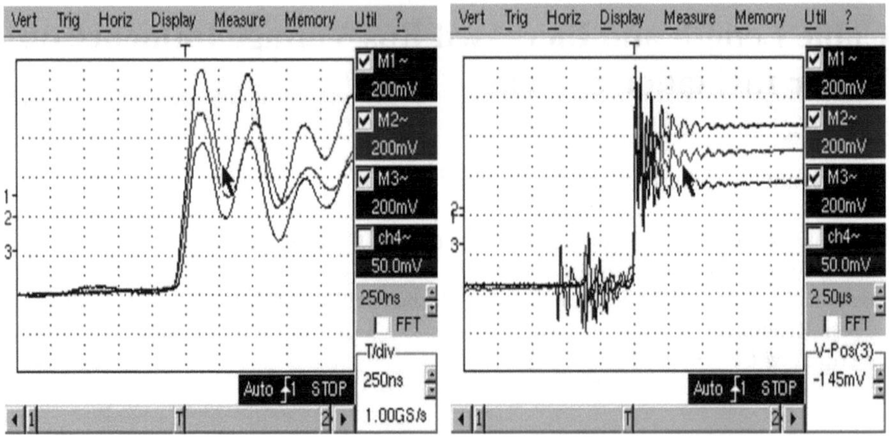

Fig. 1 and **2.** Wave shapes at different drive power supply voltages (400, 500, 580V AC)

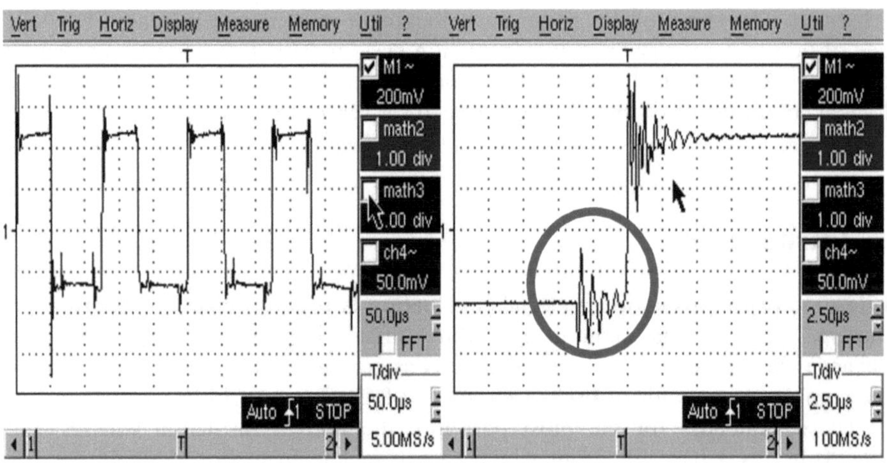

Fig. 3 and **4.** Wave shapes (825V DC BUS) at 50μs and 2,5 μs scope time base

Fig. 1. and 2. show wave shapes at highest BUS Voltage 825V DC at different scope time base. Usually there is the wave undulation also on low state of the signal before rise (red circle on Fig. 4.).

To check the impact of these voltage spikes to the insulation life time is a test proposal - by using a drive connected normally to a motor (no common mode choke). The connection between the motor and the drive is by cable of variable length to show its influence on peak voltages and dv/dt ratio.

The drive setting is standard setting for used motor with standard 8 kHz of the PWM frequency. It is important to keep the motor temperature at the same level and make periodical measurement all key factors.

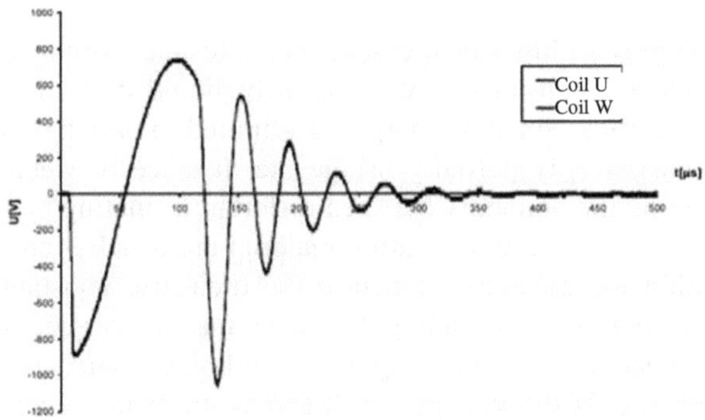

Fig. 5. Surge test of on-fault winding

During the testing key performance indicators have to be monitored. Checking of insulation resistance is one of the basic methods of evaluation of insulating system. This check need to power off the measured motor, disconnection of cable at motor side.

Surge test is used for detection of short connection at winding. The shape of forward and backward surge is measured, compared and evaluated. This method is very reliable for detection of shortage within the one phase winding; even the insulation resistance to the ground is within the spec.

Partial discharge (PD) is a localised dielectric breakdown of a small portion of a solid or liquid electrical insulation system under high voltage stress. While a corona discharge is usually revealed by a relatively steady glow or brush discharge in air, partial discharges within an insulation system may or may not exhibit visible discharges, and discharge events tend to be more sporadic in nature than corona discharges.

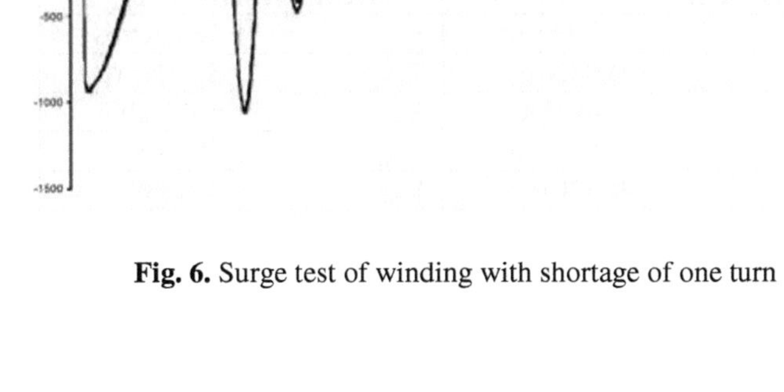

Fig. 6. Surge test of winding with shortage of one turn

PD usually begins within voids, cracks, or inclusions within a solid dielectric, at conductor-dielectric interfaces within solid or liquid dielectrics, or in bubbles within liquid dielectrics. Since discharges are limited to only a portion of the insulation, the discharges only partially bridge the distance between electrodes. PD can also occur along the boundary between different insulating materials.

Partial discharges within an insulating material are usually initiated within gas-filled voids within the dielectric. Because the dielectric constant of the void is considerably less than the surrounding dielectric, the electric field (and the voltage stress) appearing across the void is significantly higher than across an equivalent distance of dielectric. If the voltage stress across the void is increased above the corona inception voltage (CIV) for the gas within the void, then PD activity will start within the void.

Once begun, PD causes progressive deterioration of insulating materials, ultimately leading to electrical breakdown. PD can be prevented through careful design and material selection. In critical high voltage equipment, the integrity of the insulation is confirmed using PD detection equipment during the manufacturing stage as well as periodically through the equipment's useful life. PD prevention and detection are essential to ensure reliable, long-term operation of high voltage equipment used by electric power utilities.

The equivalent circuit of a dielectric incorporating a cavity can be modeled as a capacitive voltage divider in parallel with another capacitor. The upper capacitor of the divider represents the parallel combination of the capacitances in series with the void and the lower capacitor represents the capacitance of the void. The parallel capacitor represents the remaining unvoided capacitance of the sample.

3 Measurement

All of published measurements have been done with scope Tektronix TDS 3012 and high-voltage scope probe Tektronix P6015A 75MHz (line to ground) and high voltage differential probe Tektronix P5210 (line to line). Two motors of different size and rating have been measured under comparable conditions.

Table 1. AKM2 and AKM7 motors parameters

AKM 22C-AN5NR-00		AKM 74P-AN5NR-00	
I_{cs}	1.39Arms	I_{cs}	18.6Arms
T_{cs}	0.84Nm	T_{cs}	52.5Nm
V_s	640VDC	V_s	640VDC
N_{rtd}	8000 rpm	N_{rtd}	2000rpm
P_{rtd}	0.57kW	P_{rtd}	7.52kW
R_{LL}	19.98Ω	R_{LL}	0.48Ω

Table 2. AKM2 (on the left side) and AKM7 measurements

Cable length	dv (peak) [V]	dv 10-90% [V]	dt [ns]	dv/dt [kV/µs]
100	1503	1202	360	3,3
90	1600	1280	340	3,8
80	1630	1304	310	4,2
70	1650	1320	330	4,0
60	1690	1352	280	4,8
50	1650	1320	250	5,3
40	1640	1312	240	5,5
35	1680	1344	250	5,4
30	1640	1312	133	9,9
25	1600	1280	120	10,7
20	1560	1248	110	11,3
15	1440	1152	102	11,3
12	1320	1056	82	12,9
10	1220	976	65	15,0
8	1100	880	70	12,6
6	890	712	61	11,7
4	870	696	75	9,3
2	730	584	118	4,9

Cable length	dv (peak) [V]	dv 10-90% [V]	dt [ns]	dv/dt [kV/µs]
100	1670	1336	415	3.2
90	1620	1296	405	3.2
80	1610	1288	330	3.9
70	1660	1328	350	3.8
60	1670	1336	259	5.2
50	1680	1344	248	5.4
40	1650	1320	228	5.8
35	1650	1320	217	6.1
30	1670	1336	122	11
25	1620	1296	125	10.4
20	1570	1256	107	11.7
15	1470	1176	88	13.4
12	1370	1096	81	13.5
10	1270	1016	71	14.3
8	1270	1016	66	15.4
6	1210	968	75	12.9
4	1020	816	80	10.2
2	970	776	140	5.5

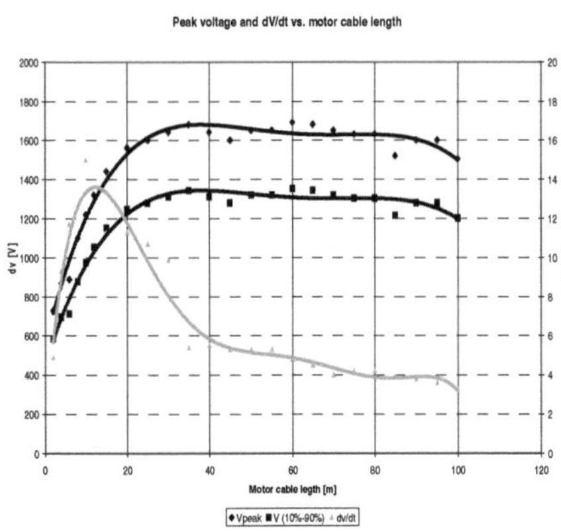

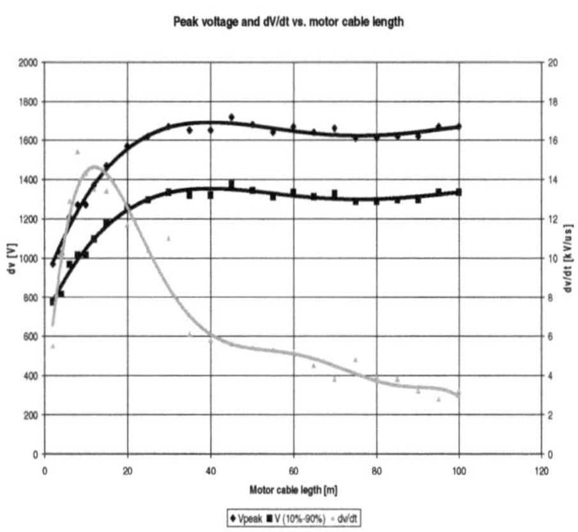

Fig. 7 and **8.** dv/dt ratio and peak voltages for AKM2 (left side) and AKM7

4 Conclusions

Figures 7. and 8. show that maximum peak voltage and maximum dv/dt ratio does not occur at the same cable length. Peak voltage culminates at 36m of cable length, while the dv/dt ratio is highest at approx. 12m of cable length. Comparing both measurements is apparent the difference of dv/dt ratio which becomes worse at AKM7 motor, while the peak voltage remained comparable to AKM2.

Acknowledgements. The research has been supported by MSM0021630518 "Simulation Modeling of Mechatronic Systems".

References

[1] Barták, A., Mravináč, L., Vařák, J.: Diagnostika poruch izolací elektrických strojů., SNTL, Praha (1984) ISBN 04-523-84
[2] IEEE 400-2001, IEEE Guide for Field Testing and Evaluation of the Insulation of Shielded Power Cable Systems
[3] IEEE 1434-2000, IEEE Trial-Use Guide to the Measurement of Partial Discharges in Rotating Machinery
[4] IEC 60270:2000/BS EN 60270:2001, High-Voltage Test Techniques - Partial Discharge Measurements
[5] Engineering Dielectrics, Volume IIA, Electrical Properties of Solid Insulating Materials: Molecular Structure and Electrical Behavior, R. Bartnikas, R. M Eichhorn, ASTM Special Technical Publication 783
[6] IEC 60085:2004 Electrical insulation. Thermal classification

Evaluation of Control Strategies for Permanent Magnet Synchronous Machines in Terms of Efficiency

E. Odvářka and Č. Ondrůšek

Brno University of Technology, Faculty of Electrical Engineering and Communication, Department of Power Electrical and Electronics Engineering, Technicka 2848/8, Brno, Czech Republic
xodvar01@stud.feec.vutbr.cz

Abstract. Permanent Magnet Synchronous Machines find a place in variety of different applications. There is a common need for enhancement of efficiency for electric drives due to general tendency of increase of the energy prices and well-known environmental issues. This paper summarizes available control strategies for an inverse-salient permanent magnet synchronous machine with emphasis on overall efficiency of the electrical drive. The discussed drive is primarily designed for vehicular technology. A mathematical model of the electrical machine including computation of power electronics losses is introduced so that steady-state efficiency at any operational point may be estimated. A part of the process is generation of power loss profiles dependent on reference d- and q- axis currents as controlled variables showing the optimization potential. Efficiency of a drive depends on a selection of control strategy. Therefore, maximum efficiency, maximum torque-per-ampere and Id=0 control strategies among voltage constrained field-weakening strategy are a part of the comprehensive study. A comparison in all considered control strategies in terms of drive efficiency is carried out so that one may track a difference in efficiency for any defined load point.

1 Introduction

Energy-efficient drives play critical role in design of hybrid or electric vehicles leading to more sustainable transportation. According to CleanCities initiative, a fleet of 82 000 of hybrid vehicles accounted in 87 US cities only in 2007 for total savings of 16 millions gallons of fuel. One percent rise of efficiency of electrical drive in series hybrid electric vehicles reduces fuel consumption of the vehicle over a drivecycle by 2.5% [1]. This paper investigates impact of control strategies for Synchronous PM machines on efficiency improvement of the electric drive.

The approach is presented on an inverse-salient synchronous machine with embedded permanent magnets able to develop 100kW of rated power at 1300rpm with constant power operation in field weakening up to 2800 rpm. This topology has been more deeply described in [2]. The machine is able to develop twice as much power in overload operation over entire speed range. The machine is designed for the minimum DC bus voltage of 500 VDC.

2 Model and Parameters of the PMSM

The machine model assumes abc to dq0 transformation convenient for calculation of reference i_d and i_q currents as controlled variables for any control strategy. Models for both d- and q- axis including representation of iron losses by a speed – dependent resistance R_c is depicted on Fig. 1a,b below.

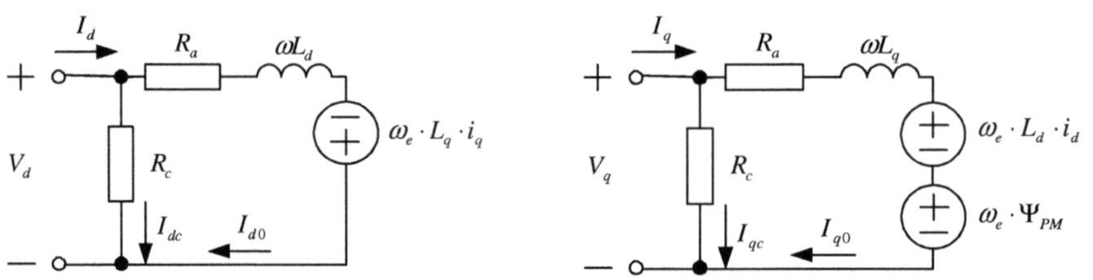

Fig. 1. Model of a PMSM in d- (a) and q (b) axis.

The symbols on Fig 1a,b represent inductance in d- axis L_d and q- axis L_q, phase resistance R_a, Ψ_{PM} as a normalized flux linkage and ω_e as electrical angular frequency. Such a model of the PMSM is sufficient for description of the machine in terms of voltage equation. The torque production of a PMSM is given by:

$$T = \frac{3}{2} p_p \cdot i_q \left(\Psi_{PM} + \left(L_d - L_q \right) \cdot i_d \right) \tag{1}$$

where p_p indicates number of polepairs. The operation of the machine is restricted by the maximum voltage constrains so that the input voltage $V_{inp} = \sqrt{\left(V_d^2 - V_q^2\right)}$ and the maximum phase current I_a constraint giving a condition so that $I_a = \sqrt{\left(i_d^2 - i_q^2\right)}$. The necessary input for performance analysis is machine's parameters according to Fig 1. With consideration of results of Finite-Element Analysis, Ψ_{PM}=0.1631Wb, p_p=9, R_a=18.2mΩ and inductances in d- axis and q- axis at nominal current L_q=0.508mH and L_d=0.419mH respectively. However, due to inherently nonlinear structure of the magnetic circuit, inductances depend on the level of loading of the machine. Therefore, a Finite-Element Analysis is needed to obtain profiles of L_d=f(I_d) and L_q=f(I_q), as is presented in Fig 2:

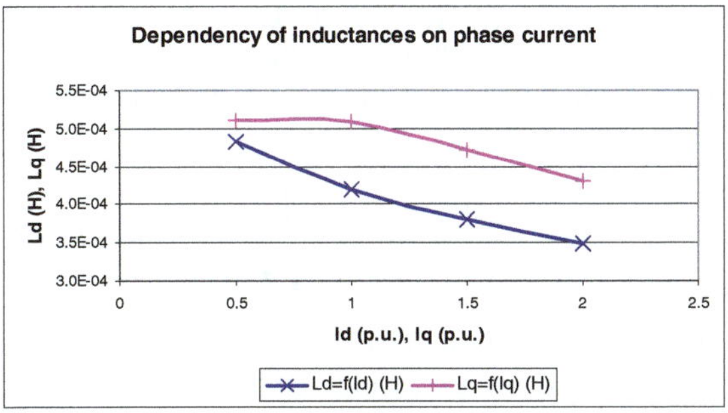

Fig. 2. Dependency of inductances Ld, Lq on current loading.

The resistor R_c, representing core losses of the machine, comprises of two resistor in parallel representing eddy current losses R_e and R_h taking into account hysteresis losses. The model assumes that hysteresis losses are speed dependent so that $R_h = R_{h,b} \cdot \omega_e / \omega_{e,b}$, where subscript '$_b$' indicates base speed of 1300 rpm. Both resistances R_e=93.18 Ω and R_h=104.56Ω are obtained from the hysteresis and eddy current losses in the iron core at base speed from the analytical model of the machine.

3 Control Strategies for Inverse-Salient PMSM

The family of inverse-salient machines differs in comparison to most common surface mounted synchronous PM machines in terms of inductances dependent on angular rotor position with L_q>L_d. That means the machine is able to produce reluctance torque. This is an advantage in field weakening operation, which occurs once the back-EMF become excessive in comparison to available input voltage as is depicted on Fig 3a.

The basic control strategy which may be applied to an inverse salient machine is I_d=0 aligning phase current entirely into q-axis so that only magnet torque is produced. The Equ 1 then reduces to $T = 3/2 \cdot p_p \cdot i_q \cdot \Psi_{PM}$. A vector diagram illustrates this case on Fig 3a. The Maximum Efficiency control strategy is based on loss minimization conditions used for obtaining expression for $i_{d,ref}$[3]:

$$\frac{\partial P_{loss}}{\partial i_d} = 0 \text{ and } \frac{\partial T}{\partial i_d} = 0 \tag{2}$$

where P_{loss} represent losses of the drive, which may be affected by a control action. With respect to Fig 1, P_{loss} comprises of copper P_{Cu} and iron P_{Fe} loss:

$$P_{Loss} = P_{Cu} + P_{Fe} = 3/2 \cdot R_a \cdot \left(I_d^2 + I_q^2\right) + 3/2 \cdot R_c \cdot \left(I_{dc}^2 + I_{qc}^2\right) \tag{3}$$

The analytical solution for loss minimization solved for i_d, which might be found in [3], appears to be to complex for practical implementation. Therefore, frequently adopted solution is Maximum Torque-per-Ampere (MTPA) control strategy considering reluctance torque too, which is basically simplification of Maximum Efficiency (ME) strategy assuming $\omega_e \rightarrow 0$ and $R_c \rightarrow \infty$ so that analytical expression for $i_{d,ref}$ is [4]:

$$i_{d,red} = \frac{\Psi_{PM}}{2 \cdot (L_q - L_d)} - \sqrt{\frac{\Psi_{PM}^2}{4 \cdot (L_d - L_q)} + i_{qs}^2} \tag{4}$$

Availability of previously mentioned control strategies reaches above the base speed their limitations in terms of available input voltage V_{inp}. The controller needs to fulfil in any case a condition so that:

$$V_{inp}^2 \geq V_d^2 + V_q^2 = (\omega_e \cdot L_q I_q + R_a I_d)^2 + (\omega_e \cdot (\Psi_{PM} + L_d I_d) + R_a I_q)^2 \tag{5}$$

Once the controller starts tracking the maximum available voltage, the machine enters into Field Weakening operational mode as is represented by a vector diagram on Fig 3b. Equ 5 needs to be solved in conjunction with Equ 1 in order to deliver desired torque and fulfill the voltage constraint in the same time. This usually requires numerical solution and storage of reference values in a table look-up.

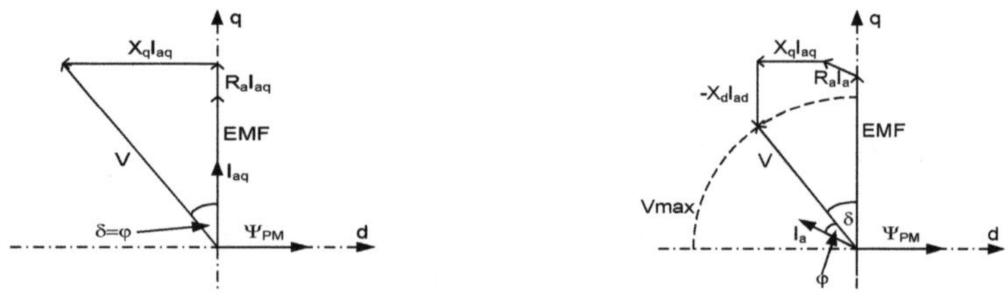

Fig. 3. Vector diagram for Id=0 control strategy (a) and field weakening operation (b).

Among the mentioned control strategies, several others may be found but with low spread in motor drives, such as Unity Power Factor Strategy.

An integral part of drive's efficiency modeling is a power electronics model in terms of assessment of its losses. Power electronics losses consist of conduction and switching losses. Conduction losses P_c may be determined out of Volt-Ampere characteristic of a device in on-state providing threshold voltage V_t and dynamic on-state resistance R_d at specific load current so that:

$$P_c = V_t \cdot I_{av} + R_d \cdot I_{rms}^2 \tag{7}$$

Switching losses P_{sw} generally depends on switching frequency f_{sw} and dissipated energy during turning-on W_{on} and –off process W_{off}:

$$P_{sw} = f_{sw}(W_{on} + W_{off}) \tag{8}$$

A convenient way of obtaining switching losses is by obtaining the switching energy as a function of current $W_{sw} = f(I)$ from a datasheet of the device by a quadratic approximation so that switching loss is function of time-variant current:

$$W_{sw}[I_a(t)] = E_{max} k^2 \sin^2 2\pi / T \cdot t \qquad (9)$$

Where E_{max} is datasheet value of the switching energy at maximum current, T is period of the current waveform and k is current utilization factor as a ratio between actual phase current and maximum current of the switching device.

The model of power electronics losses principally follows the diagram below:

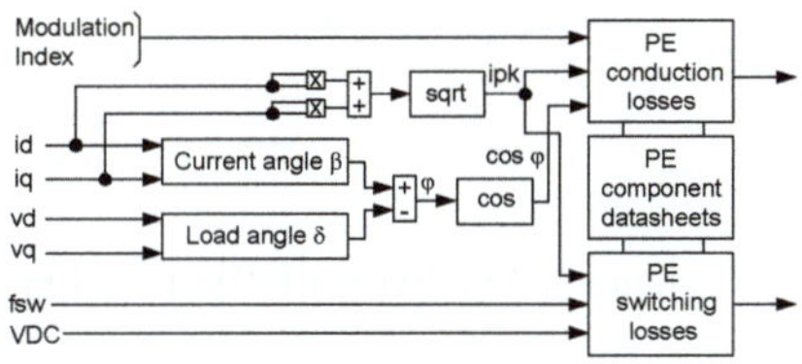

Fig. 4. Block diagram describing calculation of power electronics losses.

Once power electronics and electrical machine's losses are combined, the total power loss profile vs i_d and i_q reference currents depicts what is the optimization potential for the drive. The maximum optimization potential depict the cyan curves of total drive losses, where the minimum of the power loss function belongs to ME strategy, unlike operation with zero i_d current at y-axis of the plots below:

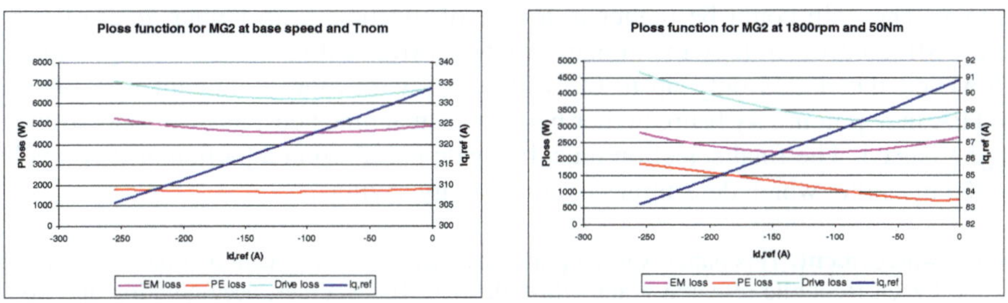

Fig. 5. Loss profiles Ploss=f(id,iq) at 1300rpm, 734.9Nm (a) and 1800rpm, 200Nm (b).

The biggest potential for loss minimization appears to be at load points, where iron losses tend to be proportionally more dominant, that means at high speed/ low torque operation as shows Fig 5b.

4 Comparison in between Control Strategies

The main objective of the comprehensive study is to highlight, what is the most feasible control strategy for a certain operational point or zone. This will be carried out by subtracting efficiency maps for particular strategies. Let assume that Id=0 is a datum for the comparison and both MTPA and ME will be able to improve efficiency η in wide torque-vs-speed range. Fig 6a shows result of subtraction η (MTPA) – η (Id=0) unlike Fig 6b presents difference η (ME) – η (MTPA):

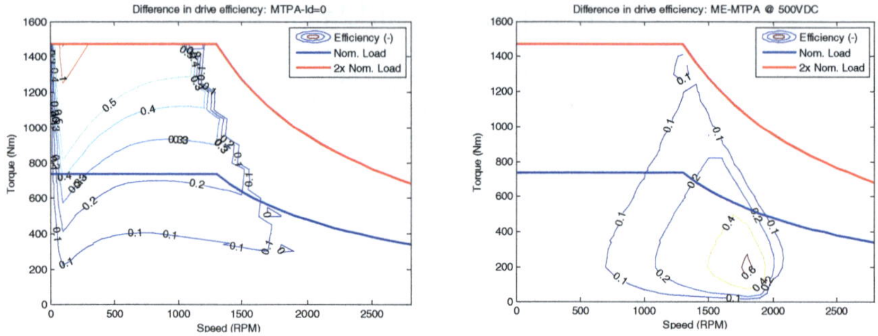

Fig. 6. Difference in efficiency η(MTPA)-η(Id=0) (a) and η(ME)-η(MTPA) (b).

The optimization potential is restricted by available voltage determining fictive line from maximum torque, nominal speed, where the machine starts weakening the field down to zero torque approximately at 2200rpm. Either or MTPA are therefore not applicable to the top-right hand side corner of the efficiency map.

5 Conclusion

MTPA shows benefit at low-speed/high torque operation of up to 1% of efficiency. The gain could be higher in a case of machine with higher level of saliency. ME control strategy maximizes efficiency over entire speed-vs-torque range. However, the rise is only up to 0.6% in zones with more dominant iron losses. MTPA may potentially bring higher benefit for machines designed for operation at higher speed. The final selection must be made based upon a particular drivecycle of the machine, which determines overall energy savings.

Acknowledgement. This paper was supported by Ministry of Education, Youth and Sports of the Czech Republic research grant MSM0021630516 "Sources, accumulation and optimization of energy exploitation in the conditions of permanently sustainable growth" and GA 102/08/1118 "Intelligent diagnostics for electric machines".

References

[1] Clean cities Annual Metrics Report (2007), `http://www.nrel.gov/docs/fy07osti/41753.pdf`

[2] Odvářka, E.: Electric Motor Generator for a Hybrid Electric Vehicle. Engineering Mechanics 16(2) (2009) ISSN 1802-1484

[3] Morimoto, S.: Loss Minimization Control of Permanent Magnet Synchronous Motor Drives. IEEE Transactions on Industrial Electronics 41(5) (October 2005)

[4] Sue, S.-M.: Voltage-Constraint Tracking Based Field Weakening Control of IPM Synchronous Motor Drives. IEEE Transact. on Industrl. Electronics 55(1) (January 2008)

A Two Layered Process for Early Design Activities Using Evolutionary Strategies

A. Albers, H.-G. Enkler, M. Frietsch, and C. Sauter

IPEK – Institute of Product Development, University of Karlsruhe,
Kaiserstraße 10, 76131 Karlsruhe, Germany
{albers,enkler,frietsch,sauter}@ipek.uni-karlsruhe.de

Abstract. Especially in system and conceptual design activities designers resort to already existing components that are combined and arranged to a new system which has to fulfill a predefined set of requirements. Designers have to deal with requirements and constraints that are changing during the development process. General goal is an automatic generation of compatible conceptual design proposals that meet the predefined requirements such as design space, EMC, etc. To this, libraries containing standardized data on components have to be developed. These libraries include component specific characteristics and data such as CAx models or efficiency factors. In an iterative process compatible systems are configured and evaluated by means of CAx based analyses. Afterwards an optimization system based on genetic algorithms accesses these data to find optimal configurations. By combining the optimization algorithm with a CAD system, design proposals are directly visualized and can be processed by the designer in the further product development process.

1 Introduction

In product development processes, computer aided software tools help to avoid expensive and time-consuming failures and iterations. The step from a functional description of a system behavior to the design of its components is a challenging task which can be supported, e.g., by the Contact and Channel Model (C&CM) [1]. Starting from a low level of detail, most simulation tools cannot be used in early activities. Additionally, there is a huge number of catalogues of different manufacturers, especially in multi domain systems. Therefore a further challenge is to find an optimal configuration. Designers have to deal with requirements and constraints that are changing during the development process. Goal of this work is a (semi-)automatic generation of compatible conceptual design proposals that meet the predefined requirements.

2 A New Framework for Computer Aided Conceptual Design

New products are often based on a combination and arrangement of already existing components. The new system has to fulfill a predefined set of requirements.

Due to this fact new mechatronic interfaces such as connectors and protocols are processed using C&CM. It is necessary for designers to be at least familiar with the involved domains and their requirements and boundary conditions. Close communication is one of the main factors to avoid suboptimal solutions. Furthermore designers have to deal with requirements and constraints that are changing during the development process. A fast and automated evaluation of the current system design regarding these changes and—if necessary—the derivation of a new system design is desired. The impact of fuzzy requirements and boundary conditions on the 'optimal' system design should also be considered. In other words the task is to map continuous requirements to the discrete world of catalogues. Finally, a complete and systematic evaluation of possible component configurations even off the beaten track could possibly lead to innovative solutions. Manual exploration of all these aspects is often not possible due to time restrictions in the product development process. Since many evaluations can be performed through structured procedures, e.g. calculation of the center of gravity for a component configuration, an automated computer aided approach seems appropriate.

Especially in the development process of mechatronic systems, the designers have to manage a lot of different catalogues to select the needed components from multiple domains like motors, controllers, brakes, sensors, gears, etc. Another challenge is to keep 'up-to-date' with the newest products and innovations in the different domains. Additionally there are diverse kinds of catalogues, e.g. book-like ones, CDs or web pages, which all have to be organized. Although each catalogue is intended to help the designer to find the best suitable product as fast as possible, one can find four different 'levels of assistance'.

Using a level 0 catalogue—often big tomes with complete product portfolios in one single book, annually updated, with text links to compatible components—will result in a very time consuming process: new requirements and boundary conditions regarding controllers or sensors can arise by selecting one component. Level 1 catalogues provide comfortable search functionalities and hyperlinks to compatible components. Thus it is faster but not easier for the designer to look for compatible configurations. The bulk of manufacturers' web pages and also most of the product CDs offer these functionalities. Level 2 catalogues include tools like component filtering to avoid selection of incompatible configurations [2]. The user does not have to look at the text- or hyperlinks to check for compatible components manually. This is done automatically by the software. By choosing one component, the number of possible configurations is reduced to assist the designer in finding the required combination. All preceding levels are limited to the products of only one particular manufacturer. This is in fact unsurprising because every company aims to distribute their own products. But from a designers point of view a manufacturer spanning solution would be much more auxiliary. A small step towards this goal is already realized in terms of CAD-models. In [3] a level 3 approach based on a database of over one thousand motors and one thousand transmissions of different manufacturers was implemented. This database contains over ten thousand possible configurations. The designer has to enter the required torque, angular velocity and optional boundary condition and the selection process, based on an automated dimensioning, is executed automatically. This facilitates the selection-process and allows faster reactions in case of changing requirements or boundary conditions.

The remaining disadvantage of this framework is its lack of flexibility regarding additional boundary conditions like design space, EMC, resonance frequencies, dynamics and other multi-domain effects. Therefore additional tools, like CAx or simulation tools, have to be integrated. Also fuzzy requirements or a weighting of different criteria against each other are not possible in the existing frameworks. Due to the situation described above, efforts in generating a greater support for the development of mechatronic systems have been made, see [3] or [4].

3 Approach

We propose a new level 4 catalogue approach that is divided into two main layers. The first layer is based on the 'conceptual verification' method developed at the IPEK and presented in [5]. The second layer supports the designer in the following optimization, or rather, evaluation step. Therefore an interaction-process of this catalogue with the CAD software Pro/Engineer (ProE) is presented. The whole framework aims to assist the designer in time-consuming and simple tasks in order to gain more time for the creative part of his work, which cannot be transferred to a computer.

When designing complex new systems, the design task is commonly segmented into smaller subtasks resp. subsystems. For complex mechatronic systems a segmentation based on the functional structure as proposed in [5] is recommended. The objective of the first layer is the generation of compatible design proposals for these subsystems. The therefore required library has to contain all component specific information like engine speed, torque, efficiency factors, CAD data but also metadata like type of motor, compatibility parameters, level of preference for a specific component, etc. To allow easy exchange and update of this library, a standardized file format is necessary. We use an XML-based file format. Based on this component library the selection process is performed using various requirements and boundary conditions. In contrast to the already existing level 3 solution, the user has easy access to every property of the components. Therefore a multitude of various criteria can be tested and taken into account when generating design proposals. The main advantage is the possibility to do so with complete compatible solutions. To generate compatible configurations that meet the requirements and boundary conditions, several steps are executed: Firstly the total number of possible configurations has to be reduced by eliminating the component combinations that do not lead to feasible configurations. If looking for an electric drive unit for example, only electric components were selected. Secondly 'don't-like' components were eliminated to realize company specific preferences. Thirdly components with parameters outside a specific range were eliminated as well. Then the generation of compatible configurations is performed. This is done by creating all possible combinations of e.g. motors and gears following the product hierarchy. Every potential solution consists of 'component primitives', that are combined using one library for each primitive. CAD-models can be used either for the selection process or for the later step. Additionally the designer does not have to look for each single model, if a configuration is chosen at the end of the process. Within the selection process, the models can be used for a multitude of analysis, e.g., design space or centre of gravity. For a correct assembling of single components in the CAD environment an approach based on C&CM is used. Each component has to be provided with working surfaces for connecting them to each other and to a possibly existing environment. In Fig. 1 an early concept of a humanoid robots elbow can be seen [6]. At the moment this time consuming step must be performed by hand.

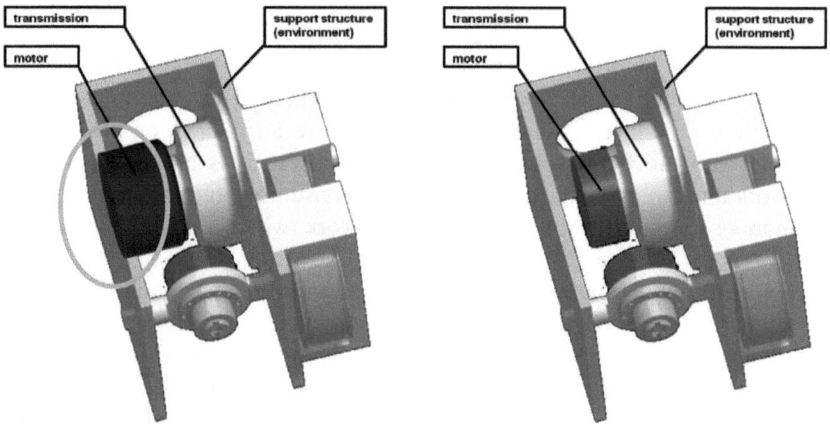

Fig. 1. Colliding and therefore rejected (left) and accepted (right) design proposals

Additionally a new concept of mechatronic interfaces is used to determine the validity of design proposals. To realize the integration of fuzzy or indistinct requirements, thresholds and ranges are used. The evaluation is accomplished iteratively for each single criterion to steadily reduce the number of possible solutions. The evaluation methods vary strongly in their computing time. Hence the optimal sequence regarding the computing time of those tests will be evaluated automatically in the future. The output of this processes are several different design proposals that fulfill the requirements. By creating a target function containing the fulfillment level of each criterion, a customized weighting of the different requirements can be realized. The final design proposals are further optimized by the optimization layer. Initially it is a time intensive process to collect all relevant datasets and to feed them into the database. But firstly this has to be done only once resp. for new data only and the longer the catalogue is used, the bigger is the benefit of this framework. Secondly the long term goal is to establish a standard catalogue system. Each company could provide their catalogues to facilitate it for the user to keep its own library up to date. This concept offers also big advantages to manufacturers: a fast distribution of a new product, resp. the knowledge that a new product is available, without having to wait for the new printed catalogue.

After having generated a set of compatible component configurations it is necessary to arrange the components spatially, i.e. to define position and orientation. During this process several restrictions such as design space, EMC, etc. have to be taken into account. In order to provide an automated process, we propose an integrated approach using a combination of CAD (e.g. PTC Pro/Engineer—ProE), CAx and genetic algorithms. There are several free and commercial software implementations. We are using the software package DAKOTA (Design Analysis Kit for Optimization Applications) [7]. In order to integrate ProE, CAx and DAKOTA, it is necessary to develop an interface (based on Perl and Java/Jlink) to link DAKOTA and ProE models of the components chosen in the iterative selection process described above and—if necessary—to further CAx analysis tools. During the process, parameterized ProE models related to the respective components are loaded, assembled and located using a parameter set.

A simple mechatronic demonstrator system consists of an energy source (component A), an engine, gear box and sensor (component B). Component B results from a variety of possible configurations selected in layer 1. Fig. 2 shows an intermediate optimization step for cable routing to connect components A and B in a suitable way.

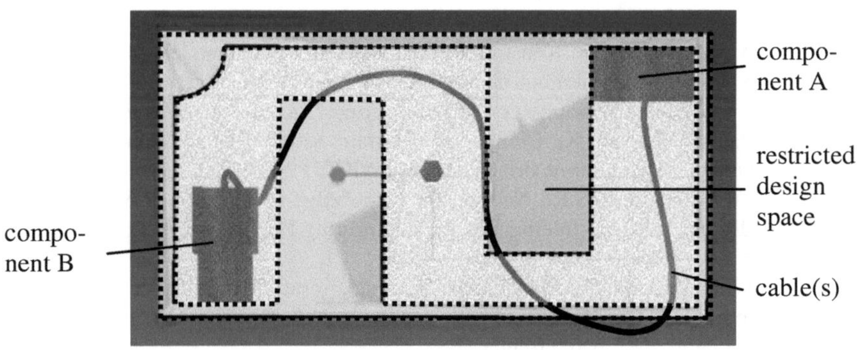

Fig. 2. Screenshot of the ProE model during the optimization of cable routing (layer 2)

The cable is modeled using a spline with supporting points that are modified iteratively by the optimization algorithm. In this demonstrator system, the cable has to be guided through a restricted design space. The configuration is analyzed by ProE with respect to collisions and radius of curvature. Data between ProE and DAKOTA are exchanged using small ASCII files including a new set of parameters for ProE or evaluation results for DAKOTA. Due to its high flexibility, the user may integrate additional analyses such as FEA. Further information on the implementation and a detailed example can be found in [8].

4 Summary and Future Work

This article introduced a novel method to support designers in conceptual design phase by means of a computer aided approach. General goal of this framework is a (semi-)automatic generation of compatible conceptual design proposals that meet the predefined requirements. Currently, the presented method is being implemented in a software tool. Amongst others the method will be evaluated during the development process of the humanoid robot ARMAR IV and V [6]. This will include extending the current elementary component library. On the one hand this extension will be done regarding components of one class (motors, sensors, etc.) and on the other hand by adding new component classes (couplings, brakes, etc.). It is also intended to integrate C&CM, detailed CAD models of manufacturers to enhance accuracy of design space analyses, masses, etc. Besides CAD models, a connection to CAx tools such as Multi Body Simulation or Finite Element Solvers to improve analysis capabilities is possible. Due to performance reasons a further goal is to feed back knowledge acquired in former development processes.

Acknowledgments. We are grateful for the support of the DFG (Deutsche Forschungsgemeinschaft) within the Collaborative Research Centers 499 'Micro Molding' and 588 'Humanoid Robots'.

References

[1] Albers, A., Matthiesen, S.: Konstruktionsmethodisches Grundmodell zum Zusammenhang von Gestalt und Funktion technischer Systeme. Konstruktion 54, 55–60 (2002)
[2] `http://shop.maxonmotor.com/ishop/app?language=en&country=INT`
[3] Devanathan, S., Ramani, K.: Combining constraint satisfaction and non-linear optimization to enable configuration driven design. In: ICED 2007 (2007)
[4] Kim, J., Will, P., Ling, S.R., Neches, B.: Knowledge-rich catalog services for engineering design. Artificial Intelligence for Engineering Design, Analysis and Manufacturing 17, 349–366 (2003)
[5] Brudniok, S.: Methodische Entwicklung hochintegrierter mechatronischer Systeme am Beispiel eines humanoiden Roboters, Karlsruhe, Universität (TH), Dissertation (2007)
[6] DFG Collaborative Research Center 588, `http://www.sfb588.uni-karlsruhe.de`
[7] `http://www.cs.sandia.gov/DAKOTA/packages.html`
[8] Albers, A., Leon, N., Aguayo, H., Maier, T.: Multi-objective system optimization of engine crankshafts using an integration approach. ASME, Boston, USA (2008)

Virtual Design of Stirling Engine Combustion Chamber

Z. Kaplan, P. Novotný, and V. Píštěk

Brno University of Technology, Faculty of Mechanical Engineering,
Institute of Automotive Engineering, Technicka 2896/2, Brno, Czech Republic
kaplan@fme.vutbr.cz

Abstract. The paper deals with the designing of a combustion chamber of the Stirling engine using the CFD approach. Virtual prototypes enabled to optimize the parameters in the combustion chamber, to increase efficiency of the energy conversion, and to decrease emissions. The presented results help to increase the Stirling engine efficiency together with a significant time decrease of development process.

1 Introduction

Efficient utilisation of biomass for energy can be achieved with a minimum of environmental impact if biomass is used for small-scale combined heat and power (CHP) production in smaller towns and villages close to biomass production sites as well as in the wood processing industries. In the range of an electric power output up to 100 kW_{el} Stirling engines are presently the only useful technology for CHP generation based on solid bio-fuels which has the potential to meet the technical end economical demands in an environmental way [1].

Stirling engines for CHP plants using gas as fuel have been developed for more than five years at the Brno University of Technology, Institute of Automotive Engineering. As a part of these activities a 500 W pilot Stirling engine was build. The pilot engine development has been supported by advanced computational modelling.

Requirements for computational modelling of different physical phenomena rise in the present time. Dynamics of Stirling engine components or dynamics of fluid processes of external heat supply engines are specific which is given by the fact that the course of observed values (force, temperature, pressure, heat transfer, etc.) is periodical [3].

Modern computational models like Multi-body System (MBS), Finite Element Method (FEM) or Computational Fluid Dynamics (CFD) deliver relatively accurate results but only if correct inputs are included. This represents a fundamental drawback of modern computational methods. The correct inputs can be greatly obtained from measurements, therefore, the measurements are continuously a fundamental part of the Stirling engine development.

The development of a Stirling engine starts with an initial design of thermodynamic engine parameters. These proposals significantly affect engine geometry (engine displacement, working medium etc). However, there are still many restrictions, one of the strictest ones is the economical aspect and this has to be carefully considered.

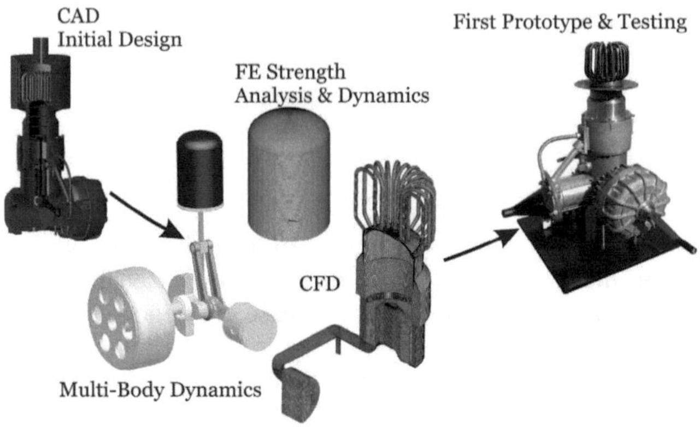

Fig. 1. CAE models and first prototype of Stirling Engine

After the main engine parameters are proposed the accent is changed to advanced CAE (Computer Aided Engineering) computational models. CAE models of the Stirling engine are presented in Figure 1. Stirling engine geometry is set to an initial cranktrain position for the start of all simulations. If flow processes are to be studied, a volume of a working medium (for CFD calculations) has to be created by a subtraction of a Stirling engine CAD model from a properly chosen volume. The design phase finishes with a real prototype of the Stirling engine. The real prototype engine also enables to validate CAE calculation results.

2 Stirling Engine Combustion Chamber

The main objective of the Stirling combustion chamber development is to ensure a uniform heat transfer to the heater tubes, beside a reduction of the internal manifold volume, and to adapt the geometry of the Stirling heater in order to allow an efficient implementation of components like a burner or an intake manifold. The CAD model of resultant version of Stirling engine combustion chamber is presented in Figure 2.

The Stirling heat exchanger includes 24 u-shaped tubes connected to the cylinder side. The burner is positioned in a combustion chamber axis and can be axially moved. The intake manifold is placed tangentially to a outer cover of the combustion chamber and enables efficient air warm-up.

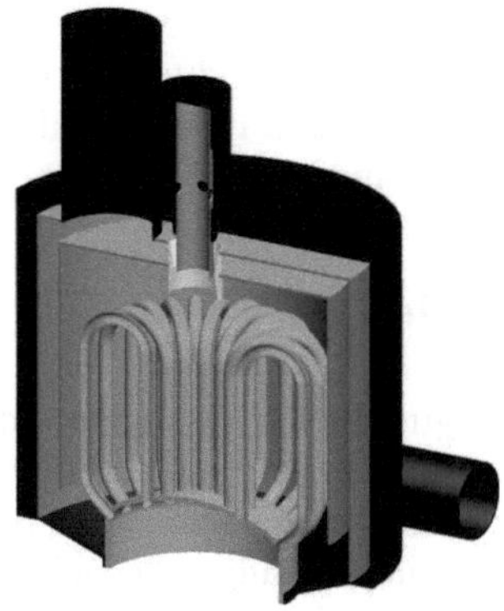

Fig. 2. CAD model of Stirling combustion chamber

3 Design of Combustion Chamber Volume

During the development phase a number of versions of air intake manifolds have been investigated. The experiences as well as CFD results confirmed the tangential position of the intake manifold.

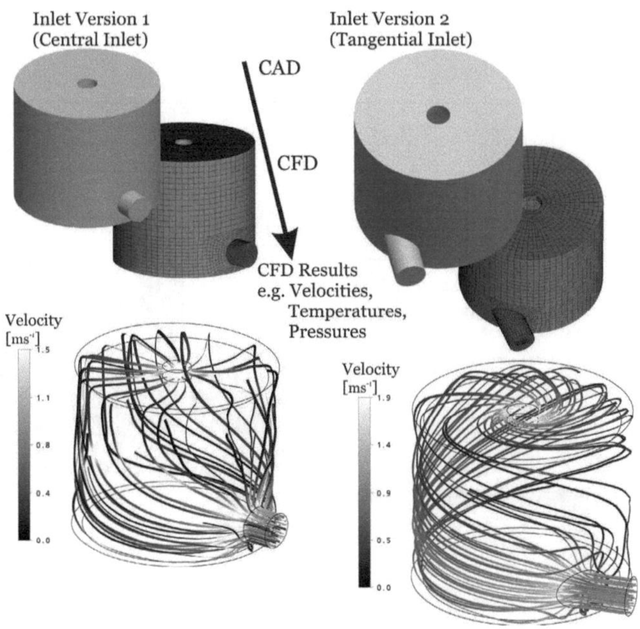

Fig. 3. Combustion chamber outer volume design versions and results

Figure 3 shows the CFD computational model and result examples of two versions - intake manifold located centrally and tangentially to the combustion chamber outer cover. The tangentially placed intake manifold reduces flow losses and enables a better warm-up of air compared to the centrally placed intake manifold.

4 Burner Position Solution

The proposed pilot Stirling engine fuel is gas supported from a gas-bottle. Gas enters the combustion chamber thru a nozzle and a burner. Air and gas mixture burns and resultant heat is distributed to the heat exchanger pipes.

A burner position can be axially changed to achieve the optimal combustion process. First burner designs showed large non-uniform temperature distribution on the heat exchanger pipes. After a few design modification the final design has been proposed and suitable temperature distributions have been achieved. The axial position of the burner pipe can be easily adjusted for the best working conditions. Figure 4 presents details of three burner pipe positions and results (pressures and velocities) on section plane for the final burner design.

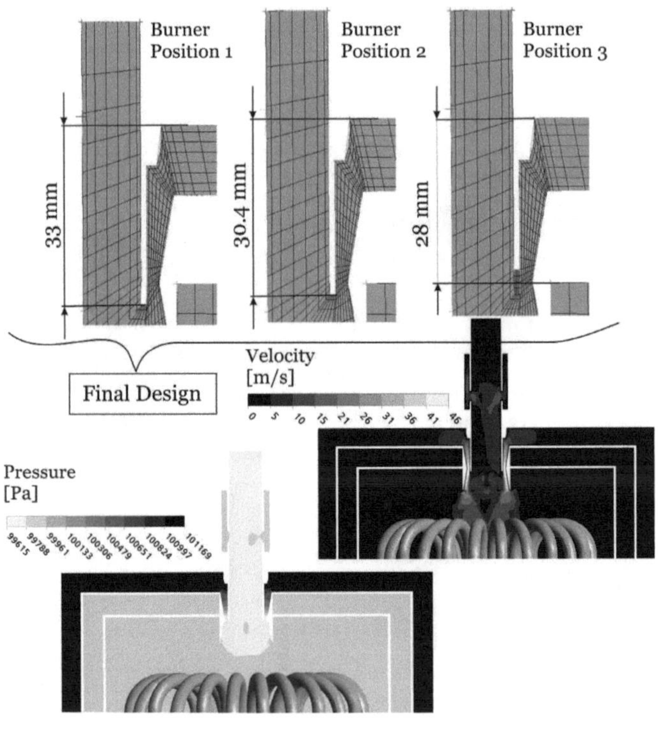

Fig. 4. Burner position description and results

5 Complete Combustion Chamber Solution

The combustion chamber is mainly made from 2 mm steel plates connected by welding technology. The combustion chamber design can be characterized as a low cost because this is one of the current Stirling engine design objectives. Therefore, all the shapes (channels, pipes etc.) are not ideally designed for fluent flow.

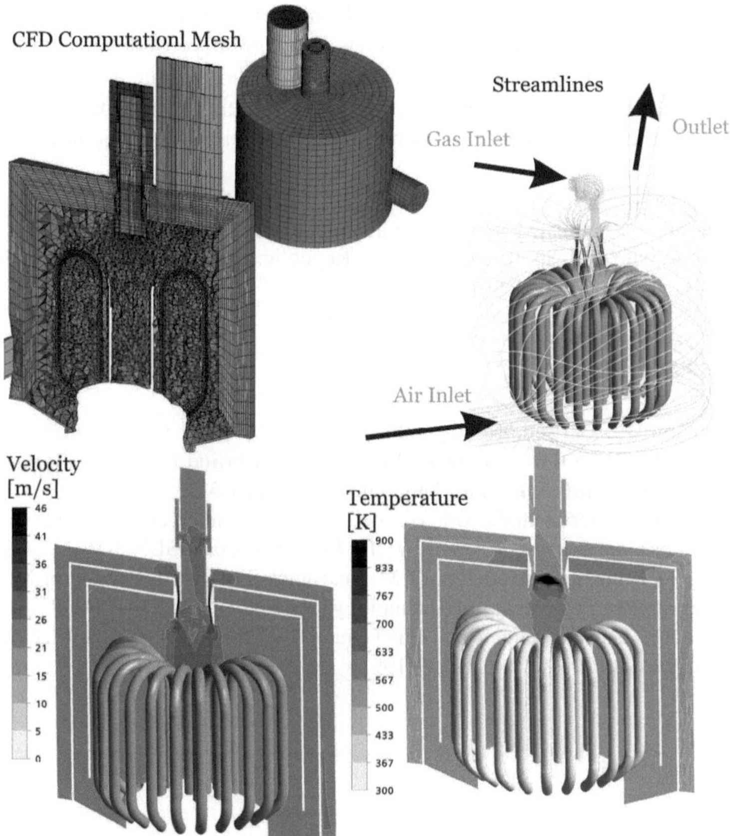

Fig. 5. Combustion chamber model and results

The complete combustion chamber is modelled by using CFD as well. The computational hybrid mesh includes hexa and tetra cells with defined boundary layers. Figure 5 shows the computational model and examples of results like flow streamlines or velocities and pressures in cut plane. The results are presented only for the final design of the combustion chamber.

6 Conclusion

The combustion chamber of the Stirling engine with electric power output 500W has been successfully resolved and the experience enable to produce a new combustion chamber for more powerful Stirling engines prepared by our team of researchers. Computational models of thermodynamic processes in the combustion chamber and in the Stirling engine respectively have been created as higher-level computational models based on CFD models of physical processes occurring in real units, using only the minimum simplifying assumptions. Developed and successfully verified computational models will be greatly used before the future, more powerful Stirling engine is made. All performed technical experiments will also offer advances in the future. Computational models and technical experiments will speed up the development of Stirling engines with better technical and economic parameters.

Acknowledgments. Published results were acquired using the subsidization of the Ministry of Education, Youth and Sports of the Czech Republic, research plan MSM 0021630518 "Simulation modelling of mechatronic systems".

References

[1] Biedermann, F., Carlsen, H.: Operating Experiences with a Small-scale CHP Pilot Plant Based on a 35 KW_{EL} Hermetic Four Cylinder Stirlind Engine for Biomass Fuels. In: International Stirling Engine Conference, Rome (2003)

[2] Siegel, A.W.: Experimentalle Untersuchen zum Warmeubergang und Druckverlust im Regenerator einer Stirling-Machine. VDI VERLAG (2000) ISBN 3-18-312419-X

[3] Brezina, T.: Simulation Modeling of Mechatronics Systems II. Faculty of Mechanical Engineering, Brno University of Technology (2006) ISBN 80-214-3341-8

[4] Organ, A.J.: Thermodynamics and Gas Dynamics of the Stirling Cycle Machine. Cambridge University Press, Cambridge (1992)

500W Stirling Engine Development

P. Novotný and V. Píštěk

Brno University of Technology, Faculty of Mechanical Engineering,
Institute of Automotive Engineering, Technicka 2896/2, Brno, Czech Republic
novotny@fme.vutbr.cz

Abstract. A successful realization of Stirling engines is conditioned by its correct conceptual design and optimal constructional and technological mode of all parts. Initial information should provide computation of real cycles of the engines. The paper presents calculation models of engine part strength, dynamics and thermodynamic cycles of the external heat supply engines. High-level FE (Finite Element), MBS (Multi Body System) or CFD (Computational Fluid Dynamics) models arising from the description of real processes which run in an external heat supply engine are used for virtual prototype of Stirling engine.

1 Introduction

The goal of the research team's work was to create a modern Stirling engine. A virtual prototype of the Stirling engine as a complex computational model of a higher level is an instrument for a successful Stirling engine design. This virtual prototype shall enable solving issues from the area of dynamics and fluid mechanics of a mechanism with high accuracy (by as much as one order) as required for further solution of control and diagnostics of a micro plant for combined production of heat and generation of power with the Stirling engine. The solution dealt with γ-modification Stirling engine with the useful output up-to 500 W [4].

2 Development Process of Stirling Engine

The development process of Stirling engine starts with an initial design of thermodynamic engine parameters. These proposals determine basic engine properties (an engine displacement, working medium etc.). However, there are still many restrictions. One of the most strictest ones is the economical aspect which has to be carefully considered.

After the main engine parameters are proposed the accent is changed to advanced CAE computational models. The development process and computational methods are presented in Figure 1. The process continues with the first prototype of the engine. The real prototype enables to validate calculation results.

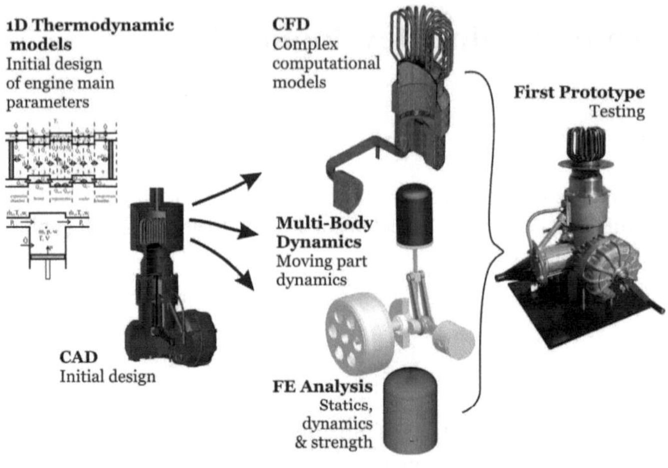

Fig. 1. Stirling engine development process

3 Stirling Engine Driving Mechanism Dynamics

Various alternatives of the crank mechanism or the rhombic mechanism are usually applied as the driving mechanism of the Stirling engine. It can also be realised as an engine with free pistons.

A centric crank mechanism has been applied for the piston and a centric crank mechanism with a crosshead has been applied for the displacer in the Stirling engine. While the issue of the dynamic balance of the driving mechanisms has been well elaborated and is based on the fact that both the geometrical and mass parameters of individual cylinder units are identical in internal combustion engines, these prerequisites are usually not met in case of the Stirling engine.

Multi-body systems can be applied as effective tools for solving Stirling engine dynamics. Multi-body systems enable solving different dynamic issues of complex systems combining rigid and flexible bodies. In the case of Stirling engine mechanisms, they can be used to find the optimum alternative for balancing the driving mechanism.

When the mechanism moves, inertial forces of different moving parts take effect, causing vibrations and must be "captured" by means of the machine seating. A virtual mechanism prototype has been created in the multi-body system and the end-point trajectory of inertial force resultant vector traced during one engine revolution running at the unit angular velocity of the crankshaft.

4 FE Analysis of Main Parts

FE computational tools enable calculations of engine part strength including fatigue calculations. Loading (forces, pressures, temperatures etc.) can be obtained from MBS or CFD results. For example, Figure 2 shows FE analysis results of displacer piston deformations for two design variants. The first version of the displacer piston includes ribs and the second one is without any ribs.

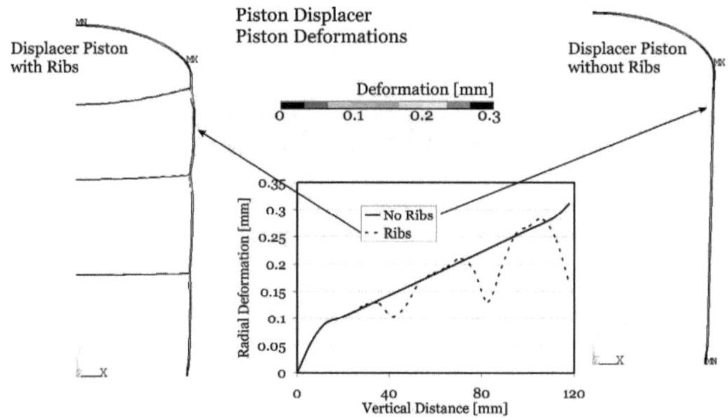

Fig. 2. FE analysis of displacer piston deformation for two variants

5 Regenerator of Stirling Engine

To use the positive effect of regenerative heat accumulation on the thermal efficiency of Stirling engine, the real regenerator should meet the following requirements:

- Heat capacity of regenerator material should be as high as possible.
- Heat conductivity of regenerator material should be as high as possible.
- For growing values of heat transfer coefficient, a temperature difference between the working gas and the regenerator material decreases; this then decreases energy losses through the heat transfer. To reach higher values of heat transfer coefficient, it is necessary to increase the velocity of working gas flow. This can be obtained through decreased sectional areas, and consequently a decreased regenerator free volume.
- The surface area for heat transfer in the regenerator should be as large as possible.
- On the other hand, a free volume of regenerator decreases pressure amplitudes and thus also the output of Stirling engine.
- Pressure losses in the regenerator should be as little as possible.

Based on the comparison of the above, mostly contradictory requirements, it is evident that at present there is no real material to meet these requirements. A search for appropriate regenerator material is therefore an optimization problem that always has to be re-solved for each particular Stirling engine.

Operational revolutions of Stirling engines are usually in the range of 900 - 3000 rpm so that time intervals for accumulation and removal of heat in the regenerator are relatively very short. Therefore the appropriate structures of regenerator should have the largest possible ratio of the surface area to the volume, and also high heat conductivity. Examples of appropriate material structures for the regenerator are shown in Figure 3.

Fig. 3. Examples of appropriate regenerator material structures

Metal materials (e.g. high-quality steel, copper or bronze) have higher heat capacity and better heat conductivity than ceramics. Therefore, the layers of fine wire netting often form the regenerators of Stirling engines. For optimization of regenerator, it is necessary to construct a sufficiently accurate computational model for determination of its heat and pressure losses.

The empirical Darcy's law has been applied for flows through porous media when the Reynolds number based on the pore size d_p is very small. The momentum equation for fluid flows passing through an isotropic media is given by

$$-\Delta p = \frac{\nu \mathbf{U}}{K}, \tag{1}$$

where p is the pore pressure, ν the fluid viscosity, and $\mathbf{U}$ the Darcy velocity. Here, Darcy velocity is taken as a superficial velocity by regarding the media as a continuum and ignoring the details of porous structures. The permeability, K, is defined as

$$K = \frac{\phi^3 d_p^2}{a(1-\phi)^2}, \tag{2}$$

where Φ is the porosity of porous material and a is a constant to parameterize the microscopic geometry of the porous materials.

Engineering practice requires the operation of flows in porous materials at high Reynolds number, such as those in Stirling engine regenerators. Experimental evidence showed that equation (1) was unable to describe the flows at high Reynolds number.

On the basis of experimental data, the equation (1) was gradually supplied with further members (Ergun, 1952) that extend its validity to a larger range of Reynolds numbers under a steady flow

$$-\Delta p = \frac{\nu \mathbf{U}}{K} + \frac{H\sqrt{\rho\nu|\mathbf{U}|}\mathbf{U}}{K^{3/4}} + \frac{F\rho|\mathbf{U}|\mathbf{U}}{\sqrt{K}}, \tag{3}$$

where ρ is the fluid density and the Forchheimer coefficient F is given by $F = b/\sqrt{a\phi^3}$ where b is again a constant to parameterize the microscopic geometry of the material. The dimensionless coefficient H, like F, is a function of porosity and microscopic solid geometry. There remains a task to construct a model for unsteady flows through porous materials, which to the first-order approximation is valid over the entire ranges of time scale and Reynolds number.

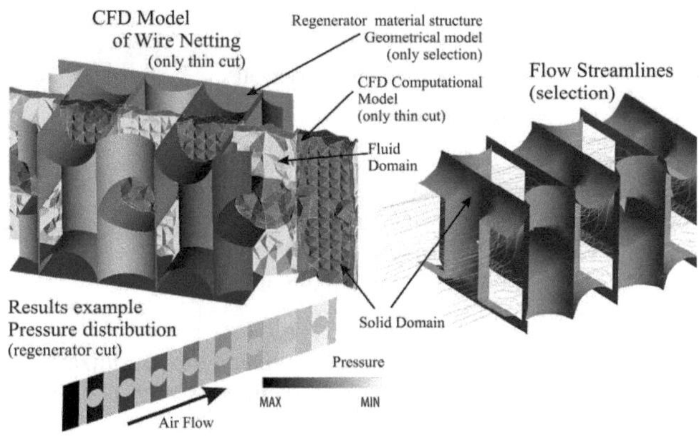

Fig. 4. CFD model and result examples of Wire Netting-type regenerator

Contemporary tools of Computational Fluid Dynamics (CFD Software) offer new possibilities of solving the problems of heat flow and transfer in Stirling engine regenerators. Figure 4 illustrates the example of wire netting-type regenerator material structure, its CFD model and examples of CFD results. Initial and boundary conditions (fluid flow velocities, temperatures etc.) for solution are obtained from CFD global model of Stirling engine.

6 Conclusion

The virtual prototypes including thermodynamic cycles or dynamics and strength of components of the Stirling engine are being created as higher-level computational models based on CFD, MBS of FEM models of physical processes occurring in real units, using only the minimum simplifying assumptions. New computational models are created after the necessary number of technical experiments is made. They will speed up the development of Stirling engines with better technical and economic parameters.

These computational models enable to compare different regenerator types like Wire Netting, Tubes, Stuffed wire mesh and Honeycomb ceramics relatively quickly. Different balancing of driving mechanism or designs of displacer pistons can be analyzed by these models as well.

Figure 5 presents the real prototype of Stirling engine with computed and measured brake power and total efficiency.

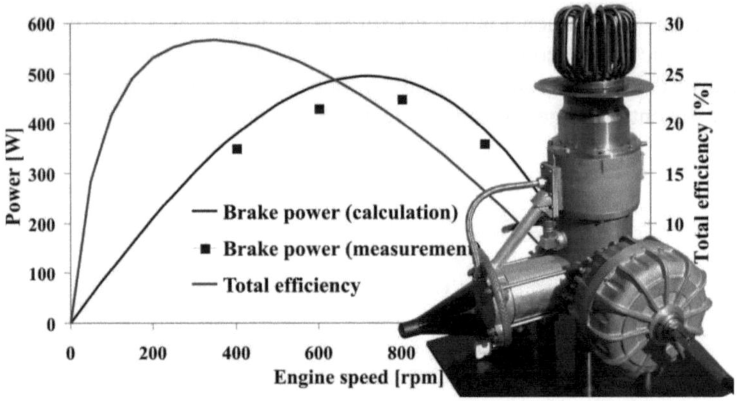

Fig. 5. Stirling engine prototype

Acknowledgments. Published results were acquired using the subsidization of the Ministry of Education, Youth and Sports of the Czech Republic, research plan MSM 0021630518 "Simulation modelling of mechatronic systems".

References

[1] Siegel, A.W.: Experimentalle Untersuchen zum Warmeubergang und Druckverlust im Regenerator einer Stirling-Machine. VDI VERLAG (2000) ISBN 3-18-312419-X

[2] Brezina, T.: Simulation Modeling of Mechatronics Systems II. Faculty of Mechanical Engineering, Brno University of Technology (2006) ISBN 80-214-3341-8

[3] Kaplan, Z., Novotný, P.: Thermodynamics of Stirling Engine, Engineering Mechanics, Brno (2005)

[4] Píštěk, V., Novotný, P.: CFD Tools in Stirling Engine Virtual Design. In: Mechatronics 2007, Poland (2007)

[5] Novotný, P., Píštěk, V.: Stirling Engine Development Using Virtual Prototypes. Acta Mechanica Slovaca (2008)

The Design of an Insulin Pump – Preliminary Requirements

H.J. Hawłas and K. Lewenstein

Warsaw University of Technology, Faculty of Mechatronics,
Institute of Metrology and Biomedical Engineering,
Św. Andrzeja Boboli 8, 02-525 Warszawa, Poland
h.hawlas@mchtr.pw.edu.pl

Abstract. The material presented in this paper is an attempt to lay down requirements for the planned design of an insulin pump. It seems important to establish proper requirements for a device before starting developing any design for an insulin dispensing device. This paper is a compilation of medical requirements along with user suggestions of presently offered insulin pumps. For a scientific point of view we are going to propose the closed loop insulin dispensing device with control built on the base of predictive neural network system. The blocks of such a systems are listed and main problems to overcome defined.

1 Introduction

An insulin pump is a device used for continuous dosage of insulin at a selected rate. An ordinary insulin pump is automatic drug-dispensing device that works in open loop mode. This means it exactly follows programmed drug dispending instructions with no any feedback form the patient body. Even with such easy job to do this device significantly facilitates treatment and improves the lives of diabetic patients.

In conventional insulin therapy [1], two types of insulin are needed: long-term and short-term insulin. Long-term type is used for daily insulin demand. Short-term type is fast acting insulin that is needed to reduce postprandial blood glucose (BG) level and its dose mainly depends on carbohydrates contents in meals.

An insulin pump uses only short-term insulin, which is distributed evenly during the day, so it works on a long-term basis (called basal), and a dose to counteract food (called bolus) that works on a short-term basis. Basic information along with an example of an insulin pump with an infusion set attached to its user may be seen in [2]. Pictures and parameters of the most popular insulin pumps may be found on the website of Diabetes Mall [3].

This small mechatronics device makes the life of patients more normal. It is especially important in the case of children. The youngest patients need to educate and socialize with their peers and to grow up in better conditions, which is decisive in their future life.

Unfortunately, insulin pumps are not easily affordable because of their price, as high as several monthly incomes of an average family in Poland. There is no manufacturer of such pumps in Poland, which is one of the reasons of their high price. Starting production in Poland would lower the price and make the device more affordable, which is one of the reasons why we started our design.

Our work is the result of the analysis of the insulin pump market [3] [4] [5] [6] [7] [8], various publications [9] [10] and user suggestions. The Internet diabetes forums were used as a source of insulin pump user opinions [11] [12]. The preliminary requirements are the compilation of a medical and user requirements along with existing product parameters.

There were two (as far as we know) attempts to design an insulin pomp in Poland, but so far there is no working model or prototype available.

From a scientific point of view we are going to propose an original and intelligent system of insulin administration with BG sensor closing the loop. This closed loop insulin dispensing device with control algorithm built on the basis of a predictable neural network system will fully or semi-automatically choose required drug dose.

2 Insulin Pumps

There are three types of such devices (a) hospital insulin pump, (b) personal insulin pump, (c) implantable insulin pump. This paper regards insulin pump of personal type, which is a pager-size unit. The whole unit and each of its blocks must fulfill some requirements, which may be classified as:

- Clinical / medical requirements, i.e. treatment parameters, biocompatibility, etc.
- Technical requirements, i.e. accuracy, power consumption, reliability, lifetime, etc.
- User requirements, i.e. usage, shape, color, etc.

Treatment requires that the person be connected to the insulin pump for nearly 24 hours a day, 7 days a week. Thus patient carries the pump with himself (or herself) mostly all the time and the pump acts almost like a part of his (or her) body. This indicates the importance of a well-thought out design.

Prior to the functional specifications there is mandatory requirement for reliable and safe operation. Health or even human life depends on the proper functioning of the device, so durability and reliability of an insulin pump is of the greatest importance. The device must have a monitoring system that ensures safe operation in case of hardware failure or improper usage. It is obvious that it must comply with various legal standards for a group of similar medical devices. Not only electrical and mechanical components have to be reliable but working algorithm has to be safe too. It is also important that the pump can undergo in-depth tests under various usage conditions.

Another requirement is miniaturization – an insulin pump must be a small compact device that may be carried all the time with the patient, so its size and weight are limited. Miniaturization also implies energy-efficient design that is dictated on one hand by lightweight and on the other hand by a long battery life and/or recharge cycle.

2.1 Main Requirements

Insulin pump is compact mechatronics device, which may be split into blocks as show on Fig. 1. Each of blocks or modules needs to be well defined for medical, technical and user requirements.

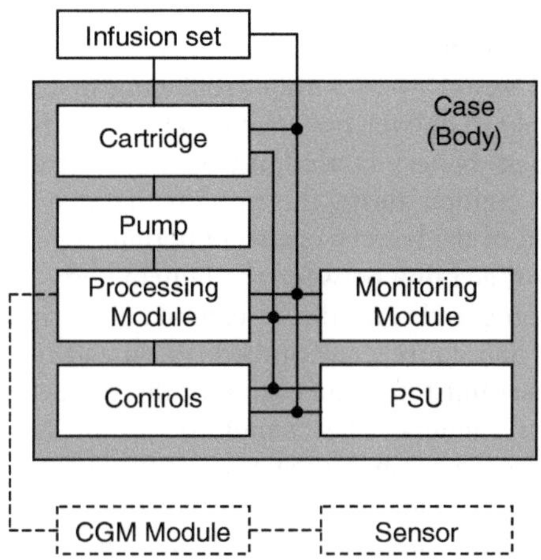

Fig. 1. Functional diagram of an insulin pump.

All of the requirements need to be established in consultation with medical specialists and users of insulin pumps.

Each functional block must be described with its technical, medical and user requirements; some of them are mandatory type but some may be treated as optional.

- Controls - The user interface must be easy to operate and must allow for both ways of communication with the user. The insulin pump is to be provided with a screen and keyboard, but also a sound and a vibration signalization are desirable. A wide range of user's age requires different pictogram menus for children and big symbols for elder persons. The keyboard itself should have a small number (about 4) of large keys. Keys have to be easily accessible and recognizable by touch even through clothing layers. Another requirement is that the screen must be readable in full sunlight as well as in night-light. The insulin pump requires some means of communication with a computer provided with necessary programming.
- Cartridge, i.e. a reservoir for insulin. Insulin from the cartridge goes into the blood system and that is why it is important that it be aseptic and sterile. The best option is that cartridges come preloaded with the required type and amount of insulin from pharmacist companies. However, most pumps require refilling

by users. There is no standard for insulin reservoirs used in the pumps so it may be good idea to establish such a standard. If refilling a cartridge is for some reason unavoidable, we need to supply the user with an easy to use and sterile disposable set.

- The Processing Module is a "brain" of the whole unit and is a master device for all other functional blocks. It is responsible for insulin pump options, dosage and communication with the user, programming or storing data. It may also accomplish some extra functions of an alarm clock or note taking. The stored data and configuration must remain undeleted when power is disconnected (when replacing the battery).
- The Power Supply Unit with batteries has to provide power to the insulin pump for at least several weeks (4-6) without replacing or recharging the battery. It also needs to be able to deliver information about the battery charge level. If a rechargeable (storage) battery is used, it has to be recharged by an external unit because of the heat emitted during this process that may affect loaded medicine. The size and weight of the battery is also important.
- The Infusion Set – It is a link between the human body and the insulin pump. It consists of thin tubing and a needle or cannula. The material from which it is made must be soft and not react with medication and blood or cause any allergies. The infusion set must have an option of disconnecting the tubing from the pomp and from the cannula. The cannula or needle has to be secured to the body by glue or tape. There should be a possibility of using an automatic insertion device that helps to place a needle in position and reducing the stress for some people who do not like to do it by hand. The design and materials used for the insertion set should allow for the tubing and the cannula to last as long as possible, because of the discomfort when connecting it and limited number of places on the body to insert it. A longlife infusion set eliminates the need to install a new set and refilling tubing and needle with insulin.
- The Pump – has to dose out medication from the cartridge with a preselected accuracy (about 5μl). It also has to work silently. The actuator used in the pump module must be energy efficient.
- The Monitoring Module may be a function of the Processing Module instead of being a device by itself. It has to monitor proper work of all blocks of the pump. It must detect any malfunction of the whole unit and also has to inform the user about it, e.g. about a disconnected infusion set. Another function of the Monitoring Module is to respond to circumstances that may impair the pump or affect the proper operation such as dropping the pump, overheating or exposing to radiation, etc.
- The Case, which due to the need for miniaturization is also a body of a device. It has to be esthetical, functional, ergonomic, waterproof and easy to clean and practical to fasten.
- The Continuous Glucose Monitoring (CGM) Module with a Sensor – an external device for measuring level of blood glucose. Our plan is that the glucose sensor would allow creating a closed loop control of insulin dosage making the insulin pump work almost as an artificial pancreas.

There are also many other requirements such as basal range, duration, frequency, or types of bolus. All the requirements will be discussed with medical specialists to select a proper range.

2.2 Patient Wish List

Some users would like to have some extra features in their pumps [13] among with the feature often mentioned is the remote control. Most patients prefer to have their pumps concealed on them. Pumps are quite often hidden under clothing and secured to underwear or taped to the body, so using a remote control would help to take a bolus or check an alarm when in public or when in haste. Remote control in the form of a hand watch is a solution suggested by some of them.

Another feature that users of the insulin pumps would like to have is the possibility of scanning meal for contents of carbohydrates leading to the automatic insulin dosage. Unfortunately, this feature will not be implemented in the first model because it still looks sci-fi. However, we can create a user-edited library with meals and their contents of carbohydrates and knowing approximate body response to insulin and carbohydrates than calculate the insulin bolus.

3 Conclusion

All the above-specified requirements will make it possible to design an insulin pump with all the necessary functions. This pump will be also easy to use and to maintain and most importantly safe to use. We hope that this work will result in a working model of an insulin pump. This model will be functionally and medically tested, so in the future we can build a final working device that will be commercially available alternatively to existing pumps. Our insulin pump will fill a hole in the market. The Polish market, not to mention the international market, is ready to absorb some 20 000 devices enough to start a large-scale production.

We have already started consultation with medical specialists to meet medical requirements of an insulin pump. In the future our design will enable us to intercompare typical insulin treatments to meet individual requirements using a planned closed loop insulin control based on a glucose sensor and a neural network system.

References

[1] Tatoń, J.: Cukrzyca wymagająca wstrzyknięć insuliny – Poradnik dla pacjentów, 1st edn., p. 424. Wydawnictwo Lekarskie PZWL, Warszawa (1998)
[2] Insulin pump, http://en.wikipedia.org/wiki/Insulin_pump
[3] Diabetes Mall - Comparison of Current Pumps, http://www.diabetesnet.com/diabetes_technology/insulin_pump_models.php
[4] Accu-Chek Polska Word Wide Web, http://www.accu-chek.pl
[5] Medtronic Word Wide Web, http://www.medtronic.pl
[6] Roche Diagnostics Word Wide Web, http://www.rochediagnostics.pl/

[7] Smith Medical Word Wide Web, http://www.smiths-medical.com/
[8] Soil Word Wide Web, http://www.sooilusa.com
[9] Lewenstein, K., Grabowski, M., Chojnacki, M.: A family of computer controlled infusion pumps – technology transfer from university to the small industry. Eng. Mechanics 12(3), 179–185 (2005)
[10] Lewenstein, K., Chojnacki, M.: Cooperation of the university and the small firm – a family of computer controlled infusion pumps. In: 4th Int. Conf. inter-academia 2005, Wuppertal, vol. 1, pp. 49–56 (September 2005)
[11] Cukrzyca i pompa insulinowa, http://www.pompiarze.pl
[12] The Diabetes Mall Forums, http://www.diabetesnet.com/forum
[13] Your Insulin Pump Proposals, http://www.diabeteshealth.com/read/2007/12/13/5586.html

Some Notes to the Design and Implementation of the Device for Cord Implants Tuning

T. Březina[1], O. Andrš[2], P. Houška[3], and L. Březina[4]

[1] Brno University of Technology, Faculty of Mechanical Engineering, Institute of Automation and Computer Science, Technicka 2896/2, Brno, Czech Republic
brezina@fme.vutbr.cz
[2] Brno University of Technology, Faculty of Mechanical Engineering, Institute of Automation and Computer Science, Technicka 2896/2, Brno, Czech Republic
yandrs01@stud.fme.vutbr.cz
[3] Brno University of Technology, Faculty of Mechanical Engineering, Institute of Automation and Computer Science, Technicka 2896/2, Brno, Czech Republic
houska.p@fme.vutbr.cz
[4] Brno University of Technology, Faculty of Mechanical Engineering, Institute of Solid Mechanics, Mechatronics and Biomechanics, Technicka 2896/2, Brno, Czech Republic
ybrezi00@stud.fme.vutbr.cz

Abstract. This contribution deals with a position control analysis of the biomechanical testing device based on the Stewart Platform. The constructed mechanism represents a six-degree of freedom positioning manipulator. Its closed kinematic chain and parallel linkage structure give it great rigidity and large load-to-weight ratio. That makes it suitable for testing of the backbone segments and hip joints. The mechanism constructed on FME BUT contains two plates (base and platform) which are connected with six linear mechanical actuators to the each other. The device control task is based on inverse position kinematics. A model of the mechanism inverse kinematics was built for the control purposes in MATLAB software. Finally the controller algorithm was designed and implemented into LabVIEW environment which is suitable for a real-time control.

1 Introduction

The cord implants tuning leads to numerous problems. It is necessary to optimize its engineering design (geometry, materials) as well as to verify the prototype function. This is of course done outside the human body. The proposed mechanism makes possible to simulate the physiological movements of the human body and observe on long–term horizon how the cord implant affect the spinal element.

2 Requirements for the Device

The following requirements have been defined with respect to previous analysis of relevant tasks [1]:

- The device must be able to load up the test specimen with the assigned loading cycle in six DOF with frequency of 0.5 Hz, the accuracy of positioning is without special demands.
- Load forces and moments were estimated, with values approx. 2000 N and 10 Nm.
- The device dimensions should be as small as possible.
- Due to assumed clinical application of the device an exploitation of electro – mechanical transmission is advised.

3 System Conception

The device must be capable of simulation of the physiological movements as close as possible. The demand for fastening the spinal segments determines the size of particular segment of the device. The device encircles the test specimens in layers. Single layer of the device is created of couple of toroidal bodies (base and platform) linked to each other. Both plates are coupled with linear controlled actuators which simulate movement and load of specimen. Delineate conception is inspired by mechanical functions of human organism. The concept of parallel mechanism called Stewart platform naturally corresponds to a single layer of such device.

4 System Design Methodology

The constructed device represents fully mechatronic system. The intended design solution for the development of mechatronic products is presented as V model with the industrial guideline – VDI 2206 [2]. There are three main design stages (Fig. 1).

4.1 Preliminary Design Phase

The device design was started with mechanical parts modeling and assembling in CAD software. Then the kinematic model based on the Inventor model was created in MATLAB. The base and platform are modeled as semiregular hexagons. The basic model of system dynamics was created in Simulink/SimMechanics. The model is suitable for basic verification of the machine behavior. There was designed a position control using a loop shaping synthesis for this simple model. According to numerous simulations it was confirmed the feasibility of the device, it was determined the workspace and derived preliminary loading of the links.

4.2 Middle Design Phase

This phase was characteristic with improving of the mechanism model. There was created detail construction of the base and platform shape and detail construction of links. The links consist of a ball screw with a nut which transforms the rotational movement to the translational, a spur gearing, Maxon motor RE 35 with a planetary gearbox and IRC sensor. This was followed by improving of the dynamic model in SimMechanics and dividing the control into two layers. The lower layer provides the control of linear actuators (links) using method of loop shaping synthesis and pole placement. The upper layer provides the actuators synchronization which is based on inverse kinematic solution. There was constructed a prototype of the linear actuator with control electronics [3] at the end of the phase. The improved position and torque control was designed for the prototype. These were implemented for PC and real-time PC in LabVIEW environment where were intensively tested.

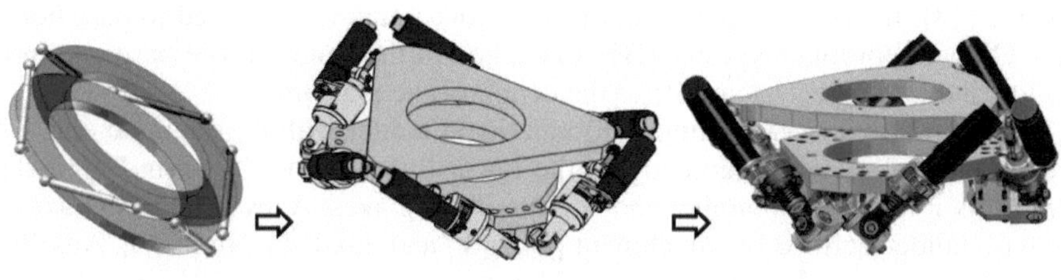

Fig. 1. Progressive design stages

4.3 Final Design Phase

There were made small construction adjustments. Important changes in the construction were caused by both the results of the experimental verification of the linear actuator prototypes behavior and its stress – strain analysis. The most important changes were made in the lower part of the actuator casing. The lower joint structure was modified as well as the whole part shaping because of the future assembling of the force sensors (strain gauges) and absolute position sensors.

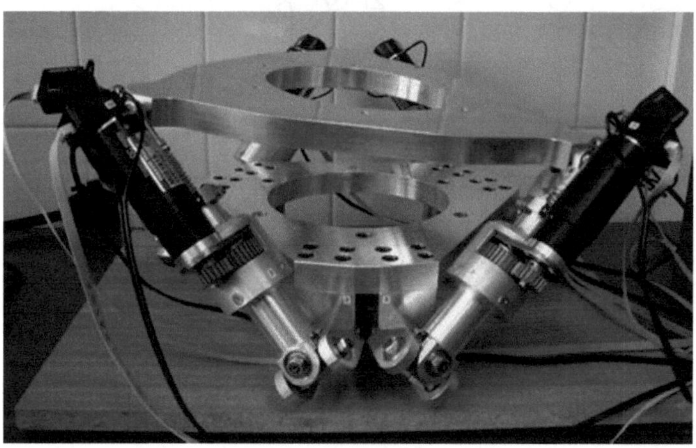

Fig. 2. The Stewart platform prototype

Several changes were also made in the control design – a state space controller with compensation of the disturbance seems to be the most suitable for the implementation in the lower control layer.

Finally the device prototype was assembled. There was necessary to conformed kinematic model to the real device (Fig. 2). Thus a new kinematic model was made.

5 Kinematic Models

The machine shown in Fig. 2 represents a full 6-DOF parallel mechanism. Because of computing complexity of an accurate model a simplified model was first designed [4]. Intersection points $\mathbf{B}_i$ of lower link joints were moved to base holder axis. Global coordinate system {B} is attached to the centre of the base and moving lo-cal coordinate system {P} to the centre of the platform.

It is necessary to verify simplified kinematic model with the accurate one. The accurate model assumes geometry of the lower link joint as shown in Fig. 3. The lower link joint has nonparallel and nonintersecting axes. Axes o_i represent axes of the base holder defined by attachment points $\mathbf{P}_i$ and auxiliary points $\mathbf{M}_i$. Auxiliary points $\mathbf{M}_i$ are derived from existing holder geometry. Intersection points $\mathbf{D}_i$ represent the axes of the second bearing of the lower link joints. Their positions in {B} are temporary unknown. Dimensions $\boldsymbol{c}_i$ are derived from the joints geometry. It is profitable to fix new local coordinate systems to attachment points $\mathbf{P}_i$. Orientation of new coordinate axes is shown in Fig. 3 according to (1).

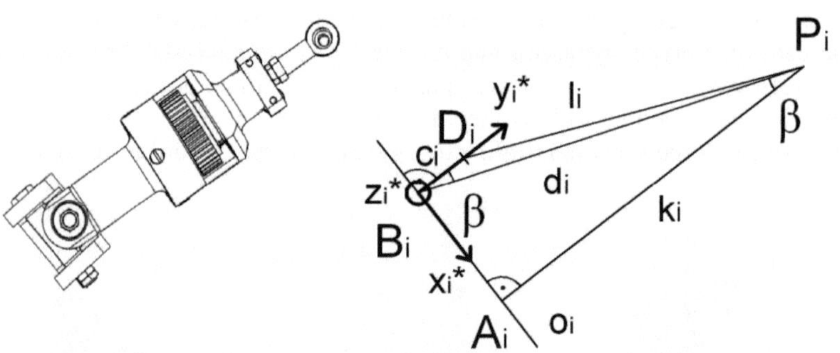

Fig. 3. Detail of the linear actuator

$$\begin{aligned} \mathbf{x}_i^* &= \mathbf{B}_i - \mathbf{M}_i \\ \mathbf{z}_i^* &= \mathbf{x}_i.\mathbf{B}_i\mathbf{P}_i \quad , \quad i = 1\ldots6 \\ \mathbf{y}_i^* &= \mathbf{z}_i^*.\mathbf{x}_i^* \end{aligned} \tag{1}$$

The position of intersection points $\mathbf{D}_i$ can be expressed in {B} by $\mathbf{T}_i$* the position vector of the origin of new local coordinate system relative to {B}, and rotation matrix $\mathbf{R}_i$*

$$\mathbf{R} = \begin{bmatrix} \frac{x_{xi}*}{|\mathbf{x}_i*|} & \frac{y_{xi}*}{|\mathbf{y}_i*|} & \frac{z_{xi}*}{|\mathbf{z}_i*|} \\ \frac{x_{yi}*}{|\mathbf{x}_i*|} & \frac{y_{yi}*}{|\mathbf{y}_i*|} & \frac{z_{yi}*}{|\mathbf{z}_i*|} \\ \frac{x_{zi}*}{|\mathbf{x}_i*|} & \frac{y_{zi}*}{|\mathbf{y}_i*|} & \frac{z_{zi}*}{|\mathbf{z}_i*|} \end{bmatrix}, i = 1...6 \quad (2)$$

The positions of the points $\mathbf{D}_i$ are then

$$\mathbf{D}_{i\{B\}} = \mathbf{D}_i*.\mathbf{R}_i*+\mathbf{T}_i*, i = 1...6 \quad (3)$$

Finally lengths of every link l_i can be expressed as

$$l_i = |P_iD_i| = \sqrt{(x_{Di} - x_{Pi})^2 + (y_{Di} - y_{Pi})^2 + (z_{Di} - z_{Pi})^2}, \ i = 1...6 \quad (4)$$

Finally simplified kinematics model was verified with the accurate model in Simulink. Combination of harmonic excitations was used as referential time course of the platform, see Fig. 4.

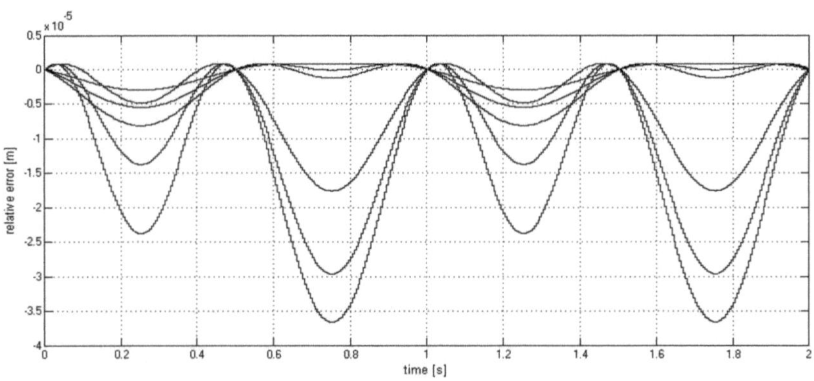

Fig. 4. Position deflection error

Note that the accuracy of the simplified kinematic model is acceptable and it is appropriate for testing on real machine.

6 Conclusion

There were shown main design stages of the biomechanical testing device based on the Stewart Platform. Then according to the simulation results was chosen the simplified model for the implementation in the mechanism position control system. The control system itself was implemented for PC and real-time PC in LabVIEW environment. Desired position and orientation of the platform is adjusted by six position controllers with an integrator on the input and with a disturbance (torque) observer. Nowadays the presented control system is tested and benchmarked on a real machine.

Acknowledgements. This work is supported from research plan MSM 0021630518 "Simulation modeling of mechatronic systems" and project 1P05ME7891 "Simulation of mechanical function of selected segments of human body".

References

[1] Brezina, T., Florian, Z., Caballero, A.A.: The design of the device for cord implants tuning. Recent Advances in Mechatronics (2007) ISBN: 978-3-540-73955-5
[2] VDI 2206. Design methodology for mechatronic systems. Beuth Verlag, Berlin (2004)
[3] Andrs, O., Brezina, L., Vetiska, J.: Position control implementation of a linear mechanical actuator. In: The proceedings of conference 12th International Symposium Mechatronika 2009, Trencianske Teplice, Slovakia (2009)
[4] Brezina, T., et al.: Simulation modelling of mechatronics systems I, BUT, Brno (2005) ISBN 80-214-3144-X

Controller Design of the Stewart Platform Linear Actuator

T. Březina and L. Březina

Brno University of Technology, Faculty of Mechanical Engineering,
Technicka 2896/2, Brno, Czech Republic
brezina@fme.vutbr.cz

Abstract. The article presents approach to the control of a linear actuator of the Stewart platform. The proposed method is based on the state – space control with disturbance compensation. The controller itself has a structure with an input integrator and the disturbance compensation is provided by a torque observer. The whole controller design is based on a linear state-space model of the actuator with the disturbance.

1 Introduction

A Stewart platform is very versatile mechanism which can be used in a large number of applications where its high stiffness and positioning preciseness are crucial apart from the limited workspace. Thereby the mechanism satisfies requirements on a biomechanical components testing machine which is its planned utilization. The construction of the presented mechanism is based on the time proven 6 – UPS topology [1]. Its movement is caused by six linear actuators where each of them contains a DC motor actuating a ball screw.

There was built a SimMechanics model [2] and designed a control which was simulated before the prototypes construction (a single linear actuator and the Stewart platform). The control was designed as a two-layer torque control and it was based on the linearized SimMechanics model [3]. The lower layer controls each of the linear actuators and upper layer synchronizes all of DC motors together. The control suitability was successfully verified by simulations of the control of the nonlinear model.

Nevertheless the original torque control was later replaced by presented position control with the torque observer. The reason is current impossibility of precise direct measurement of the torque produced by the linear actuator movement. Direct strain-gauge measurement of the torque implemented and tested does not guarantee obtaining of the value for the small platform loads yet.

2 State Space Representation of the Linear Actuator

The torque control of the Stewart platform presented in [3] was not developed because of the current impossibility of the produced torque measurement.

That was the reason for designing of the lower layer control with a torque observer. A state – space control with compensation of the disturbance seems to be the most suitable for the implementation in the lower control layer. Concretely it was used a combination of the control structure with an input integrator and compensation of the torque (disturbance) in the steady state, i.e. with the torque observer. Simplified linear state – space model with a disturbance (1) was used for the control design purpose.

$$\begin{aligned} \mathbf{x}' &= \mathbf{Ax} + \mathbf{Bu} + \mathbf{Ez} \\ \mathbf{y} &= \mathbf{Cx} + \mathbf{Du} \end{aligned} \tag{1}$$

A, **B**, **C**, **D**, **E** are in sequence the state matrix, the input matrix, the output matrix, the feedforward matrix and the disturbance matrix.

There was used usual model of the DC motor RE 35 with statically modeled linear actuator mechanism (only gear ratio motor shaft/ball screw) as the initial model (2).

$$\begin{aligned} \begin{bmatrix} \varphi' \\ \omega' \\ i' \end{bmatrix} &= \begin{bmatrix} 0 & 1 & 0 \\ 0 & -k_f/J & k_M/J \\ 0 & -k_b/L & -R/L \end{bmatrix} \begin{bmatrix} \varphi \\ \omega \\ i \end{bmatrix} + \begin{bmatrix} 0 \\ 0 \\ 1/L \end{bmatrix} [u_M] + \begin{bmatrix} 0 \\ 1/(nJ) \\ 0 \end{bmatrix} [M_{ZD}] \\ \begin{bmatrix} \varphi_M \\ \omega_M \\ i_M \\ M_M \\ \varphi_D \\ M_D \end{bmatrix} &= \begin{bmatrix} 1 & 0 & 0 \\ 0 & 1 & 0 \\ 0 & 0 & 1 \\ 0 & -k_f & k_M \\ 1/n & 0 & 0 \\ 0 & -nk_f & nk_M \end{bmatrix} \begin{bmatrix} \varphi \\ \omega \\ i \end{bmatrix} + \begin{bmatrix} 0 \\ 0 \\ 0 \end{bmatrix} [u_M] \end{aligned} \tag{2}$$

Where k_f, k_m, k_b, R, L, J are the motor parameters and n is the gear ratio between rotation movement of the motor shaft and the ball screw. The state vector **x** represents angular displacement of the rotor φ, angular velocity ω and electrical current in the stator i, $\mathbf{x} = [\ \varphi\ \omega\ i]^{\mathrm{T}}$, the vector of inputs $\mathbf{u} = [u_M]$ presents motor driving voltage u_M. The vector of model outputs contains angular displacement of the rotor φ_M, angular velocity ω_M, electrical current in the stator i_M, motor shaft torque M_M, angular displacement of the ball screw φ_D and its torque M_D. The disturbance vector $\mathbf{z} = [M_{ZD}]$ presents the ball screw loading torque and the compensator of the disturbance works with the only gain k_Z ($\mathbf{K_Z} = [k_Z]$).

3 The Controller Design

Implementation of the structure with the integrator on the input combined with the steady state disturbance compensation leads to the following control law

$$\begin{aligned} \mathbf{u} &= -\mathbf{Rx} + \mathbf{r}_i \mathbf{v} - \mathbf{K}_Z \mathbf{z} \\ \mathbf{v}' &= \mathbf{w} - \mathbf{y} \end{aligned}, \tag{3}$$

where $\mathbf{R} = [r_\varphi\ r_\omega\ r_i]$ is the vector of gains of the system's state vector.

Presented control law modifies the state – space model of the system to the state – space equation:

$$\begin{bmatrix}\mathbf{x}'\\ v'\end{bmatrix}=\left(\begin{bmatrix}\mathbf{A} & \mathbf{0}\\ -\mathbf{C} & 0\end{bmatrix}-\begin{bmatrix}\mathbf{B}\\ -\mathbf{D}\end{bmatrix}\begin{bmatrix}\mathbf{R} & -\mathbf{r}_i\end{bmatrix}\right)\begin{bmatrix}\mathbf{x}\\ \mathbf{v}\end{bmatrix}+\begin{bmatrix}\mathbf{0}\\ 1\end{bmatrix}[\mathbf{w}]+\left(\begin{bmatrix}\mathbf{E}\\ \mathbf{0}\end{bmatrix}-\begin{bmatrix}\mathbf{B}\\ -\mathbf{D}\end{bmatrix}\mathbf{K}_z\right)[\mathbf{z}] \tag{4}$$

The system with the integrator of the regulation deviation of the rotor/ball screw angular displacement, i.e. for $\mathbf{y} = [\varphi_M]$ or $\mathbf{y} = [\varphi_D]$, is controllable. The controller gains $\mathbf{R}$ and the integrator gains $\mathbf{r_i} = [r_{i\varphi}]$ are adjusted in such way that state matrix of the system with the integrator

$$\begin{bmatrix}\mathbf{A} & \mathbf{0}\\ -\mathbf{C} & 0\end{bmatrix}-\begin{bmatrix}\mathbf{B}\\ -\mathbf{D}\end{bmatrix}\begin{bmatrix}\mathbf{R} & -\mathbf{r}_i\end{bmatrix} \tag{5}$$

has specified eigenvalues p_1, p_2, p_3 and p_4, providing stability of the system for the sufficient velocity, in allowed control values range and with given power limitations.

It must be valid for the state equation without the integrator (i.e. with the control law $\mathbf{u} = \mathbf{-Rx - K_z z}$)

$$\mathbf{x}'=(\mathbf{A}-\mathbf{BR})\mathbf{x}+\mathbf{Bw}+(\mathbf{E}-\mathbf{BK}_z)\mathbf{z} \tag{6}$$

$$(\mathbf{A}-\mathbf{BR})\mathbf{x}=-(\mathbf{E}-\mathbf{BK}_z)\mathbf{z}, \tag{7}$$

if it is desired that disturbance should not move the system from the steady state $\mathbf{x}' = \mathbf{0}$ for the zero referential value $\mathbf{w} = \mathbf{0}$.

This is accomplished in our case for

$$\omega=0, i=-\frac{M_Z}{k_M}, \varphi=-\frac{1}{r_1}\left(k_Z-\frac{R+r_i}{k_M}\right)M_Z. \tag{8}$$

Thus for $k_Z=\dfrac{R+r_i}{k_M}$ is $\varphi = \omega = 0$ and the disturbance will not influence the steady angular displacement and the system steady state. It is necessary to take into account that in practice it is valid for the maximal loading torque that $|M_{Z\max}| \le k_M |i_{\max}|$.

It was already written that the disturbance observer was used for the torque M_{ZD} estimation. Its structure is based on the system model which simulates the system behavior. Eventual variances of the system outputs $\mathbf{y}$ from the model outputs $\hat{\mathbf{y}}$ are corrected in the model via the variance of its states $\mathbf{\Delta x}$ which linearly depends on the variance of the outputs $\mathbf{y}-\hat{\mathbf{y}}$, i.e. $\Delta\mathbf{x}=\mathbf{H}(\mathbf{y}-\hat{\mathbf{y}})$. The structure of the observer is extended with the part for the disturbance estimation for the construction of the disturbance observer. This part works on the similar principle as the structure with the integrator on the input. The output of the system's model is corrected inside the observer to correspond with the real output of the system. The correction is based on the variances between outputs, on the integration of variances between outputs and the last contribution is the disturbance estimation.

Than it is obtained

$$\begin{aligned}\hat{\mathbf{x}}' &= \mathbf{A}\hat{\mathbf{x}} + \mathbf{B}\mathbf{u} + \Delta\mathbf{x} + \mathbf{E}\hat{\mathbf{z}}\\ \hat{\mathbf{y}} &= \mathbf{C}\hat{\mathbf{x}} + \mathbf{D}\mathbf{u}\\ \Delta\mathbf{x} &= \mathbf{H}(\mathbf{y} - \hat{\mathbf{y}})\\ \hat{\mathbf{z}}' &= \mathbf{K}_i(\mathbf{y} - \hat{\mathbf{y}})\\ \hat{\mathbf{y}}_z &= \mathbf{I}\hat{\mathbf{z}}\end{aligned} \tag{9}$$

A, **B**, **C** and **D** are matrices of the state – space model of the observed system, $\hat{\mathbf{x}}'$ is the vector of the estimated system states, y (used $\mathbf{y} = [\varphi_M \quad i_M]$), or $\hat{\mathbf{y}}$ is the vector of the observed or estimated system outputs, $\hat{\mathbf{z}}$ is the estimated vector of the disturbance states and finally $\hat{\mathbf{y}}_z = [M_{ZD}]$ is the vector of the estimated disturbance.

The state – space representation of the disturbance observer is then

$$\begin{aligned}\begin{bmatrix}\hat{\mathbf{x}}'\\ \hat{\mathbf{z}}'\end{bmatrix} &= \left(\begin{bmatrix}\mathbf{A} & \mathbf{E}\\ \mathbf{0} & \mathbf{0}\end{bmatrix}^T - [\mathbf{C} \quad \mathbf{0}]^T \begin{bmatrix}\mathbf{H}\\ \mathbf{K}_i\end{bmatrix}^T\right)^T \begin{bmatrix}\hat{\mathbf{x}}\\ \hat{\mathbf{z}}\end{bmatrix} + \left(\begin{bmatrix}\mathbf{B} & \mathbf{0}\\ \mathbf{0} & \mathbf{0}\end{bmatrix} - \begin{bmatrix}\mathbf{H}\\ \mathbf{K}_i\end{bmatrix}[\mathbf{D} \quad \mathbf{I}]\right)\begin{bmatrix}\mathbf{u}\\ \mathbf{y}\end{bmatrix}.\\ \hat{\mathbf{y}}_Z &= [\mathbf{0} \quad \mathbf{I}]\begin{bmatrix}\hat{\mathbf{x}}\\ \hat{\mathbf{z}}\end{bmatrix} + [\mathbf{0} \quad \mathbf{0}]\begin{bmatrix}\mathbf{u}\\ \mathbf{y}\end{bmatrix}\end{aligned} \tag{10}$$

Values of **H**, $\mathbf{K_i}$ matrices were established in such way that state matrix

$$\begin{bmatrix}\mathbf{A} & \mathbf{E}\\ \mathbf{0} & \mathbf{0}\end{bmatrix}^T - [\mathbf{C} \quad \mathbf{0}]^T \begin{bmatrix}\mathbf{H}\\ \mathbf{K}_i\end{bmatrix}^T \tag{11}$$

has specified eigenvalues p_1, p_2, p_3 and p_4, providing stability of the system for the sufficient velocity.

The control law is then

$$u = -r_\varphi\left(\varphi_{ref} - \varphi_D\right) - r\hat{\omega}_D - r_i i_M + r_{i\varphi}\int\left(\varphi_{ref} - \varphi_D\right)d\tau + k_Z\hat{M}_{DZ} \tag{12}$$

where the loading torque $\hat{M}_{DZ}$ is estimated by the disturbance observer and the ball screw angular velocity estimation $\hat{\omega}_D$ is obtained from two – points approximation of the ball screw angular displacement derivation φ_D. The proposed control behaves better with this estimation than with the $\hat{\omega}_D$ estimation using the state – space observer.

The position and the electrical current sensor were modeled as first order systems (13), (14) for the examination of the sensors influence over the behavior of the controlled system.

$$\varphi'_{Ds} = -1/T_\varphi \, \varphi_{Ds} + K_\varphi/T_\varphi \, \varphi_D \tag{13}$$

$$i'_{Ds} = -1/T_i \, i_{Ds} + K_i/T_i \, i_D \tag{14}$$

Where φ_{Ds}, i_{Ds} are values generated by sensors.

Its influence appeared as unimportant and therefore it was not necessary to compensate it. There was used a converter model in the numerical simulation. Its influence over the controlled system behavior appeared as more important.

The lower control layer simulation with the torque (disturbance) observer was implemented in Matlab – Simulink. Fig. 1 and Fig.2 documents the impact of the step change in M_Z on the position response φ_D.

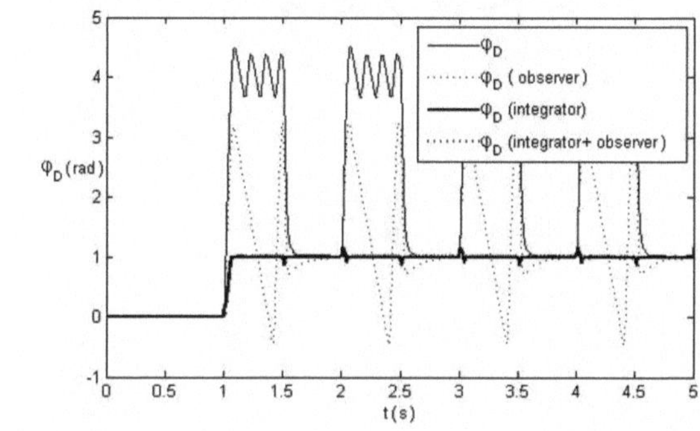

Fig. 1. The impact of the step change in M_Z on the position response φ_D

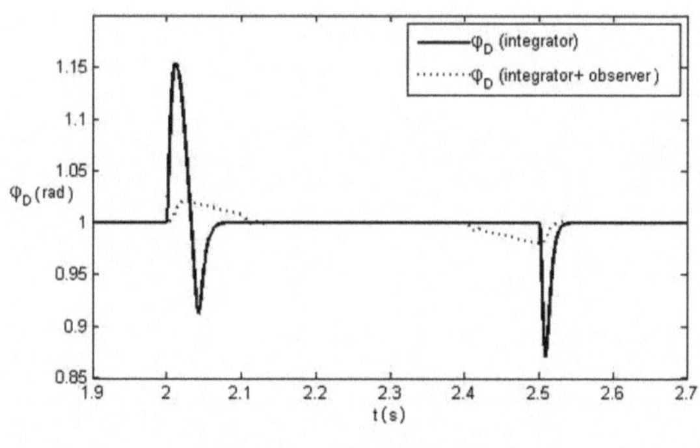

Fig. 2. Detail of the Fig. 1

4 Conclusion

The proposed method is based on a control structure with an input integrator and with a torque observer. The simulation proved that the controller is able to compensate the additional dynamics of the sensors and the converter. The controller is ready for implementation on the prototype of the Stewart platform.

Acknowledgments. This work is supported from research plan MSM 0021630518 "Simulation modelling of mechatronic systems" and project 1P05ME789 "Simulation of mechanical function of selected segments of human body".

References

[1] Merlet, J.P.: Parallel robots. Springer, Heidelberg (2006)
[2] Brezina, L., Andrs, O., Brezina, T.: NI LabView – Matlab SimMechanics Stewart platform design. Applied and Computational Mechanics 2, 235–242 (2008)
[3] Brezina, L., Brezina, T.: The device for implants testing: The control. In: Proceedings of Engineering mechanics, Svratka, pp. 84–89 (2008)

Design and Implementation of the Absolute Linear Position Sensor for the Stewart Platform

P. Houška, T. Březina, and L. Březina

Brno University of Technology, Faculty of Mechanical Engineering,
Institute of Automation and Computer Science, Technicka 2896/2, Brno, Czech Republic
houska.p@fme.vutbr.cz

Abstract. The article deals with design and implementation of absolute linear position sensor for sensing of a ball screw position. The ball screw is one of the main parts in a linear actuator of the "Device for mechanical assessment of human body functional segments". The proposed sensor is based on principle of sensing a position of a permanent magnet of family Hall IC (integrated circuit). The magnet is attached to the endstop of the moving ball screw. PCB (printed circuit board) with Hall IC are mounted to the casing of the linear actuator. Data acquisition is executed by an 8-bit microcontroller which provides the actual position of the ball screw via serial communication interface.

1 Introduction

Experimental devices for different tasks in field of biomechanics are developed in laboratories of Brno University of Technology, Faculty of Mechanical Engineering. An universal experimental device has been developed here for last four years [1]. The advantage of the device should be the possibility of usage in wide rank of problems. Previous devices built on BUT were developed for specific task, e.g. for loading tests of two spinal segments. Problems solved in biomechanics area are more and more complex thus it is necessary to build a device which will be more universal. That was the main reason for starting the development of the "Device for mechanical assessment of human body functional segments". The device (Fig. 1) is made of two plates linked together with six linear actuators. The actuators must be compact, high-performance and sufficiently accurate to meet the demands on the whole device. There was found no commercially available linear actuator satisfying the requests. Hence the development of own linear actuator began. The mechatronical approach of design of the actuator allowed us to design and implement the actuator exactly meeting our demands.

2 The Linear Actuator

The linear actuator is basically solved as a ball screw with rotating screw nut actuated by a DC motor. The sensory part of the actuator is represented by an incremental encoder connected to the DC motor shaft. This is used for the position control of the actuator. The second sensor is the absolute linear position sensor which is necessary for the control of reached position when starting the device and also during the operation. Finally the third sensor is the force sensor which is made of four strain-gauges connected to the full bridge. The sensor is glued-on the body of the actuator.

The used incremental encoder is HEDM 5540B type. The two remaining sensors were developed simultaneously with the actuator in such way to reach the high compactness of the whole actuator and at the same time to keep the actuator dimensions in reasonable area.

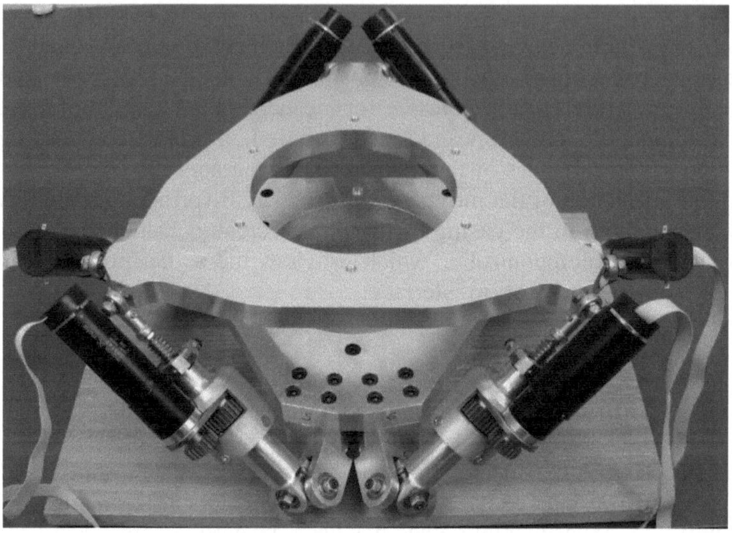

Fig. 1. Realized Stewart platform

3 Selection of the Linear Sensor

The required parameters of the absolute linear sensor (table 1) were obtained by successive iteration of the design steps of the linear actuator. A market research was accomplished according to the requests and no suitable commercial sensor was found.

Hence the own approach to design the absolute sensor was chosen. It was decided to use the principle of sensing the permanent magnet position by a magnetic field of Hall IC. The magnet is attached to the endstop of the moving ball screw. As was already mentioned the PCB with Hall IC are mounted to the casing of the linear actuator (Fig. 2). The data acquisition is provided by an 8-bit microcontroller. The measured actual position is obtained via the serial communication interface.

Table 1. Required parameters of the absolute linear sensor

Operating range	32mm
Measurement accuracy	0.02mm
Maximal dimensions of the sensor	34x28x20mm
Minimal life-cycle	$1.5\,10^8$ cycles
Working temperature range	10 to 50° C

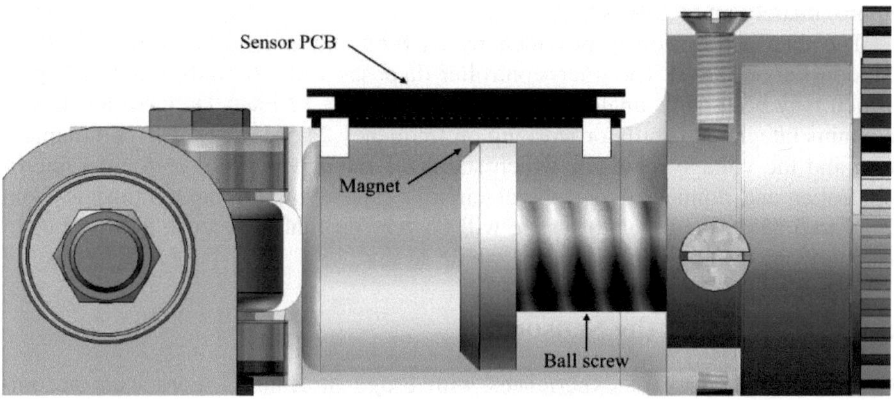

Fig. 2. The sensor placement on the linear actuator of the Stewart platform

4 Selection of the Sensor Elements

The magnet dimensions are chosen in such way to satisfy three following requirements. The magnet must be as close as possible to the inner surface of the actuator casing, it must be as small as possible, and finally it must influence by its magnetic field the ball screw and other iron parts of the actuator as few as possible. The width of the actuator casing in place of the sensor attachment is 1.5mm. The minimal distance between the magnet and sensor can not be smaller. Hence the material of the magnet should have the magnetic induction as high as possible and the working temperature higher than 50° C. Consequently a NdFeB magnet from N35 material quality class was selected [2]. Dimensions of the magnet are Ø3x4mm. Other parameters of the material are in the table 2.

A1301 type by Allegro MicroSystems Inc. was selected as a Hall IC [2]. The A1301 has a linear voltage output with sensitivity 0.25mV/mT. The IC sensitivity is temperature and driving voltage V_{cc} dependent. It is 0.248mV/mT for the desired temperature range (table 1). The temperature dependence is negligible in this case (approximately 0.4%). The output voltage V_{out} of the sensor is in direct ratio with the driving voltage V_{cc}. Zero value of the output, corresponding with measured inductivity 0T, equals the half of the driving voltage $V_{cc}/2$. The sensitivity

Table 2. Properties of NdFeB magnet N35

Magnetic remanence	1180 to 1250 mT
Maximal working temperature	80° C
Temperature coefficient Br	$-1.2\,10^{-3}\ {}^{\circ}C^{-1}$
Driving voltage V_{cc}	4.5 to 5.5V

is also driving voltage dependent. The change caused by this dependency might be approximately up to 25%/V.

The data acquisition is provided by an 8-bit microcontroller C8051F530 by Silicon Laboratories. The microcontroller disposes with 16 configurable I/O pins which may be used as analog switchable inputs of a 12-bit A/D converter. It also contains integrated oscillator working on frequency 24.5MHz, UART communication and the voltage regulator which stabilizes the driving voltage for the microcontroller and connected peripheral circuits. The regulator input range is from 2.7V to 5.25V. The regulator usage will be very profitable because it allows us to use only one driving voltage for the whole sensor.

5 Realization of the Sensor

Considering the lack of experiences with usage of Hall IC there were accomplished several experiments for the verification of the sensor feasibility. There were realized three testing PCB complemented with four Hall ICs. The first PCB was complemented with Hall ICs with distance 2mm between each other, the second 3mm between each other and the third 4 mm between each other. Complemented PCBs were in sequence mounted to the test jig with linear slide-way Alu-Rol AD208R. A holder with the selected magnet was attached to the cart of the linear slide-way. The position of the cart was changed and measured by a micrometer head Schur with resolution 0.002mm and accuracy 0.003mm. The data acquisition was provided by multifunctional DAQ card NI PCIe 6251 and for the software implementation it was NI LabVIEW.

There were accomplished several experiments with this configuration and with all of PCBs mounted in distance of 0.5, 1.5 a 2mm from the magnet and in distance of 2mm from a 1.5mm thick dural plate inserted between the magnet and sensors. This corresponds with the real configuration of the sensor and actuator. The real characteristics of the Hall IC and used magnet were obtained from the experiment. The characteristics were used for the modeling of the sensor in NI LabVIEW.

The location of Hall IC on the PCB was designed according to the model in such way to keep the sensor accuracy the highest possible. The result is usage of 11 circuits with spacing 3mm placed in a line. The sensor was devided into two PCBs because of requirements on its dimensions. The first PCB with Hall IC has dimensions 18x34mm. The second PCB with the microcontroller, supporting circuits, communication and programming connector (Fig. 3) is connected in parallel to the first one.

Fig. 3. Realized linear absolute position sensor

The data acquisition proceeds continuously. The maximum sampling period of the 12-bit A/D converter is 170kHz and corresponds with the recommendations of the manufacturer. The Hall IC output voltage is done eight times immediately in sequence. A 15-bit value where the lowest four bits contains the noise is obtained. Measuring of the grounded input GND is accomplished before switching the multiplex of the A/D converter to the next input. The reason is to discharge the electric charge accumulated in the input circuits of the converter and suppress a "mirror effect". The data acquisition form all of Hall ICs proceeds with frequency 1.7kHz and the data from each Hall IC are saved to a field for the further storage.

The calculation of the resulting position and data processing is in parallel with the measuring.

6 Algorithm of the Sensor Data Processing

The algorithm has to be most efficient because the fast 8-bit microcontroller with no mathematical coprocessor must provide sufficient computational power. That is the reason of writing the most of operation in terms using the fixed-point mathematics which is on such devices much faster than floating-point mathematics. The implementation is in the C language.

The algorithm proceeds in following steps. The circuit with the highest and the lowest excitation are selected from the collected data In the first step. These circuits are the closest to the north and south pole of the magnet. The one with the most or the least excited neighbor is selected then. This leads to the particular elimination of the method noise sensitivity because the chosen values are the most distant from Hall IC output zero value. It is possible to execute the method without calibrating because the Hall IC output zero value is exactly the half of the circuit voltage which is same for all elements and only source voltage dependent.

The calibration of the selected pair of the values is made in the second step. The magnet position against the Hall IC pair is solved in the third step. The absolute value of the position against the beginning of the sensor is computed in the fourth step. The beginning of the sensor is set during the calibration.

The calibration is the most important step of the algorithm. It was accomplished linearly for the first realized sensor and the reached repeated accuracy is ±0.03mm. According to experiments the polynomial of the third order should be sufficient for the desired accuracy and computational time.

7 Conclusion

The article describes the design and implementation of the absolute linear position sensor of the linear actuator. The realized sensor is compact and integrated to the linear actuator. The sensor accuracy is given by the accuracy of the calibration algorithm of the each Hall IC outputs. The advantage of using the ICs is their insignificant temperature dependency in range of working temperatures. Only the permanent magnet is temperature dependent in this sensor now. However the used algorithm is immune against this influence. The biggest challenge is the improving of the preciseness of the sensor calibration and automation of the process.

Acknowledgments. This work is supported from research plan MSM 0021630518 "Simulation modelling of mechatronic systems" and project 1P05ME789 "Simulation of mechanical function of selected segments of human body".

References

[1] Březina, T., Houška, P., Březina, L., Andrš, O.: Device for Experimental Modelling of Biomechanical System Properties. Simulation modelling of mechatronic systems III 3, 231–242 (2007)

[2] Gilbert, J., Dewey, R.: Linear Hall-effect sensors, Application note 27702A, Allegro MicroSystems, Inc. (1998)

A Touch Panel with the Editing Software and Multimedia Data Base

M. Skotnicki[1], K. Lewenstein[1], and M. Bodnicki[2]

[1] Warsaw University of Technology, Faculty of Mechatronics,
Institute of Metrology and Bioengineering, 8 Sw. A. Boboli Str., 02-525 Warszawa, Poland
m.skotnicki@mchtr.pw.edu.pl, k.lewenstein@mchtr.pw.edu.pl
[2] Warsaw University of Technology, Faculty of Mechatronics,
Institute of Micromechanics and Photonics, 8 Sw. A. Boboli Str., 02-525 Warszawa, Poland
m.bodnicki@mchtr.pw.edu.pl

Abstract. The project described in the paper concern the construction of the information Touch panel with the panel's software and the data base. At first we formulate the assumptions for the project as follows: developing the data base, as well as the universal application, which in an attractive way present information and multimedia objects from the data base, creating a software which enables adding and editing information as well as integrating the formed applications with the touch panel. There is planned to use the panel as the element of the information center in the hall of the Faculty building. The most valuable feature of the panel is that there's no need of any reception help. All information which our guest needs is available in self service.

1 Touch Panel – Design Forms

First of all we tried to buy such a system but the market offer was difficult to accept. There were a lot of examples of some factory models with prices from about 2500 Euro to 20000 Euro [1,2]. Some of them are presented on Fig. 1; there is also the cheapest model Alltrim Monolith 17' (with the smallest 17' screen). Taking above into account we concluded that the best solution is to design own casing, because in this case we can obtain the cut of the final cost and get the solution witch exactly fits to our needs.

We analyzed three possible forms of the touch panel: free standing, wall mounted and desk. The last one can be used only in the case of gastronomy and hotel trade, and is not useful in advertising (due to the defects of the two other forms), so we exclude it. In the Table 1 below we present advantages and disadvantages of the standing and wall design.

Table 1. Advantages and disadvantages of the standing and wall design

Adv/Disadv (+/–)	Free standing panel	Wall mounted panel
+	the most popular form	takes little space
+	diversity of shapes	simple casing (cheap production)
+	can be located in any place	large screens (42'' and more)
–	mostly quite small screen	flat surface is required
–	expensive in production	large screen increases the costs

Fig. 1. Forms of the touch panel: free standing, desk and wall mounted.

On base the above characteristics we chose that the best solution will be the construction which can be used as a standing kiosk as well as a wall panel.

Touch screen technologies

The most expensive and sophisticated panel's element is the touch screen, which could be done on several technologies. Below we'd like to give a short review of the functioning principles of the most popular screen technologies.

Resistive technology

Resistive technology works as follows: at the glass surface there is located a conductive layer separated by an isolator from the second conductive layer (put on the polyester membrane). Polyester membrane due to its durability protects at the same time the screen from the damages. The isolator and the conductive layers are transparent. Touching the screen results in shorting of the two conductive layers and voltage created in this way is the analogue representation of the place of touching. The touch screen controller changes the analogue signal into the digital information about the position of touch and sends it to the computer.

Capacitive technology

On capacitive technology the working area of the screen is made of two layers of glass, containing a network of sensors between them, which react to the changes in the electrical capacity (content). Touching the screen results in the change of the electrical capacity and the information about this is provided by the sensors to the screen controllers. The touch screen controller conveys data about the position of touching on the basis of the achieved information.

Infrared technology

On infrared technology the working area of the screen is made of the pure glass. In the edges of the screen there are placed IR diodes, vis-à-vis sensors reacting to infrared. Touching the screen causes a blockade/blocking of the rays in the axis x and y, which results in giving information about the place of touching.

Surface Acoustic Wave

The surface acoustic wave needs the working area of the screen made of the pure glass. In the edges of the screen there are put converters emitting and receiving ultrasounds and series of the ultrasound reflection generators, which are combined with the glass. The converters emit and receive ultrasound waves, which are dissipated on the screen via the generators. It results in creating the two axes: X and Y. Touching the screen causes absorption of a part of waves and in this way a shadow of the ultrasound wave is created. A touch screen controller reads the information passed by the converters and compares them with the digital map of the reflections programmed in the controller.

Acoustic Pulse Recognition

In Acoustic Pulse Recognition method the screen is also made of the pure glass. The edges of the screen contain four piezoelectric converters, which process the acoustic waves into the digital signal, understood by the computer. Touching the screen causes appearing of the acoustic wave, which approach the converters with different intensity. The converters change the received waves into the digital signal and send it to the tactile screen controllers. The information gained are then compared with/to the matrix of the sounds programmed in the controller during production.

Dispersive Signal Technology

Dispersive Signal Technology needs the screen consisting on the chemically amplified/ strengthen glass. In every corner of the screen there are piezoelectric sensors connected to/with the controller. Sensors measure the oscillatory energy of the surface vibrations caused by touch.

Strain Gauge

Strain Gauge is similar to the above described. the screen is fixed elastically in the corners, strain gauges have to be used in order to detect strain.

Optical Imaging

Optical Imaging system has two miniature cameras are put in the corners, they are supposed to "see" the finger or scriber with the use of the scanning surface along the screen. The frame is enlighten by the IR light. Touch creates a shadow on the basis of which its position can be defined.

The touch screen, which is exposed to use by many people, has to be resistant to dirt – its work cannot be disturbed by e.g. liquid drops on the glass. Due to the data being projected (large amount of text, which has to be rewind) the static objects have to be detectable such as a motionless finger on the rewind strip/button. These required features are in different way realized by the technologies of the touch screens, what is illustrated in Table 2.

Table 2. Comparison between different techniques used in touch panels design

Technology	Resistance to dirt	Detecting static objects
Resistive	YES	YES
Capacitive	YES	YES
Infrared	NO	YES
Surface Acoustic Wave	NO	YES
Acoustic Pulse Recognition	YES	NO
Dispersive Signal	YES	NO
Strain Gauge	YES	YES
Optical Imaging	YES	YES

Analysis of the Table 2. eliminates four technologies so we need some additional advantages for choosing the final technology of the screen. This is shown in Table 3. However working through different surfaces can easily replace high durability – covering the surface by a thick glass would be enough.

Table 3. Detailed analysis of the selected techniques

Technology	High sensitivity	Working through different surfaces	Durability	Low cost
Resistive	YES			YES
Capacitive	YES	YES		
Strain Gauge			YES	
Optical Imaging		YES		YES

Taking above into account we decided to use cheap and sensitive resistive technology with minimum 20 inches diagonal of the touch screen.

2 The Software of Touch Panel

The software of the touch panel consists on two main parts; firstly the database and editing software and the secondly the panel's software.

The data base we'd like to use is based on the .xml and .txt files. There should be a possibility of opening the application directly from the pen drive, without installing the extra software of data base.

The content of the database should include:

- the list of the attached multimedia files (photos, films) together with their descriptions and position in the catalogue;
- a set of all the informative texts about the Faculty (specialization descriptions, information about recruitment, scientific organizations etc.);
- connections between information from points 1 and 2 (assigning photos to the particular texts, table of contents, specializations led by particular institutes etc.) and already existing patterns of connections (e.g. during introducing the list of specializations, program should ask whether to put the links with their description too or not)

Editing software should provide that the program enables looking through the already existing multimedia data base, its edition and adding new content, as well as marking the content which is supposed to be put in the current presentation and creating menu. As a result of the working program a .xml file will be generated and it will contain all texts and connections as well as the copies of all the files connected with these texts.

We would like to use C++ programming language or applet in flex/air technology (in order to guarantee compatibility) and intuitive interface making the software easy in use. The interface has also to be convenient in use what means that it has to be served by the touch screen as well as by the use of mouse. All the buttons have to be big enough and distant from each other in order to eliminate the risk of pressing the wrong button because of the wrong read-out of finger position on the screen.

The presentation should be visually attractive making that the panel should in some way encourage people to come and read something.

Format of the panel's software ensure presentation created in Flash, could be put in the older format of Flash (cs2) in order to guarantee compatibility on the Unix systems and to the full screen projector (exe). Except installing in the kiosk the software will also be distributed on the compact discs, and we have to ensure activity in different operational systems.

After changing the content there would be no need in repeating the compilation. With every start the application reads in the whole content and menu structure from the .xml files generated earlier.

3 The Final Project

Describing project is the subject of the master thesis on the faculty of Mechatronics Warsaw University of Technology. Recently there have been a tendency to prepare strictly theoretical diplomas, not exceeding the software or computer model part. The touch panel is a project which is likely to bring a specific result – not only a design, scheme, but also a working system, very useful device.

As it was written earlier we decided to design our own touch panel system which should be cheaper and much better then mentioned Alltrim Monolith 17' model. We have designed our touch panel on base following bought parts:

- Touch screen 20''
- Mini-ITX + Atom panel 1,6 GHz
- 2GB RAM DDR2
- 450 W power supplier
- 500 GB SATA2 disc

All the above mentioned materials are easy to get and their whole price doesn't cost more then 1000 Euro). At the moment we have designed a special case which could be use as a free standing or as well as a wall mounted panel. Till the end of June we are going to built the prototype of the above described touch panel and start to organize the information database. The above describing software is now under development and tests of functionality; finish is planned on summer. We hope that till the conference beginning we could have one panel's prototype in the hall near the Faculty of Mechatronics Information Desk, what we'd like to show on a short film.

References

[1] http://www.monitouch.com
[2] http://www.advantech.com

How to Compensate Tool Request Position Error at Horizontal Boring Milling Machines

M. Dosedla

Brno University of Technology, Faculty of Mechanical Engineering,
Institute of Production Machines Sytems and Robotics, Technicka 2896/2,
Brno, Czech Republic
ydosed00@stud.fme.vutbr.cz

Abstract. Horizontal boring and milling machines are one of the most used machines for machining of the shape-complicated and precise workpieces with weight about several tens of tons. Nowadays these machines are frequently used in energy industry where the high demands on the machine accuracy occur. The position accuracy of the tool placed at the end of the headstock is particularly the weakest point of these machines. The position error is caused by gravitation on the horizontal machines conception. Thus this paper introduces the main methods for compensation of this position error.

1 Horizontal Boring and Milling Machines

These machines belong to middle-sized and big-sized machines. They were particularly used for machining of deep holes at big-sized workpieces in past, whereas nowadays they are also often used for milling operation. For these reasons many producers provide machines without boring spindle which means that these machines must be equip by some milling head. Fig. 1 shows the horizontal boring and milling machine by Fermat Company. This floor type machine has boring spindle and it is equip with a rotary table.

The main cutting motion is rotational movement in this kind of machines and it is provided by a tool such as drill, miller etc. The secondary motion is linear movement and it is also provided by tool or workpieces that depends on a machine conception.

The tool position accuracy is particularly the weakest point of these machines as mentioned above. The tool is clamped in a spindle at the end of a headstock. The boring spindle and whole headstock can usually horizontally travel out of machine then all these parts are deformed by gravity. For these reasons the tool request position is achieved with an error.

Fig. 1. Horizontal boring and milling machine - floor type [Fermat CZ]

The value of the tool position error depends particularly on:

- rigidity of the headstock and the boring spindle,
- rigidity of the headstock slider,
- weight of the headstock,
- weight of the tool,
- type of an used accessory.

There are used several compensation systems of the tool position error in practice. Most of these approaches are based on mechanical, electro-mechanical and hydraulics-mechanical systems.

The compensation of the headstock curvature is a rather complicate design issue. To find the best possible design solution is suitable to use some optimization method or design all parts according to systematic designing.

1.1 Systematic Designing

Systematic approach for designing belongs to academic approach for designing. It is the most demanding designing method. It is a rational designing based on theoretical knowledge. The basis of this theory was defined in the seventies of Twentieth Century. [1]

1.2 Machine Tool Optimization

At technical optimization the effort is to create a machining centre on the highest possible technical level, fulfilling the maximum demands on performance, rigidity and precision parameters. If the highest emphasis is placed on the total costs necessary to manufacture a machining centre, we speak about economic optimization. If the target is to design a machining centre on a relatively high technical level, keeping the adequate costs of development and manufacture, technical-economic optimization is selected. [2]

The optimization at building a machining centre cannot be understood as a purely mathematic task; therefore it is not easy to use mathematic methods to select its optimum building. The optimization at designing a machining centre is a task, whose target is most of all to find the best possible functional machine structure to select suitable components and to perform the right selection of a proposed design solution. The diagram of particular optimization types is shown in Fig. 2. [2].

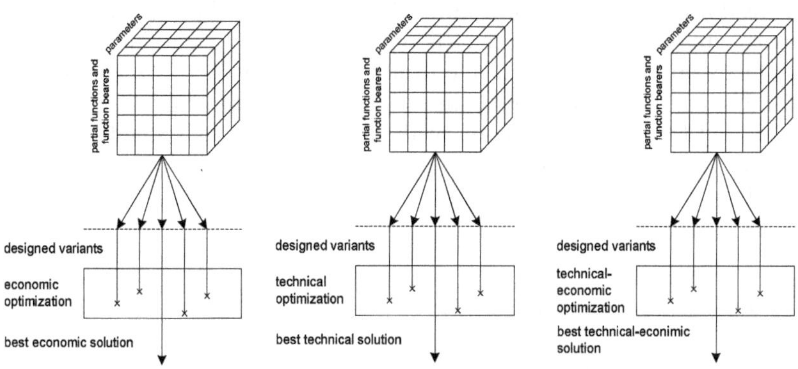

Fig. 2. Diagram of economic, technical and technical-economic optimization [2]

2 The Compensations Approaches

The first designing step of a new headstock should be a specification of headstock precision and headstock accessory in this case. Requirements for the tool position accuracy will be lower particularly for roughing machines. We usually expect the high rigidity and performance of these machines. Whereas the high accuracy and lower machine performance are expected of the finishing and universal machines.

The used machine accessory has also a high influence on final machine geometry and precision. There is a big difference between deformation of the headstock without head and with some milling head which can weight about 1000 kg. Thus for this type of big horizontal machines which are equipped by several milling heads is necessary to compensate the headstock curvature in order to achieve the high accuracy of final tool position.

2.1 The Most Frequently Used Approaches for the Compensation of the Headstock Curvature in Practice

Cam mechanism: The headstock is placed on a special bracket in this system. There are two cam mechanisms which are driven by two servo-motors and gearboxes between the cross-carriage and the headstock bracket. These cam mechanisms can lift with headstock bracket and compensate the headstock curvature by this system. This compensation type use e.g. Tos Varnsdorf Company see Fig. 3.

Hydrostatic guide way: Some producers of bigger horizontal milling machines use hydrostatic guide ways at linear and rotary machine axes. This hydrostatic system can also use for compensation of the headstock curvature. We can supply more pressure oil at the headstock front side than at rear side. The compensation of the tool position error is achieved by this system. This compensation type uses several producers e.g. Union, Pama, FPT, Waldrich Coburkg etc.

Compensation bars: The application of compensation bars is quite frequently used system by machine producers. These bars are placed within the ram at the top side and they are pushed towards the front headstock surface. We can pull bars in backward direction by hydraulic cylinders if we need any compensatio. This system also used for the compensation of the headstock curvature and it is provided by several companies e.g. MAG Gidding & Lewis, Skoda Machine Tool, Union etc see Fig. 4.

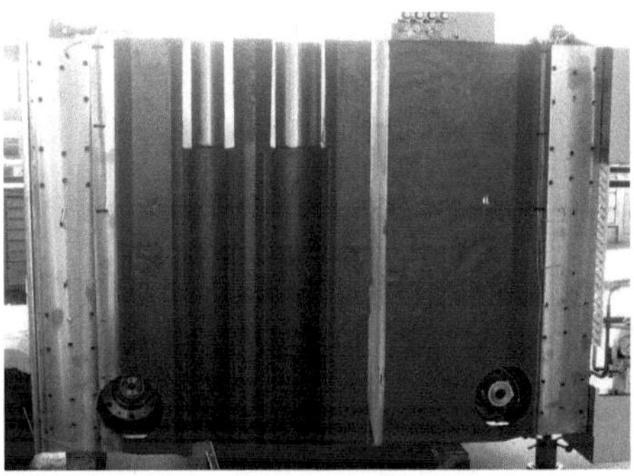

Fig. 3. Rear side of headstock slider [Tos Varnsdorf]

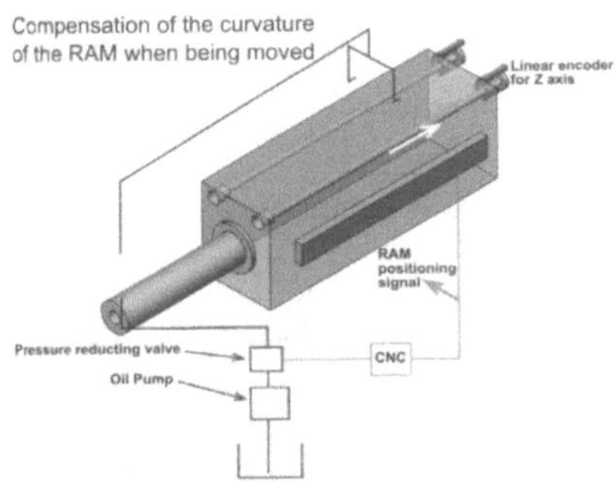

Fig. 4. Compensation bars [JUARISTI]

2.2 Headstock Curvature Compensation by Servomotor and Ball Screw

Nowadays servo-motors belong in the most used type of drive-unit in the field of the machine tool construction. We can reach easily manageable electro-mechanic system by connection the servo-motor and ball screw.

We have created digital model of the ram with cross-section 400x400 mm and length 2100 mm for simulation of the ram curvature compensation. It was simplified model without main spindle, bearings, and other parts. We have calculated that headstock front part falls down by 0,016 mm by FEM see Fig. 5.

This deformation was compensated by two forces that pushed on the front surface of ram at the top side. Deformation was almost compensated by forces F_c=20 000 N see Fig. 6.

We have decided to use a ball screw and a servomotor system for the compensation of the ram curvature see Fig. 7. The main features of designed system are shown at Table 1.

Fig. 5. Curvature of the ram without any compensation

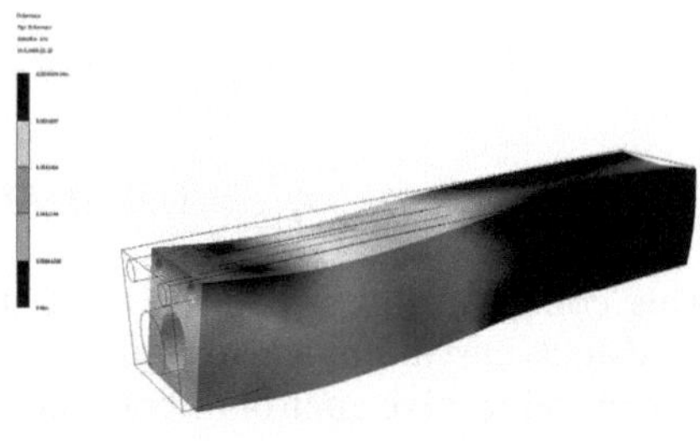

Fig. 6. Curvature of the ram with compensation

Table 1. Compensation system features

ball-screw K 40x15	diameter	40 mm
	screw-pitch	15 mm
	lenght	2250 mm
Servomotor Heidenhain QSD 155D EcoDyn	speed	0 – 3000/min
	power	5,7 kW
	torque	18,1 Nm
gearbox Stöber P521	ratio	4

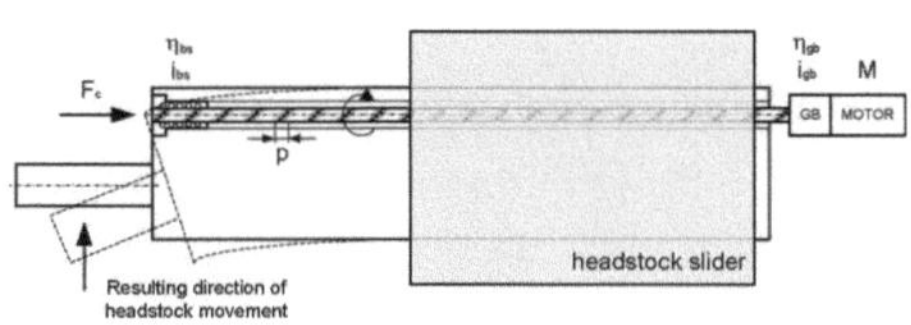

Fig. 7. Diagram of the compensation system with two ball screws

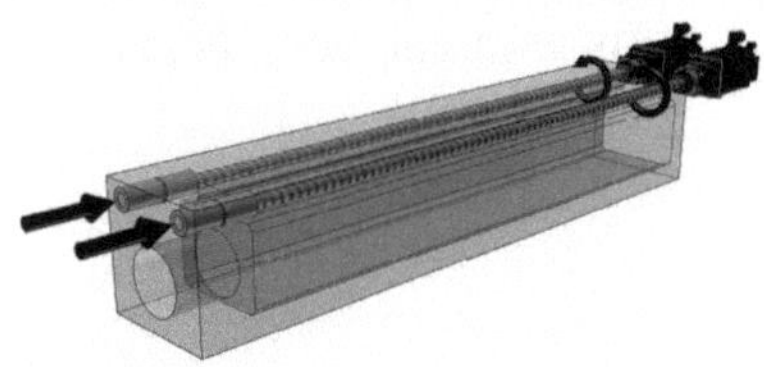

Fig. 8. 3D CAD model of the compensation system

The calculation of the compensation system:

$$F_c = M \cdot i_t \cdot \eta_t = 18{,}1 \cdot 1675{,}48 \cdot 0{,}828 = 25110N \quad (1)$$

$$i_t = i_{bs} \cdot i_{gb} = 418{,}87 \cdot 4 = 1675{,}4m^{-1} \quad (2)$$

$$i_{bs} = \frac{2 \cdot \pi}{p} = \frac{2 \cdot \pi}{0{,}015} = 418{,}87 m^{-1} \quad (3)$$

$$\eta_t = \eta_{bs} \cdot \eta_{gb} = 0{,}9 \cdot 0{,}92 = 0{,}828 \quad (4)$$

This compensation system (Fig. 8) should be suitable for compensation of the headstock curvature. There will be necessary to prepare more exact mathematic model and resolve control of both servomotors.

3 Conclusion

One of the ways how to increase a horizontal machines precision is using any compensation system of the headstock curvature. This compensation can be realized by many systems. Concurrent machine tool producers do not have the same opinion at this problem. Almost every producer uses his oven compensation system. Easy controlling of the servo-motors and high accuracy of ball screws probably will lead to frequent using of this system.

The final machine accuracy also depends at deformation of headstock slider, column, bed and machine fundament. It is necessary to keep the same level of whole machine and fundament quality as the headstock quality.

References

[1] Dosedla, M.: Concurrent and modern approaches for designing. International Journal of Applied Engineering Research (2008) ISSN 0973-4562

[2] Dosedla, M., Marek, J.: Design of a machining centre – criterion optimization. In: AEDS Workshop 2008, Pilsen (2008) ISBN: 978-80-7043-685-1

Verification of the Simulation Model for C Axis Drive in the Control System Master-Slave by the Turning Centre

J. Křepela and V. Singule

Brno University of Technology, Faculty of Mechanical Engineering,
Institute of Production Machines, Systems and Robotics, Technická 2,
Brno 616 69, Czech Republic
ykrepe05@stud.fme.vutbr.cz, singule@fme.vutbr.cz

Abstract. The paper describes the verification of the simulation model for the C axis on the multifunction turning centre. The drive works in the positional feedback and his simulation model is specified for detection of the dynamical behaviours. The C axis is designed as accurate angle position axis and with big dynamic stability. The C axis drive is constructed with help of a connected worm gearing on a spindle. Worm wheel is solved as one part with two gears. Servomotors are controlled with the mode speed/torque coupling (MASTER-SLAVE), which guarantees the constant torque preloading. The difference of a torque's guarantees the operation without backlash. In the dynamic model are involved the friction on the worm gears, torsion stiffness located with help of the FEM and moment of inertia for all parts. Results of this paper will be comparison of the outputs from simulation model of the C axis with measuring on the prototype of turning centre. It will be to compare the results of eigen frequencies, position responses on the torque step.

1 Introduction

Main Spindle of the machine, on the witch is implemented the C axis, is for turning operations driven by asynchronous motor with power 71kW. For high torque moment necessity is gearing reduced through two steps planetary gearbox and constantly belt gear. For milling and drilling operation is this main motor uncoupled through neutral position in this gear box and spindle is hydraulic coupled with worm gears, where are geared with two synchronous servomotors controlled in mode Master – Slave (e.g., Fig 1). This mode assures to change the parameters of electrical preloading between both servomotors from the machine control. Preloading of servomotors holds positions of coupling eliminated the production backlash of worm gearing. This is arranged with leaned teeth flanks against both worms opposite teeth of worm gears react in opposite direction of torque moments both servomotors [1].

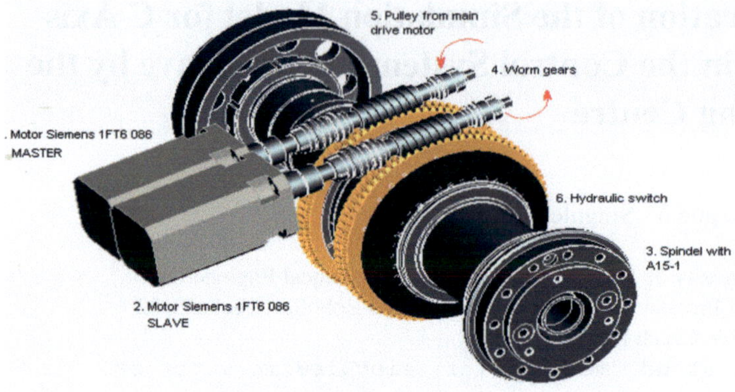

Fig. 1. Design of the C axis [1].

Simulation model of the C axis drive is created mainly for optimization of the parameters setting for both drivers and for size of preloading with torque between the both motors. Simulation model includes this generally parts (e.g., Fig 2):

- angle position and speed feedback of the motor Master
- speed feedback of the motor Slave
- PI regulation of the preloading torque for both motors
- model of friction on the worm gear
- multi-body mass model of dynamic system

For successful verification of the simulation model of C axis drive on the prototype of machine TT75 was necessary to determine optimal values for parameters of the both motors in the control system Siemens Sinumeric 840D.

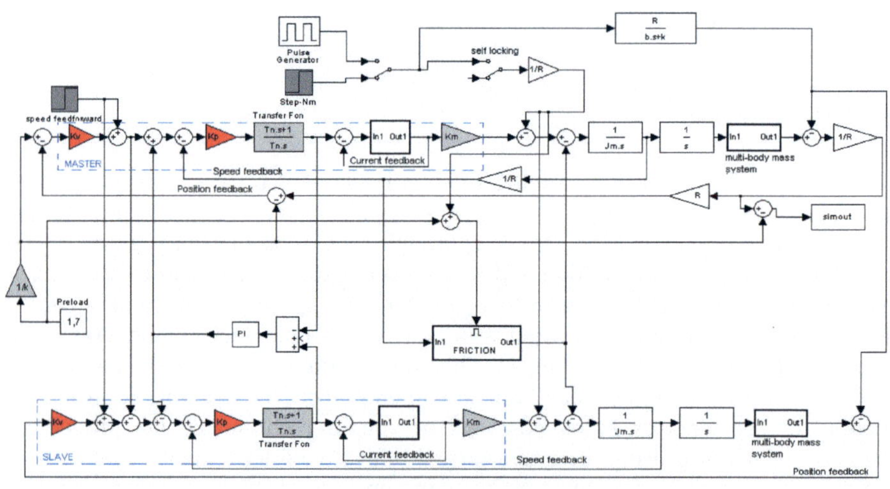

Fig. 2. Simulation model of the C axis [2].

Setting of the drives for measuring:

- Amplification of the speed feedback K_v= 33.3/s
- Integrated time constant for motor Master I_M=3ms
- Integrated time constant for motor Master Slave I_S= 3ms
- Integrated time constant for PI regulator of the torque I_R=10ms
- Proportional amplification for motor Master K_P= 2100Nms/rad
- Proportional amplification for PI regulator of the torque K_P= 15%

2 Multi-Body System

Structure of the multi-body mass system is showed on the figure 3. Worm wheel is designed as self-locking; therefore the torsion oscillation of the load is influenced solely of the torsion stiffness of the spindle in the area between the direct angle position gauge and worm wheel. Conversely the oscillation both motors by blocked load is influenced of the whole mechanical system. Moments of inertia are determined from 3D models for manufactured parts and for the traded parts are determined in the catalogue. Torsion stiffness are modelled in surrounding of FEM. This dynamic model is included in simulation model for complete C axis drive in the control system Master-Slave as subsystem. Verification this dynamic model is made with help of the measuring amplitude-phase frequency characteristic direct in speed feedback of the motor Master or the motor Slave in the tools of

Table 1. Calculated values of the reverberation frequencies of the motor.

	1.rev. frequency	2.rev.frequency	3.rev. frequency	4.rev.frequency
Calculated value [Hz]	22,5	56	65	78
Measured value [Hz]	27	52	64	80

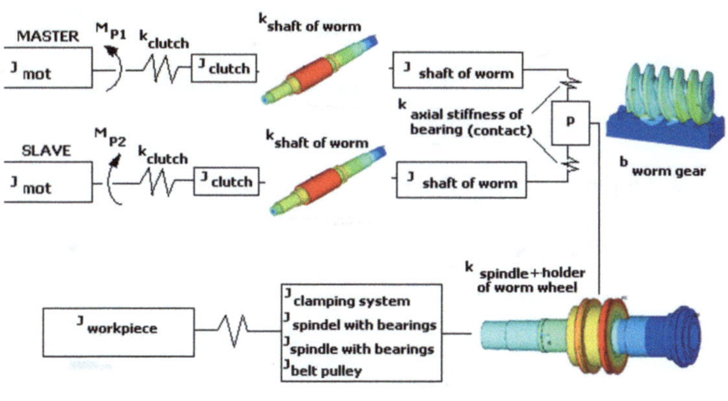

Fig. 3. Diagram of the mechanical system by the preloading Master-Slave [2].

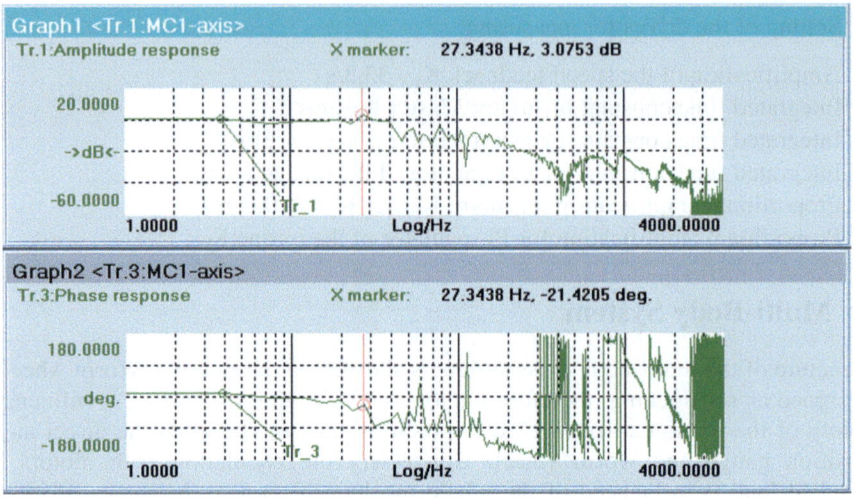

Fig. 4. Amplitude-phase characteristics for speed feedback of the motor

the control system Sinumerik 840D. For this measuring are removed all obliteration filters (e.g. Fig 4).

For precision speed feedback hasn't be amplitude amplification in the reverberation greater then 3 dB. This is fulfilled with help of large damping in the worm wheel.

3 Impact Elasticity of the Position

Impact compliance of the position is in one from preferably testified measuring of the setting quality of the whole control system of the C axis. Impact compliance testifies to whole stiffness of C axis and to dynamic behaviour by machining. The measuring (e.g. Fig 5) was made with help of the rigid bar; witch was fixed on the

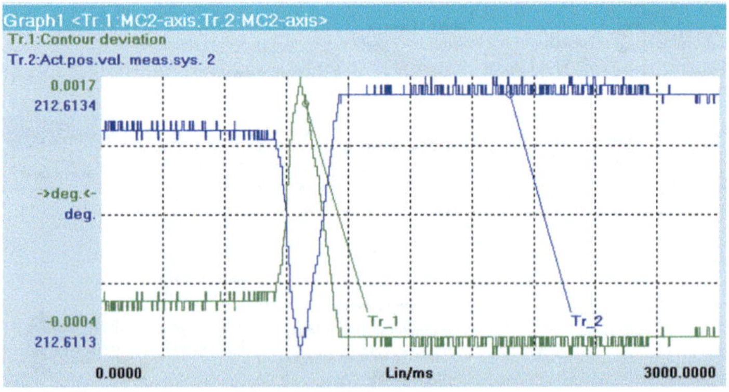

Fig. 5. Measured position response on the step change of toque 200Nm

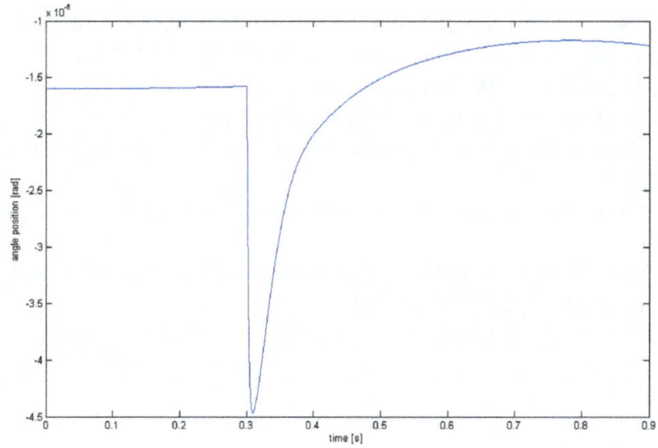

Fig. 6. Simulated position response on the step change of torque 200Nm

clamping system. With help of the accuracy gauge of force was the bar preloaded and fast loosen. The preloading of torque is possible to develop very accuracy, bat speed of the loosing is worse to repeat by more measuring.

The different of the size of the dramatic decline of position and time of decline by step change of torque is by measured and calculated course 15% (e.g. Fig 6). This different testify to difference of the calculated and real stiffness. Dynamic course of the response after the dramatic decline to the stabilization is more difference; bat for the speed is the problem adheres to the same conditions by measuring and simulation.

4 Manuscript Submission

Measuring of the C axis on the prototype of machine TT75 helps for verification of the simulation model. The difference between the measured values and calculated values are in most cases explained by incomplete measuring and at the same time of the error, which was arisen from calculation of stiffness and damping on the worm gear. Simulation model was simplified in any considerations by small nonlinearity in dynamical system. General is possible to say, that the behaviours of the C axis is for the requirements for machining very good and also was proved rather the advantage of friction on the worm gear as damping by reverberation.

Acknowledgments. This work is supported by project MSM 0021630518 and company TOS, a.s.

References

[1] Křepela, J., Singule, V.: Dynamical behaviours of the C axis multi-body mass system with the worm gear. In: Recent Advances in Mechatronics. Springer, Heidelberg (2007)

[2] Křepela, J., Singule, V.: Simulation of the dynamic behaviours of the C axis drive by the turning center. Solid state Phenomena (2009) ISSN 1012-0394
[3] Weck, M., Brecher, Ch.: Werkzeugmaschinen, Mechatronische Systeme, Vorschobantriebe, Prozessdiagnose. Springer, Heidelberg (2004)
[4] Mišun, V.: Dynamika výrobních strojů. Vysoké učení technické v Brně (1991) ISBN 80-214-0297-0
[5] Souček, P.: Servomechanismy ve výrobních strojích, ČVUT Praha (2004) ISBN 80-01-0292-6
[6] Siemens: Speed/Torque Coupling, Master-Slave (TE3). Function Manual, Siemens, 03/2006 edn. (2006) 6FC5397-2BP10-1

Compensation of Axes at Vertical Lathes

J. Marek[1] and P. Blecha[2]

[1] Technical director, TOSHULIN, a.s, Wolkerova 845, 768 24 Hulin, Czech Republic
jiri.marek@toshulin.cz
[2] Brno University of Technology, FME, Institute of Production Machines, Systems and Robotics, Technicka 2896/2, Brno, Czech Republic
blecha@fme.vutbr.cz

Abstract. If exacting and complicated technological operations are applied at titanium machining, it is not acceptable to set positions of the particular motion axes at the machine tool with too big position uncertainty. For this reason, it is necessary to manufacture all components of motion axes very precisely. However, manufacturing precision has some economic and technical limits. After the optimum limits are exceeded, it is useful to perform electronic compensation of mechanical motion elements. This paper describes the essential principles of mechatronic compensation and it also shows some practical results.

1 Introduction

Uncertainty at manufacture of important parts, especially for such industrial branches as aviation, power engineering and transport, represents a great problem leading to technical and financial damages. There are several possibilities how to minimize the working and manufacturing uncertainty of the machine tools used for manufacture of exacting workpieces. The first one is to design a very precise machine tool. However, this interferes with technical and economic properties, as excessive demands put on working uncertainty of a machine tool can make it unsellable. The second possibility is to design a machine with a certain fixed limit of working uncertainty that can be eliminated by means of mechatronic principles.

2 Manufacturing and Working Uncertainty

Dimensional variations occurring at machining of workpieces give the first hand information about manufacturing scatter or manufacturing precision. Manufacturing uncertainty is a degree of precision which can be reached when machining a workpiece using a particular machine at a defined operation status. It includes deviations conditioned by the machine as well as deviations which are not conditioned by the machine [VDI 3441]. According to the definition, all machining deviations conditioned by the machine are summarized under the term "working uncertainty", including systematical as well as random errors (Fig. 1).

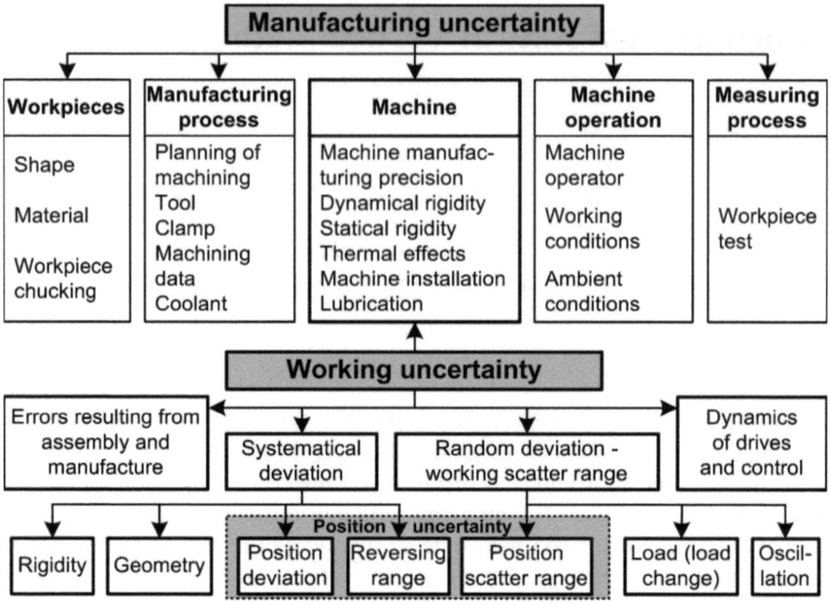

Fig. 1. Manufacturing uncertainty and working uncertainty

Systematical errors are such influences as geometric deviations, thermal effects, static and dynamic rigidity, etc. They can be partly detected e. g. by geometric examinations in accordance with DIN standards or ISO recommendations or by special independent examinations.

Random deviations determine the range of a machine tool working scatter. They can be determined by mathematic-statistic methods from dimensional variations of test workpieces (machined workpiece) under the specified machining conditions.

3 Compensation of Manufacturing and Working Uncertainty

Many authors examine the individual causes of manufacturing and working uncertainty separately [2, 3, 4, 5], especially the geometric accuracy of a machine tool in relation to the thermal machine condition [6, 7]. A great effort is put to exploration of geometric accuracy, primarily with the use of volumetry. The determined deviations are compensated either by means of hardware or software.

The following is mostly adapted by means of hardware:

- motion mechanisms of a machine including servo systems;
- geometric deviations in shape and position of machine tool frame structural parts which are intentionally and purposely shape optimized during manufacture, in order to achieve the required effect after mounting them;

- geometric deviations in shape and position of guideways at motion axes so that the resulting motion reaches the required parameters;
- heat radiation sources.

It shall be mentioned that hardware compensation is very demanding and requires very good technical and physical knowledge.

Software compensation lies in implementation of a compensation table into the digital control systems. The tables are prepared either in advance or at real time and pre-activated during a working cycle – machining. In the industrial practice the compensation tables are usually prepared in advance in the machine control system. It is mainly due to worse reliability and increased failure rate at obtaining the real-time feedback signals and also by the economy of the whole operation.

Geometric accuracy is measured at unloaded machine (without cutting forces) and the axes are evaluated independently, unless assessed by volumetry. We want to stress the fact that comprehensive consideration of all factors causing the manufacturing and working uncertainty of a machine tool is essential. Geometric accuracy and thermal conditions (deformations) represent only one "little" part.

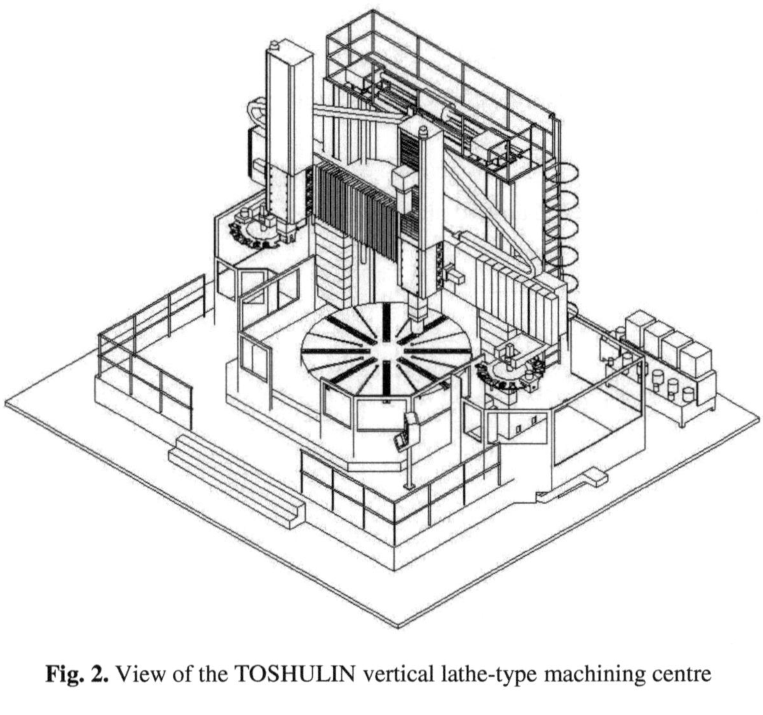

Fig. 2. View of the TOSHULIN vertical lathe-type machining centre

4 Vertical Lathe-Type Machining Centre

A vertical lathe-type machining centre is presented in Fig. 2. From the kinematic point of view, one motion is performed by a workpiece (rotation) and two motions are performed by a tool (linear motion).

The following compensations can be used as the control system options:

- straightness compensation;
- interpolation type of straightness compensation;
- compensation of the vertical axis deflection;
- increased bidirectional compensation of the screw lead;
- compensation of the cross-rail deflection;
- three-dimensional space compensation.

These compensation principles can be applied in SIEMENS as well as FANUC control systems.

Besides, the hardware compensations described above are applied at the machine. Fig. 3 shows the coordinate systems of the workpiece and of the tool at a vertical lathe-type machining centre. Workpiece compensation is necessary in case of high requirements on minimization of the manufacturing and working uncertainty:

- stroke in the Z-axis (plane XY);
- rotation of the plane XY around all axes;
- stroke in the Y-axis (plane XZ);
- stroke in the X-axis (plane YZ).

It is also necessary to perform tool compensation:

- stroke in the Z-axis;
- stroke in the X-axis;
- rotation of the plane XZ around all axes.

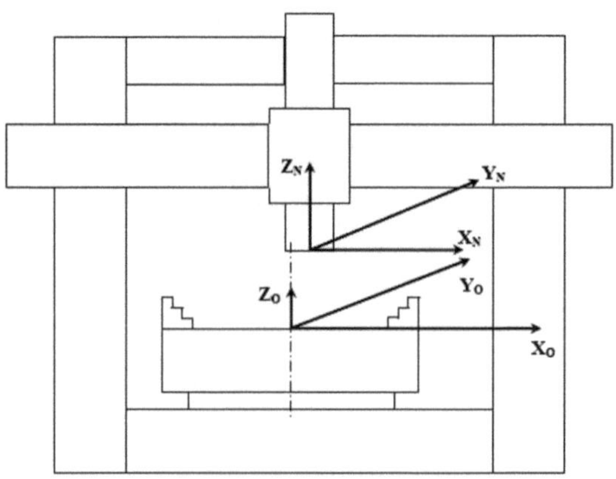

Fig. 3. Tool and workpiece coordinate systems at a vertical lathe-type machining centre

5 Practical Example

It was necessary to perform compensations in the individual axes for one important customer of TOSHULIN, a. s. company, in order to reach the required working accuracy; the price of the machined workpieces reached hundreds of thousands EUR. Another interesting fact is that 90% of all material removed from the semi product was changed to chips. The workpiece wall thickness did not exceed 2 mm, its diameter was approximately the same as its height.

Machining with such accuracy would not be possible without the following steps taken in cooperation with the machine user:

- optimization of the workpiece shape and its chucking;
- optimization of the technological process, the tool shape and of the tool edge cooling method;
- training of the highly-qualified and dutiful machine operating staff;
- optimization of the measuring process.

It is evident from what was written above that achievement of high working accuracy (minimization of manufacturing uncertainty and working uncertainty – see Fig. 1) is not only the matter of the machine tool itself, but also of its user.

The machine design was optimized by hardware as well as software adjustments to decrease the working uncertainty. The machine manufacturer performed the following hardware modifications:

- geometric shape and position deviations of frame structural parts and their guideways;
- elimination of heat radiation sources.

Software compensation consisted in this process: the actual course of the table displacement and rotation (Fig. 3) was approximated by a polynomial and the resulting compensation value was suitably used during the manufacturing process. The results obtained in practice are shown in Fig. 4.

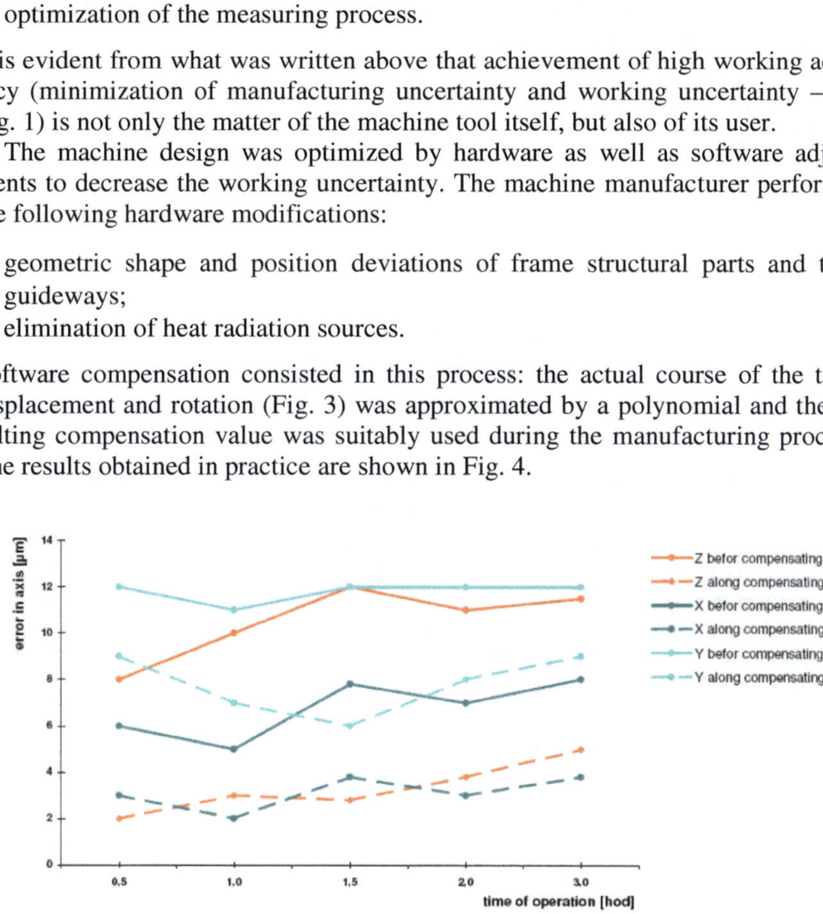

Fig. 4. Displacement compensation at the axes X, Y, Z

It is clear from the presented graph that the errors, caused by systematical and random deviations during the machine operation, have been considerably decreased in the X-axis, in the Y-axis as well as in the Z-axis (Fig. 1). This has resulted in a considerable decrease of working uncertainty and the required results have been obtained.

6 Conclusion

Elimination of manufacturing uncertainty and working uncertainty (Fig. 1) is a very important part of a working process if any machine tool is operated. Mechatronic principles are used to decrease errors, combining hardware and software means. However, it is necessary to keep in mind that it is not very suitable to explore the possibilities how to compensate the individual quantities separately, but it is necessary to consider the comprehensive effects of all factors affecting the manufacturing and working uncertainty.

Acknowledgments. Research work for this contribution was financed by TOSHULIN, a.s. world leading producers of vertical lathes and by the Ministry of Education, Youth and Sports of the Czech Republic (project 1M0507 "Research of production techniques and technologies").

References

[1] Weck, M., Brecher, C.: Werkzeugmaschinen 3: mechatronische Systeme, Vorschubantriebe, Prozessdiagnose, 6th edn., p. 442. Springer, Berlin (2006)

[2] Shen, Y.L., Duffle, N.A.: Uncertainties in the acquisition and utilization of coordinate frames in manufacturing systems. CIRP Annals – Manufacturing Technology 40(1), 527–530 (1991)

[3] Knapp, W.: Measurement Uncertainty and Machine Tool Testing. CIRP Annals – Manufacturing Technology 51(1), 459–462 (2002)

[4] Kurtoglu, A., Sohleniu, G.: The accuracy improvement of machine tools. CIRP Annals – Manufacturing Technology 39(1), 417–419 (1990)

[5] Portman, V., Shuster, V., Rubenchik, Y., Shneor, Y.: Substitute Geometry of Multidimensional Features. CIRP Annals – Manufacturing Technology 53(1), 443–446 (2004)

[6] Brecher, C., Hirsch, P., Weck, M.: Compensation of Thermo-elastic Machine Tool Deformation Based on Control internal Data. CIRP Annals – Manufacturing Technology 53(1), 299–302 (2004)

[7] Svoboda, O.: Prostorová přesnost frézovacích center při tepelně neustálených stavech, PhD Thesis, ČVUT v Praze, FS, p. 132 (2007)

Mechatronic Backlash-Free System for Planar Positioning

P. Matějka, J. Pavlík, M. Opl, Z. Kolíbal, and R. Knoflíček

Brno University of Technology, Faculty of Mechanical Engineering,
Institute of Solid Mechanics, Mechatronics and Biomechanics,
Technicka 2896/2, Brno, Czech Republic
ymatej04@stud.fme.vutbr.cz

Abstract. The application of the parallel mechanism in robotics can still be considered a valid and modern concept. The disadvantage of traditional designs of the mechanism is the low rigidity of the joint connections and the resulting hysteresis in their behaviour. This is due to the combined effect of tensile and compressive stress. The principle of the presented solution consists in realizing movement through the use of actuators that are only exposed to tensile stress.

1 Introduction

The option of employing parallel kinematic structures both in machine tool design and in robotics has been recently in great demand. In technical practice there is a host of concepts of the arrangement of parallel kinematic mechanisms available for either practical application or at least theoretically prepared. The principal requirements on their final structure include simple construction, easy control and limited reconfigurability to allow for possible modifications of the geometrical arrangement of the components during the testing of the mechanism. In addition to meeting the above demands it is also necessary that we ensure sufficient rigidity and accuracy of the mechanism under load. The design should also consider the possibility of integrating a measurement system with optional verification of the presumed characteristics and/or using this subsystem as a feedback-providing element.

For illustration, below we show some basic types of practical applications of parallel kinematics the arrangement of which clearly indicates the direction of the development. The first example is the traditional Hexapod concept (Fig. 1). Obviously, this concept fails to meet the construction simplicity requirement and its control is difficult to grasp and uneasy. Yet, it serves as the basis for a number of machines deployed in industry, such as P800M by Metrom (Fig. 2).

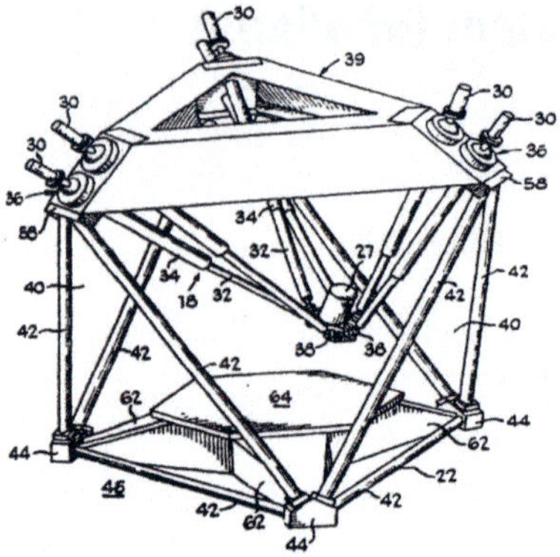

Fig. 1. Hexapod

Fig. 2. Hexapod

The parallel structure principle is also often used in the constructions of various manipulators both in the form of planar 2D or 3D workspace. The planar arrangement of parallel kinematics is dominant in machine tools (Fig. 3).

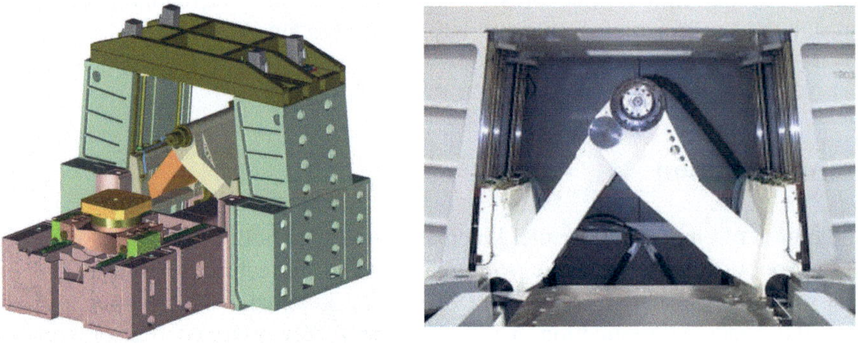

Fig. 3. TriJoint by KOVOSVIT MAS

2 Implementation of the Experimental Device

From the above examples of mechanisms with parallel kinematics it is clear that most of the devices of this type actually used in practice employ joints with combined tensile and compressive stress. These loading characteristics together with manufacturing imperfections in the links result in considerable hysteresis in the behaviour of the actuators as a whole. This is undesirable with regards to the accuracy of the machine.

The shortcomings can be eliminated in several ways, such as increasing the manufacturing precision, increasing the overall rigidity, etc. However, these solutions are very unfavourable in terms of the effect/cost ratio. Another option which seems possible is the elimination of combined stress. It seems suitable to cancel out the compressive stress of the actuators, mainly due to their strutting stability. As the loading capacity in pull is usually significantly better than the strutting stability, it is possible to achieve higher loading of the mechanism. Another step leading to greater accuracy of the parallel mechanism is the application of prestress in the actuators that results in delimiting the backlash in the individual elements and links and thus its elimination in the positioning of the mechanism.

2.1 Hardware Design

The constructional design of the mechanism employs 4 kit servomotors and is based on the mechanism using linear actuators stressed by simple pull (Robert L. Williams II, 2003). The rotary motion of the servomotor is then transformed by means of a pulley and a cable rod to translational motion. The motion is further transmitted by a rod to a platform that performs a movement predefined by us.

The advantages of cable rods include low cost, low weight, simplicity and ease of inspection to see whether the rod is permanently pre-stressed.

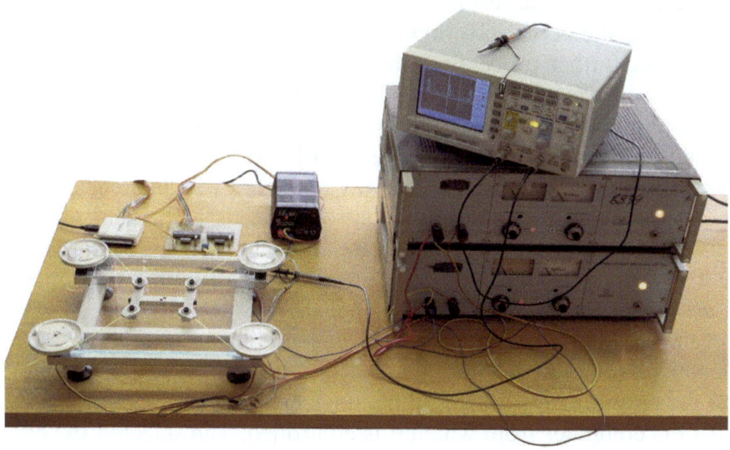

Fig. 4. Constructional design of the mechanism

3 Control

The actual path of the platform's movement was created using the LabView 8.5 software. Within a defined XY plane the path of the platform's movement can be random. It can be determined in a number of ways. On the virtual control panel (fig. 5) we can choose between manual mode or predefined movement. Manual mode offers the option of testing the correct generation of width-modulated pulses in a 10-bit word and the proper turning and engaging of the individual servomotors.

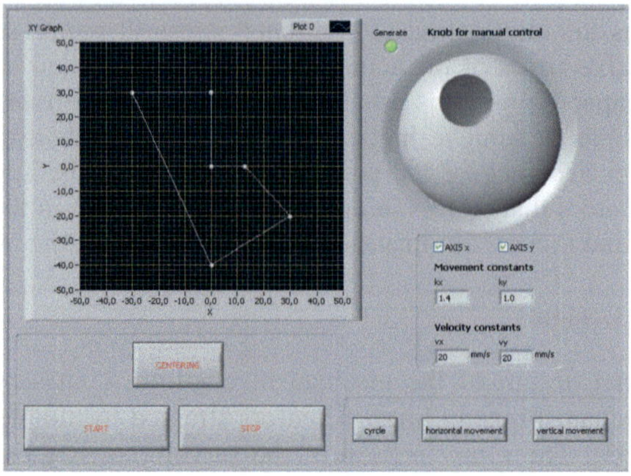

Fig. 5. Virtual control model developed in the LabView environment

The actual programme consists of several parts. The first one contains the mathematical apparatus describing the mechanism proper. By changing the constants and variables we can reach any trajectory in the plane. The output is a set of values describing the actual position of the servomotor, which is of the unsigned integer type. For further processing, these values need to be converted to 10-bits. The conversion is effected by another block of functions, which also provide for the writing on the output ports of the I/O cards. While the control of the four servomotors is accomplished using only one card with a 12-bit interface its outputs must by cyclically overwritten. In this way a single cycle is created for each servomotor. At the same time the microcontroller needs to be sent information on which servomotors is to be controlled at the moment. For this reason each cycle is assigned a servomotor identifier consisting of a pair of high-bits. Control is there effected using in total a 12-bit word where the first 10-bits specify the required turning of the servomotors and the last two added bits ensure proper addressing to the particular servomotor.

One of the requirements for the whole device was its mobility and easy connectibility to a PC computer from which it is controlled. The selected I/O card by National Instruments type USB 6009 which creates the interface between the PC and the ATMEGA microcontroller is limiting by its number of 12 I/O ports, but given its low cost and reasonable speed of overwriting the digital outputs, i.e. 150Hz, make it satisfactory to our purposes. More precise positioning of the servomotors would require opting for other peripheries available on the market, not just by National Instruments.

4 Measurement and Evaluation

Verification had to be performed to prove the above assumptions. It has been done by means of measuring the real trajectory, using machine vision. The original alternative expected using the Siemens Simatic VS722 industrial camera. This solution, however, proved insufficient for practical reasons, especially due to low camera sensor resolution.

An EOS 40D still camera with a resolution of 3888 x 2592 was then used to acquired more precise information on the real platform position. From an accuracy point of view, this resolution provides approx. 5× more accurate position data than the camera. The disadvantage of this solution, however, is the lower recording frequency. This has partly been eliminated by reducing the platform motion speed. This speed has been optimised as to the minimum measured values during one cycle. We have set a criterion of at least 30 measurements.

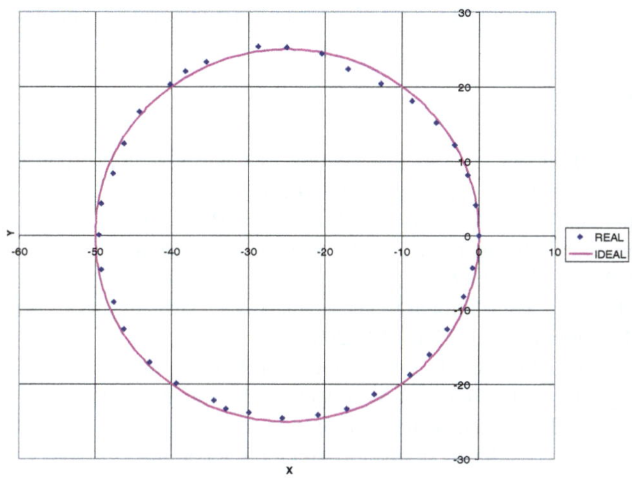

Fig. 6. Position measuring in circle interpolation [mm]

Fig 6. shows the ideal platform trajectory and the real positions detected by measuring “snapshots”. Based on repeated measuring and statistic result processing, it has been determined that the final accuracy corresponds with the above specified assumptions, which are limited mainly by the hardware component used.

5 Resume

The prime objective of this work was to verify the input assumptions. These included mainly the repeatability of accuracy of the parallel mechanism, which uses the principle of pre-stressed joints. This principle provides both a significant increase of mechanism’s overall stiffness and elimination of backlash in individual joints, which results in more accurate mechanism positioning. The problems of a reduced work area and bigger overall mechanism dimensions are the main disadvantages.

The secondary objective was to design this experiment as a low-cost solution – including the demonstration equipment – with a possibility of implementing it in the teaching process.

Both of these objectives have been achieved: verification of the mechanism behaviour and implementation in the teaching process including the involvement of students in the experiment preparation and realisation.

Implementation of this presented principle in practical applications, especially in machine tools with parallel kinematics would increase their overall stiffness and accuracy. For this reason, our further efforts will be aimed at finding partner(s) among industries in order to utilise these principles in a commercially produced machining tool.

Acknowledgments. This project was supported by Ministry of Education, Youth and Sports of the Czech Republic (Research Centre for Manufacturing Technology 1M0507).

References

[1] Valášek, M., Bauman, V., Šika, Z.: Study of Concepts of Parallel Kinematics Machines for Advanced Manufacturing. Acta Polytechnica 45(3) (2005)

[2] Neugebauer, R.: Parallel Kinematic Machines in Research and Practice. In: Proceedings of the 2nd - 5th Chemnitz Parallel Kinematics Seminar PKS 2000 - 2006, Chemnitz 2000, 2002, 2004, 2006 (2006)

[3] Williams II, R.L.: Translational Planar Cable-Direct-Driven Robots. Journal of Intelligent and Robotic Systems 37, 69–96 (2003)

[4] Kolíbal, Z.: Industrial robots for manipulation with parallel kinematic machines. In: Proceedings of 15th International Workshop on Robotíce in Alpe-Adria-Danube region RAAD 2006, Balatonfured (Hungary), June 15-17, 2006, pp. 69–74 (2006)

[5] Williams II, R.L., Vadia, J.: Planar translational cable-direct-driven robots: hardware implementation. In: 2003 ASME Design Engineering Technical Conferences (2003)

[6] Barette, G., Gosselin, C.: Kinematic Analysis and Design of Planar Parallel Mechanisms Actuated with Cables. In: ASME Design Technical Conferences, Baltimore, MD (2000)

Compensation of Geometric Accuracy and Working Uncertainty of Vertical Lathes

M. Michalíček

Brno University of Technology, Faculty of Mechanical Engineering,
Institute of Production Machines, Systems and Robotics,
Technicka 2896/2, Brno, Czech Republic
mmichalicek@seznam.cz

Abstract. Already when developing and optimizing new drafts of production machines, where also vertical lathes belong, we have to count with various kinds of compensation methods. This paper deals with this problem as well and focuses on compensation of errors caused by geometric inaccuracy (GI) and working insecurity (JI) of the vertical lathes. The main reason for description of compensation methods and working uncertainty is endeavour to create a mathematic model of a machine, from which it would be possible to predicate its manufacture qualities for the final customer, already in the offer phase.

1 Introduction

Geometric accuracy of used parts in construction of the working machine is one of the main criterion of every producer, who is obliged to decline working uncertainty from customers´ requirements all the time. Total accuracy of the working machine is affected by a large number of factors such as working conditions, a tool, rigidity of machine, a workpiece, etc. Nevertheless, an important task of the working machine is fine positioning of the tool in the face of the workpiece all over the workplace. Achievement of low positioning uncertainty is affected by errors included in the machine construction (geometric error), temperature change of a work environment, parts of the machine or the change in a machine load. The geometric errors are the main cause of the working inaccuracy of vertical lathes [1]. According to Evans, Kocken [2], quasi-static errors are even 70% of the total error of the machine. This paper is focused on compensation of geometric errors of the vertical lathes originated from finishing of workpiece faces. Errors, which are made during machining, are of a volumetric format. For face milling, it is necessary to keep minimum deviation in the axis Z. Consideration how these errors minimalize will be specify there.

2 Sources of Geometric Errors

It is necessary to define meanings of the errors and accuracy prior to suggestions of errors compensation. Accuracy can be defined as a concordance rate of a finished workpiece with required sizes and geometric accuracy [3]. Error can be understood as any deviation of a position, for example, a sharpness deviation of a tool from a theoretically desired value of the workpiece and the workpiece, which is specified by tolerance. Errors range gives measuring of accuracy. It is maximum of shift errors among two points in a workplace of the machine [4]. The errors can be divided into two caregories, namely quasi-static and dynamic. The quasi-static mistakes change in time and they have a main influence on the tool and the workpieces. These mistakes depend on a machine main building structure. These sources include geometric-kinematic errors, erros caused by dead weight of machine parts and stress generated from thermic load of the machine [2]. Especially in big machining machines (machining centers), which are composed from parts such as..., we can find errors for every mentioned part. All the part errors influence the total volumetric error of the machine. For concrete case of the vertical lathe, where the main cutting motion of a face turning is done by a rotational workpiece and additional motion is done by the tool, see fig. 1. There the main source of the errors is a rail moving on axis X (Fig.1.).

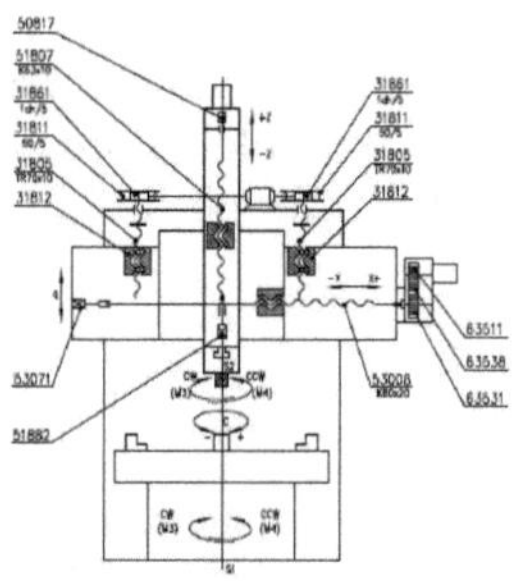

Fig. 1. Kinematic scheme of the vertical lathe [6].

In consequence of variable setout of a railhead ram towards the railhead, there are variables overturning moments. For a finishing cut, we can suppose that forces from machining are insignificant. Dominant influence on the finishing cut comes from a self-load of the machine assembly of the railhead ram, railhead and rail. In the fig.2., there are examples where the machine works in the finishing cut and in a process of general machining where force F is a load of the machine.

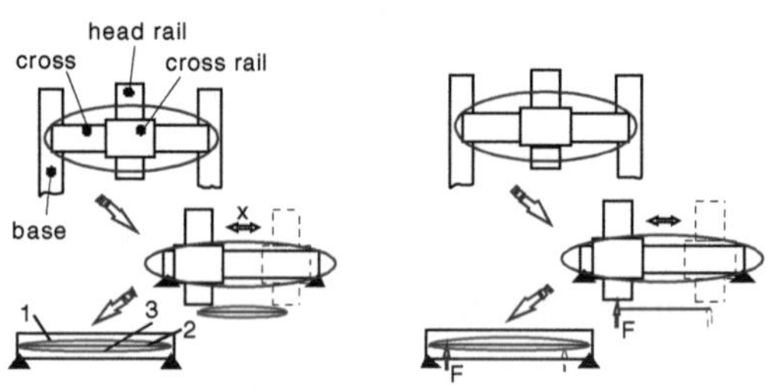

Fig. 2. Model of search of compensation according to kind of load.

In consequence of variable setout of a railhead ram towards the railhead, there are variables overturning moments. For a finishing cut, we can suppose that forces from machining are insignificant. Dominant influence on the finishing cut comes from a self-load of the machine assembly of the railhead ram, railhead and rail. In the fig.2., there are examples where the machine works in the finishing cut and in a process of general machining where force F is a load of the machine. It is possible to suppose required shape of the rail (blue curve 1) from the Fig.2 left. The tool is in its ideal position after load from the railhead and railhead ram. In contrast to fig.2. right, it is perceptible, that the force applied toward the railhead will deform it. Rail deflection will be in agreement with green curve (Fig.3). That deflection could be compensated by an appropriate force. Compensation according to the fig. 2-left is suitable for a producer, who exemplifies geometrical accuracy to a customer in unloaded state of the machine. This compensation meets requirements on the mechanical accuracy of the vertical lathes. On the contrary, compensation of ambient influences affecting the total system changing in time is a good example of mechatronic system application.

2.1 Errors of the Rail of the Vertical Lathes Excited by Static Forces

These are errors arised only from their own load of all the parts in direction from the tool applied on the rail. Considering construction and equipment of each of the machine, result end deflection curves of the railhead ram will be different. The surveyed curve of deflection will be derived from chosen discrete points all over the workplace and it will be extended by a suitable curve (fig.3.). Face turning is mainly affected by deformation in the axis Z. This deformation is eliminated by compensation of a shape of the rail face of the rail. Deflection of the rail results in tool turning around the axis Y and change in a tool position toward the workpiece. These movements are shown in the fig. 5.

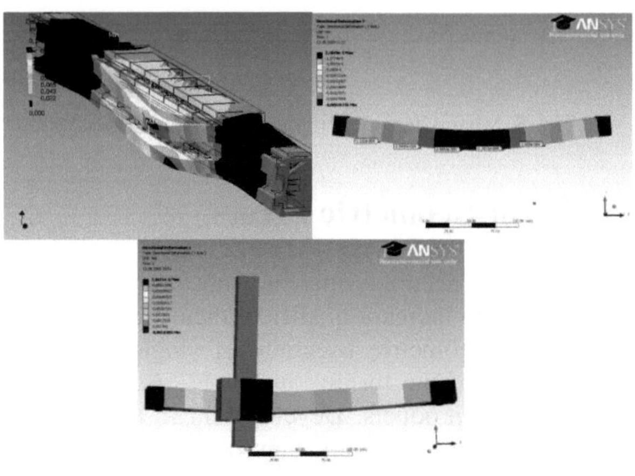

Fig. 3. Calculation models of the self weight loaded rail [6].

2.2 Errors of the Vertical Lathe Excited by General Load

Load originating from quasi-static, dynamic and turning forces influences the whole system during a cutting process. Forces effects are changeable in dependence on time, position in the workplace, used technology etc. Deformations originated at the end of the tool is possible to arrange according to their sizes in an appropriate axis like Δy>Δz>Δx [6] (the system of coordinates according to fig.1.) The optimal position of the tool in the workplace of the vertical lathes TCP,i (Tool Center Point, ideal) and TCP,r real position influenced by forces and moments applied on the tool, is displayed in the fig.4. For instance, deformation of a mathematical model of the tool during surface machining (vc=100 m/min, h=ap=0,3 mm, f=0,1 mm, workpiece - 180HB) forms 4% from the total error measured on the workpiece for axis Z.

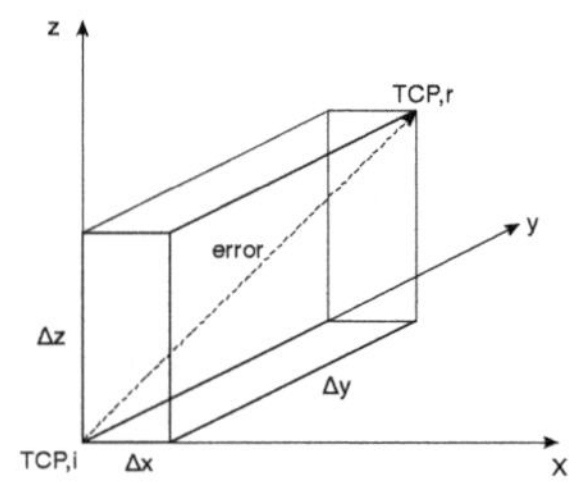

Fig. 4. Visualization of the error TCP of the vertical lathes

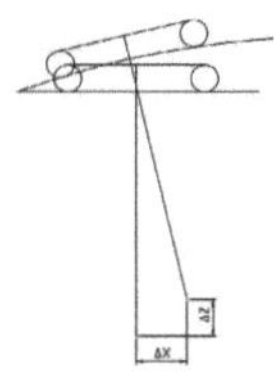

Fig. 5. Visualization of the error TCP in axises X-Z.

3 Compenzation of Geometric Errors

With requests on products accuracy increase, not only producers of turning machines have been obliged to increase quality of their products to satisfy customers. Methods of increase in geometric accuracy of single parts are given by used technology and precision of the turning machines. Accuracy increase can cause expensiveness of the final products. Development in branches of electrotechnics and informatics enabled rise of machine control. Machines coud be then software compensated. Thereby, requests on geometric and kinetic accuracy of the appropriate parts could be decreased. If 70% of the error is formed by quasi-static errors, which form deformation of the structural parts at big machines, at vertical lathes there are a column and the rail which is loaded by a constant force acting in the position of the axis X in a work length of the rail. Software method, where machine constants obtained from the measurement of the machine are introduced into the machine, is one of the methods decreasing the total working uncertainty of the machine. Today, a control system enables to import the constant values for compensation of the axis, compensation in a plane and a space. It only depends on a selected method, where criterions of use are expediency and time necessary for machine measurement and recording of the constant values into the system. Raksiri [5] describes a method established on a neuron system for combination of compensating the geometric errors and erros originated from a machining force in his paper. This system is presented in the fig. 6, where INPUT values enter the system and investigated errors are obtained on OUTPUT Layer for each axis. In the fig. 7, there are test results of four different parameters together with increase in the geometric accuracy.

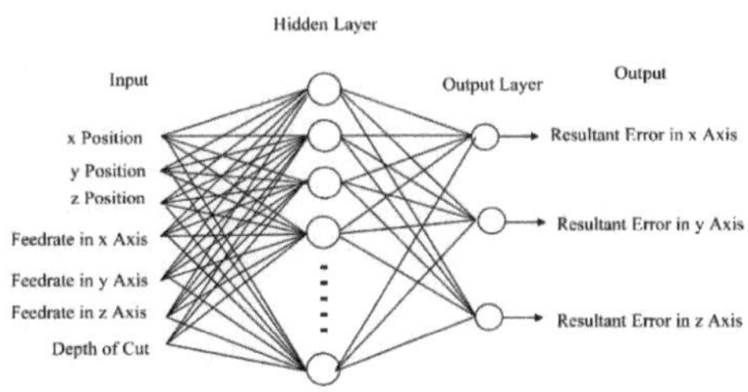

Fig. 6. 7-110 -3 Network of error model combination [5].

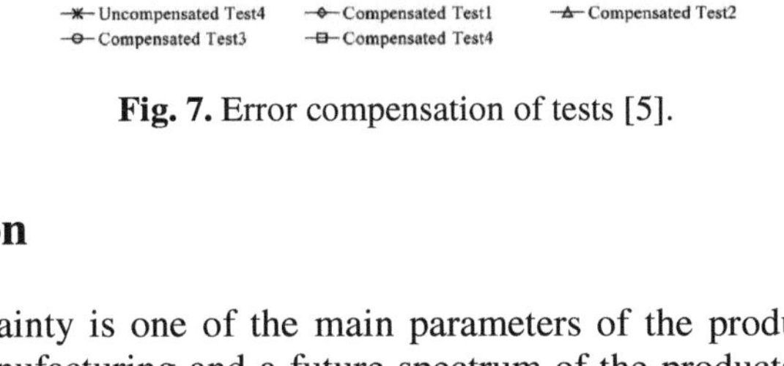

Fig. 7. Error compensation of tests [5].

4 Conclusion

Working uncertainty is one of the main parameters of the production machines, not only for manufacturing and a future spectrum of the products, but it can be a decisive item in a competitive fight, too. It urges the producers to continually develop and give new parts to construction of the production machines. It can be edited by construction (passive compensation), software compensation or usage of the mechatronic systems. Whereas a price of the whole machine decides about its marketability, it is necessary to judge the compensation methods in term of economics. Practically, we especially meet with compensation of the geometric errors by the software method. Correction of the position between the tool and the workpiece is made after the import of the correct values into the control system. Other methods are used rather in a laboratory.

References

[1] Ramesh, R., Mannan, M.A., Poo, A.N.: Error compensation in machine tools — a review Part I: geometric, cutting-force induced and fixturedependent errors. International Journal of Machine Tools & Manufacture 40, 1235–1256 (2000)
[2] Evans, C.J., Kocken, R.J.: Self-calibration: reversal, redundancy, error separation and absolute testing. Annals of the CIRP 45(2), 617–632 (1996)
[3] Jedrzejewski, J., Modrzycki, W.: Intelligent supervision of thermal deformations in high precision machine tools. In: Proc. 32nd Int. MATADOR Conf., Manchester, UK, pp. 457–462 (1997)
[4] McMurtry, D.R.: Development of the next generation of 3D probing systems for the future Coordinate MeasuringMachines and Machine Tools. In: Proc. 3rd Int. Conf on Laser Metrology and Machine Performance — LAMDAMAP, pp. 169–185 (1997)
[5] Raksiri, C., Parnichkun, M.: Geometric and force errors compensation in a 3-axis CNC milling machine. International Journal of Machine Tools & Manufacture 44, 1283–1291 (2004)
[6] TOSHULIN, company publication

Assessment of Design and Risk Analysis of a Tool Holder Manipulator

L. Novotný[1] and P. Blecha[2]

[1] Section leader, TOSHULIN, a.s, Wolkerova 845, 768 24 Hulin, Czech Republic
Lubomir.Novotny@toshulin.cz
[2] Brno University of Technology, FME, Institute of Production Machines, Systems and Robotics, Technicka 2896/2, Brno, Czech Republic
blecha@fme.vutbr.cz

Abstract. Machine tools are production facilities built of structural parts, drives, measuring devices, sensors, control systems etc. It is not common to regard a machine tool as a mechatronic system, while the service devices and peripheries of machine tools are automatically seen as mechatronic. A typical example of a mechatronic system is the manipulator with tools and tool holders of the POWERTURN 2500 machine tool. Its main task is to enable unmanned running of this CNC machine by automatic tool exchange. The paper deals with assessment of the design and with risk analysis of the tool holder manipulator of POWERTURN 2500 vertical CNC lathe. Its construction consists of a platform moving in the YZ plane of the machine tool actuated electromechanically and hydraulically as well as manually. Construction of the manipulator must meet the technical requirements and at the same time it has to be safe both for the operating staff and the machine itself. The safety issue is an essential prerequisite for placing the machine on the market and into service, with regard to the requirements of the 2006/42/EC Directive on Machinery coming into force on 29 December 2009.

1 Introduction

Machine tools are production facilities consisting of structural parts, drives, measuring systems, sensors, control systems, etc. It is not common to consider a machine tool to be a mechatronic system (a machine tool is usually understood as an independent system). However, operating devices and peripheries of machine tools are understood automatically to be mechatronic equipment. The typical example of a mechatronic system is a manipulator with tools and with tool holders (tool heads). Its main function is to provide the essential property of CNC machines, i. e. their ability to work in unmanned operation and to perform the automatic tool exchange.

2 Description of the Tool Holder Manipulator

The essential idea about the structure of the tool holder manipulator (magazine) and about its components originated during the first designing phase orientated especially on functionality, operability and serviceability of this equipment.

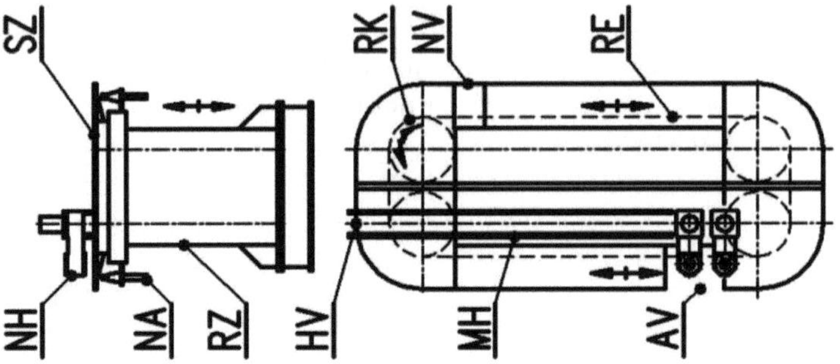

Fig. 1. Tool holder manipulator

The basis construction element of the tool magazine (Fig. 1) is a self-supporting frame (RZ). The lower part is a stationary one and the upper part moves in the YZ plane of a machine tool. Its motion is provided hydraulically. The movable part stroke is 500 mm, the height of the lifted magazine is 2600 mm. There are chain wheels (RK) located on the frame. The chain (RE) moves along these chain wheels and tool holders carrying tools (NA) are attached to the chain. The protective guard (SZ) is attached to the upper part and it is specified as a protection against dirt. There is a platform – a tool head manipulator (MH) located over the protective guard. Tool heads (TH) are located on the tool head manipulator. This manipulator moves together with the frame in the vertical direction (Z-axis). For the reason of the tool head exchange, the manipulator moves in the Y-axis. The manipulator motion is provided electromechanically along linear guiding in the range specified by technical limits and practical operation conditions. The remaining motion part is provided by an operator's manual motion so that the tool heads can be accessible for an operator during their exchange, cleaning, etc. The operator must unlock a mechanical lock before the platform can be moved. All end positions are protected by limit switches. There are three important places marked in the Fig. 1. The place for the automatic exchange of tools and tools heads (AV) serves for their automatic exchange in the ram. The place marked as NV is specified to load tools to the tool magazine. Because the tool heads must be located on the manipulator in such a place which can be reached by the machine ram, the place for the tool head exchange (HV) must be located at the opposite end of the automatic exchange area.

During the second designing phase, it is necessary to focus on the providing of the tool magazine safety.

3 Risk Analysis

It follows from the new 2006/42/EC Directive on Machinery that the machinery manufacturer or its authorised representative must provide risk assessment with the target to determine requirements put on health protection and safety of work. The machinery – in our case it is the tool magazine – must be designed and constructed considering the results of this risk assessment. The particular risk assessment process consists of the following steps:

- determination of the anticipated use and any reasonably predictable misuse of the machinery,
- determination of hazards which can follow from the machinery and hazardous situations connected with it,
- risk estimation regarding to seriousness of a possible injury or health damage and probability of their occurrence,
- risk evaluation with the target to determine if it is necessary to reduce risk in accordance with the Directive on Machinery target,
- exclusion of hazards or reduction of risks connected with these hazards by means of application of preventive measures in the following sequence: design modification, safety precautions, information of the user about persisting hazards.

The most important phase of the risk assessment process is risk analysis which consists in identification of hazards and estimation of risks connected with each particular identified hazard. A number of authors deal with various approaches to the designing process [1, 2, 3]; however, these approaches seem to be too scientific or scholastic and their application within industry is very problematic [4, 5]. This has resulted in the requirement to make the machinery designing process more transparent and to increase the support of the risk analysis realization at machineries.

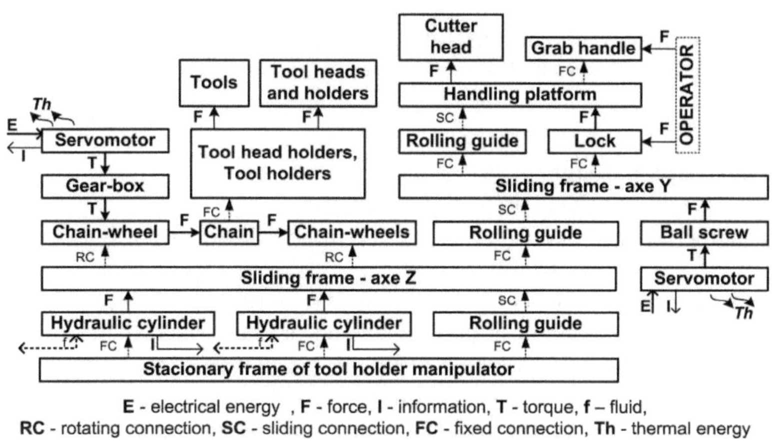

Fig. 2. Block diagram

3.1 Identification of Hazards

Identification of hazards is the most important part of risk analysis. It is necessary to identify all hazards connected with the machinery construction as well as with all life cycle phases. Doing this, it is necessary to consider especially operators' activities in the particular life cycle phases of the machinery. System analysis was performed to identify hazards connected with the tool holder manipulator design. This system analysis consists in creation of its block diagram. This diagram shows all considerable interactions of its particular elements (Fig. 2). The second step is then performed by means of this block diagram, of EN ISO 12100-1 and EN ISO 14121-1 standards and of the overview of operators' activities. This step performs analysis of significant hazards during all life cycle of the tool holder manipulator. Application of forms shown in Fig. 3 proved very well in this phase.

ANALYSIS OF SIGNIFICANT HAZARDS During transport, assembly and installation				Machine: Tool holder manipulator Model: POWERTURN 2500
No.	Phase of life cycle	Type of hazard description	ID	Description of hazardous event:
1.1	Unloading of tool holder manipulator, lifting of tool holder manipulator,	inconvenient position, overstrain, fall, overthrow, crushing	8.1, 18, 23, 27.1, 27.2, 27.3, 27.6	During lifting there is a risk of fall or overthrow of the tool holder manipulator if the gravity centre is not known or is not respected. In case of unfavourable construction the workers may also suffer overstrain or inconvenient position.
etc.				

Fig. 3. Form for analysis of significant hazards

3.2 Risk Estimation

For each identified significant hazard is necessary to design a useful preventive measure to reduce this risk. Preventive measures are designed in the following sequence:

- measures built in design;
- safeguards and additional protective measures;
- information for use.

Estimating the risk, it is necessary to consider seriousness of possible health damage as well as probability of this damage occurrence which shall be estimated with regard to frequency and length of threat, possibility of avoiding the hazard and probability of hazardous event occurrence. The form (Fig. 4) proved very well during this process. This form documents in a transparent way risk estimation as well as the iterative risk reduction process.

Brno UT FME	RISK ESTIMATION FORM responsible: (name)		machine: THM date: 10.05.2009
No. of hazard	ID	**Hazard identification** according to EN ISO 14121-1: 1. Mechanical hazards	
1.1	**3**	**Crushing**	
Phase of life cycle: *use*		**Danger zone:** *operating area*	
Exposed persons: *operator*		Operational state: *adjustment*	
Description of hazardous event:	*Crushing during moving of sliding frame of tool holder manipulator. Danger of crushing of fingers or parts of hands between stacionary and sliding frame of tool holder manipulator.*		
Initial risk	**Seriousness** of possible health damage:	*S2 - serious injury*	Size of risk **12**
	Frequency and length of threat:	*A2 - frequent to continuous*	
	Possibility of avoiding the hazard:	*E3 - hardly possible*	
	Probability of hazardous event occurrence:	*W3 - high*	
STEP 1: Measures built in design (according to EN ISO 12100-1)			
Description of measure:	*The design of the machine respects the ergonomical principles reducing the possibility of inserting fingers or parts of hands between the moving parts.*		
Reduced risk after measure	**Seriousness** of possible health damage:	*S2 - serious injury*	Size of risk **10**
	Frequency and length of threat:	*A2 - frequent to continuous*	
	Possibility of avoiding the hazard:	*E2 - possible at certain conditions*	
	Probability of hazardous event occurrence:	*W2 - medium*	
STEP 2: Safeguards and additional protective measures (according to EN ISO 12100-1)			
Description of measure:	*It is possible to activate only one dangerous movement at a time in a clearly arranged space. Frame moving with a safe speed of 2m/min. Use of enabling switch. Dangerous movements of the clamping device can be activated only by a deliberate action of the operator.*		
Reduced risk after measure	**Seriousness** of possible health damage:	*S1 - slight injury*	Size of risk **4**
	Frequency and length of threat:	*A2 - frequent to continuous*	
	Possibility of avoiding the hazard:	*E2 - possible at certain conditions*	
	Probability of hazardous event occurrence:	*W2 - medium*	
STEP 3: Information for use (according to EN ISO 12100-1)			
Description of measure:	*Warning in the instruction manual: "It is not possible to totally eliminate the hazards associated with frame moving by technical means; therefore, behave in such a way so as to avoid injury of fingers or parts of hands!"*		
Residual risk	**Seriousness** of possible health damage:	*S1 - slight injury*	Size of risk **3**
	Frequency and length of threat:	*A2 - frequent to continuous*	
	Possibility of avoiding the hazard:	*E2 - possible at certain conditions*	
	Probability of hazardous event occurrence:	*W1 - little*	
VALIDATION:	Measures are sufficient: (name)		date:

Fig. 4. Form for risk estimation

4 Conclusion

Fig. 5 shows the final plan of the machine (ST) with the already integrated tool holder manipulator (ZN). On the basis of the above-mentioned risk assessment, we have performed such measures which take into account safety requirements as well as operation requirements. The machine is operated only by one technician who is responsible for working operations and who moves only in the operation platform area (PO) and in the area on the increased operation platform specified for the tool head exchange (PH). The access to the tool magazine area is marked with arrows in Fig. 5. This access is enabled only in the tool magazine NC STOP mode. Service and maintenance are subject to a special mode and these operations can be performed only by a technician who is properly trained.

The exchange of milling heads (HV) and of tools (NV) can be performed only in two places and the operator cannot enter out of the defined area, after the door protected by an electromagnetic lock has been opened. The chain moves by safe speed and the motion can occur only by one place. The text mentions only the major precautions which have been applied to the machine on the basis of risk analysis documents and which result in safe operation of the complete machine.

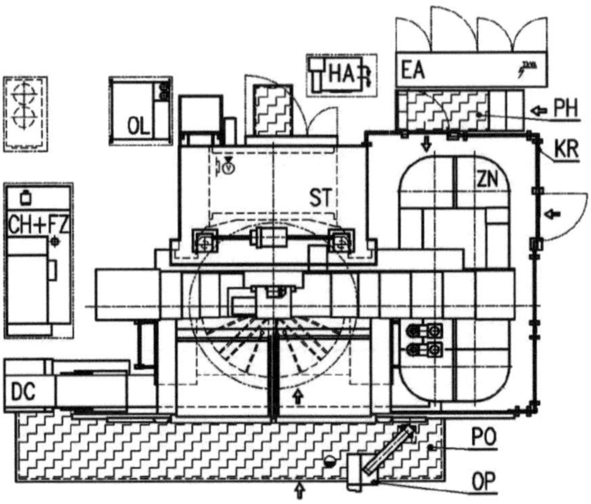

Fig. 5. Plan of the machine with the integrated tool holder manipulator

The presented approach to risks analysis has proved very well at development of machines and now it represents the company's standard at the designing process.

Acknowledgments. Research work for this contribution was financed by TOSHULIN, a.s. world leading producers of vertical lathes and by the Ministry of Education, Youth and Sports of the Czech Republic (project 1M0507 "Research of production techniques and technologies").

References

[1] Bernard, A., Hasan, R.: Working situation model for safety integration during design phase. Annals of the CIRP 51(1), 119–122 (2002)

[2] Eder, W.E., Hosnedl, S.: Design Engineering - A Manual for Enhanced Creativity. CRC Press - Taylor & Francis Group (2007) ISBN 1420047655

[3] Krause, F.-L., Kimura, F., Kjellberg, T., Lu, S.C.-Y.: Product Modelling. Annals of the CIRP 42(2), 695–706 (1993)

[4] Marek, J.: Management of risk at design of machining centres. In: Housa, J. (ed.) Proceedings of Machine Tools, Automation and Robotics in Mechanical Engineering, Praha, pp. 91–98. CVUT Praha (2004)

[5] Marek, J.: How to go further in designing methodology of machine tools? In: Hosnedl, S. (ed.) Proceedings of AEDS 2006 Workshop, Plzen, pp. 81–88. W. Bohemia University (2006)

Design of the Controller for Elimination of Self-excited Oscillations

T. Březina, J. Vetiška, P. Blecha, and P. Houška

Brno University of Technology, Faculty of Mechanical Engineering, Institute of Automation and Computer Science, Technicka 2896/2, Brno, Czech Republic
brezina@fme.vutbr.cz

Abstract. The article deals with design of the controller for elimination of self-excited oscillations during machining. These oscillations are generated between a cutting tool and a work-piece surface and lead to decreased quality of the machined surface as well as decreased geometrical precision. The secondary effect is increased level of noise emissions. In general, the self-excited oscillations negatively affect the working productivity of metal cutting machines and inhibit its growth. The design of the controller is based on a model of the cutting process which works with so-called regenerative principle of the self-excited oscillations generation and is based on the variable chip thickness. The model itself serves to derivation of the controller gains. A piezoelectric compensator of the cutting force is controlled by the designed PSD controller. The compensator changes the position of the tool against the machined surface thereby changing the chip thickness.

1 Introduction

The standard goal of a metal cutting machine construction is – (a) to get as precise work-piece as possible with quality surface, (b) in the shortest possible time. The problem is these demands stand against each other.

Machine-tool technical designers strive to obtain machine structure which is ideally stiff, lightweight and with high natural frequencies and high damping. The desired dynamical behavior or machining accuracy of the system is not guaranteed every time even with sophisticated optimization tools and new materials. The machine oscillations typically lead to decreased quality of the work-piece surface [5].

2 Self-excited Oscillations

There are three basic forms of oscillations of the machine-tools: natural oscillations, excited oscillations and self-excited oscillations. The self-excited oscillations occur in rough cutting as well as in finishing cutting. They are generated between the tool and the surface by mutual influence of the cutting process and the machine-tool without any outer periodical exciting.

Vibrations are caused by modulation of the static component of the cutting force which is influenced by relative cut (relative thickness of the chip). Chip thickness depends on the relative position of tool and the work-piece due to vibrations of the whole system machine – tool – work-piece. The consequences were already mentioned – decreased quality of the machined surface and noise.

There are several theories dealing with the problem including the reproduction principle. The wavy surface $Y_0(t)$ of the machined area originated from the previous cut is machined by the oscillating tool $Y(t)$. Waviness of the surface changes the depth of cut (the chip thickness) periodically, thus changing the cutting force which excites the whole system.

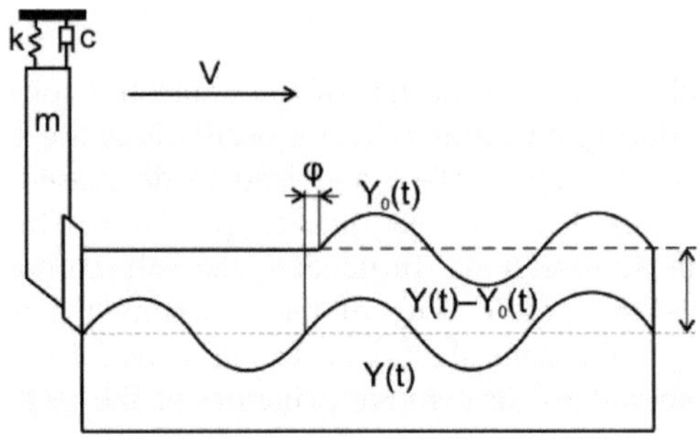

Fig. 1. The reproduction principle of self – excited oscillation

This is the principle of the reproduction of the oscillations in every turn of the spindle (Fig.1). The vibrations of the tool and the work-piece are phase-shifted. This shifting is crucial for the self-excited oscillations generation [3].

3 The Model of the Cutting Process

The model of a cutting process is based on the elementary equation [1].

$$m\ddot{x}(t)+c\dot{x}(t)+kx(t)=Fx(t) \tag{1}$$

where $Fx(t)$ is a cutting force described as

$$F_x(t) = a\left\{K_f\left[h_0-x(t)+x(t-\tau)\right]-\frac{C_i\dot{x}(t)}{V}-\frac{\alpha_i\ddot{x}(t)}{V^2}\right\} \tag{2}$$

*Fx, a, V, x(t), x(t-τ), h0, Kf, Ci,*and *αi* are in sequence: the cutting force, the cutting tool width, cutting velocity, amplitude of the inner wave, amplitude of the outer wave, the tool displacement, and three coefficients of the cutting force.

Dynamics of the process is according to Eq. 1 and Eq. 3 described as

$$a_2\ddot{x}(t)+a_1\dot{x}(t)+a_0x(t)+a_{0,\tau}x(t-\tau)=b_0 \tag{3}$$

where

$$a_2=m+a\frac{\alpha_i}{V^2}, a_1=c+a\frac{C_i}{V}, a_0=k+aK_f,\ a_{0,\tau}=aK_f,\ b_0=aK_f h_0 \quad (4)$$

4 Controller Design

The proposal is based on the cutting process model. It is assumed using of PID controller which controls generation of the compensative force $F_{a,\tau}(t)$ in such a way to keep the process dynamics on the aperiodicity margin. Thus

$$\ddot{x}(t)-3p\dot{x}(t)+3p^2x(t)-p^3\int_0^t x(\tau)d\tau=-p^3\int_0^t x_{ref}(\tau)d\tau \quad (5)$$

where $p<0$ is triple eigenvalue of the process. The controller gains are obtained for the control law

$$F_{a,\tau}(t)=K_i\int_0^t\left[x_{ref}(\tau)-x(\tau)\right]d\tau+K_p x(t)+K_d\dot{x}(t)+K_{p,\tau}x(t-\tau)+K_0 \quad (6)$$

by substituting Eq.6 to Eq.3 as

$$a_2\ddot{x}(t)+a_1\dot{x}(t)+a_0x(t)+a_{0,\tau}x(t-\tau)=b_0-F_{ax}(t) \quad (7)$$

and comparing the desired dynamics Eq.5 with Eq.7 and substituting Eq.4 to Eq.7

$$K_d=c+a\frac{C_i}{V}+3p\left(m+a\frac{\alpha_i}{V^2}\right), K_i=-p^3\left(m+a\frac{\alpha_i}{V^2}\right), K_i=-p^3a_2,$$
$$K_{p,\tau}=aK_f, K_0=aK_f h_0 \quad (8)$$

for $V\neq 0$.

Note that analytical form of the gains makes possible use the PID as adaptive controller which is adaptable to changes of the cutting force V and to the tool displacement h_0 [4].

Worse quality of the regulation is to be expected for the low cutting velocities because of appearance of the cutting velocity in the denominators of (8). It will be functionless for the zero cutting velocity.

5 Simulation Verification

The sampling period was chosen with respect to the possibilities of the hardware planned to be used for implementation of the control element as sampling period $Ts=1e\text{-}4[s]$. The achievable velocity of the controlled process was then tuned through a triple eigenvalue of the process p. First, we used simulation to find out that for $p>2700$, the response of the controlled cutting process is unstable for a step and for $p=1500$ it shows almost zero overshoot.

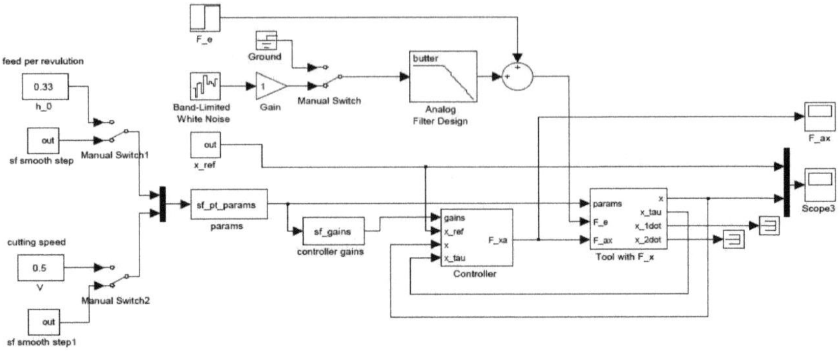

Fig. 2. Scheme of the controller connection and system model

Together with evaluation of the process dynamics, we monitored the power conditions and the amplitude of the compensation force with respect to technological feasibility of the actuating device. The character of the achieved process dynamics in relation to the size of the process eigenvalue is shown in Fig. 3.

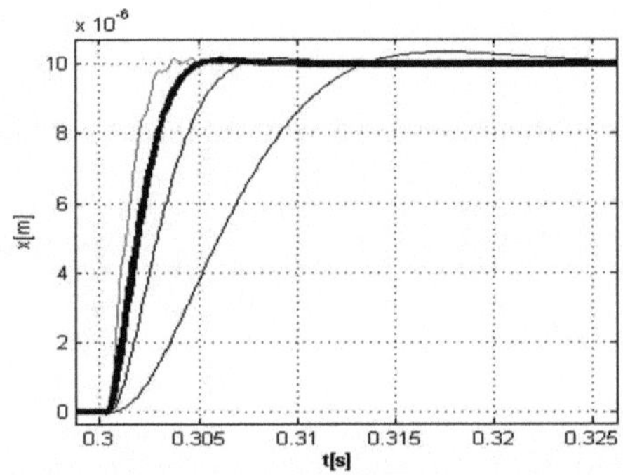

Fig. 3. Response to the jump of the desired position (p=-2200, -1500,-1000, -500)

Further testing examined the relation between the process dynamics disturbed by random failures of the cutting force and the process eigenvalue. Maximum size of the failure was set to *10N*. With increasing absolute value of the eigenvalue p, the amplitude of the tool position oscillation decreases as the compensation process is faster and the controller reacts more vigorously to the cutting force failures.

The performed simulations confirmed adequacy of the eigenvalue setting to *p=1500*, when there is still a sufficient reserve to the stability limit and at the same time the proposed control decreases approx. fourfold the self-excited oscillation of the tool position compared to the uncontrolled cutting process.

Adaptation of the controller to the changed working conditions is performed through the change of its gains. In terms of technology, the working conditions

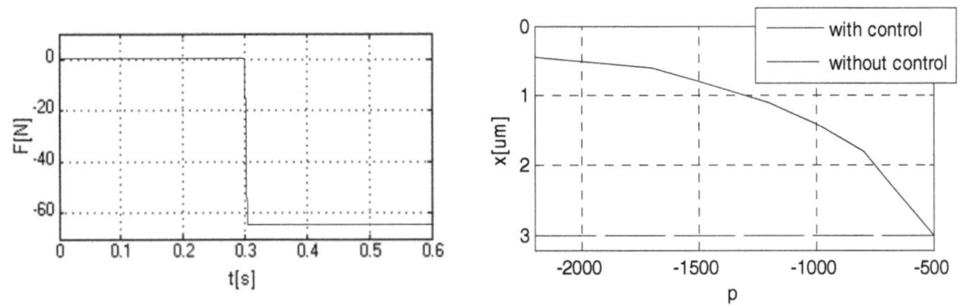

Fig. 4. Left - course of the compensation force for p= -1500, right - relation between the transmission of noise and the size of eigenvalues p

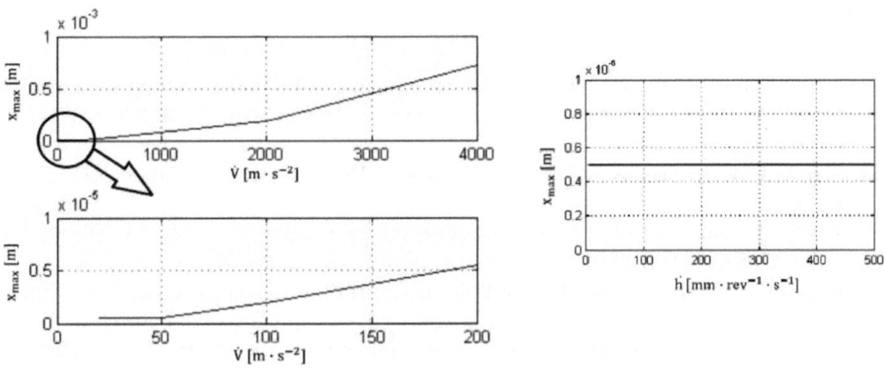

Fig. 5. Left - Influence of velocity step change on self-excited oscillation, right - Influence of step change of tool advance speed on self-excited oscillation.

vary the fastest in relation to the cutting velocity *V* and tool advance per revolution h_0. The behavior of the cutting process at step change of the cutting velocity and step change of the tool advance are demonstrated on Fig. 5. It appears that an acceptable value is the steepness of *50* ms^{-1} for V and *5 mm/rev* for h_0.

Due to the fact that at the described step changes *V* and h_0 the cutting process shows significant increase of tool oscillation amplitude and the step changes are unrealistic for technological reasons, a ramp was used for influence testing. Simulation was then used to find the greatest possible steepness of the ramp at these characteristics which did not yet affect the overshoot of the reaction to the jump in the desired tool position and did not worsen the self-excited oscillation of the tool position too much.

6 Conclusion

Use of the designed PID controller in a machining process results in marked decrease of self-excited oscillation of tool position compared to the uncontrolled cutting process. Simulation of the whole process was realized by modeling the machining with constant cutting velocity V and constant tool advance per revolution. Under the conditions of high speed machining (cutting acceleration of $\dot{V}<50\,[\mathrm{ms}^{-2}]$ and the speed of tool advance per revolution $\dot{h}_0<5\,[mm.ot^{-1}s^{-1}]$) we achieved approximately threefold decrease of self-excited oscillation. Under the standard machining conditions the achieved decrease of oscillation was approximately fourfold. Even better results, i.e. increased absolute value of the process eigenvalue p and improved noise elimination were obtained when a shorter sampling period $T_s<1e-4\,[\mathrm{s}]$ was chosen.

Acknowledgments. This work was supported by research plan MSM 0021630518 Simulation modeling of mechatronic systems.

References

[1] Altintas, Y., Eynian, M., Onozuka, H.: Identification of dynamic cutting force coefficication and chatter stability with process damping. Annals of the CIRP 57(1), 371–374 (2008)
[2] Marek, J.: Konstrukce CNC obráběcích strojů, MM průmyslové spektrum, září (2006) ISSN 1212-2572
[3] Tlusty, J.: Analysis of the State of Research in Cutting Dynamic. Annals of the CIRP 27, 583–589 (1978)
[4] Kautsky, J., Nichols, N.K.: Robust Pole Assignment in Linear State Feedback. Int. J. Control 41, 1129–1155 (1985)
[5] Novotny, L., Sveda, J.: Metody potlačování vibrací u vysoce dynamických obráběcích strojů. MMprůmyslové spektrum, listopad, pp. 16–20 (2007)

Problems of Quality of Convex Printouts for the Blind People

R. Barczyk and D. Jasińska–Choromańska

Warsaw University of Technology, Faculty of Mechatronics, Institute of Micromechanics and Photonics, Św. A. Boboli 8, Warsaw, Poland
r.barczyk@mchtr.pw.edu.pl, danuta@mchtr.pw.edu.pl

Abstract. People who have lost sight also lose the simplest ability of acquiring information about the surrounding world. Newspapers become unreadable, and writing a letter to a friend has to be done by a third person. Nevertheless, with the development of modern technologies, the number of new devices, which support the communication of blind people, is increasing (also microprocessor). Yet still Luis Braille's system, which uses point-convex signs, is most commonly used. A new problem has arisen, namely modern convex printing technology is not always in proper relation to the quality of the printouts. There are no quality standards which precisely describe the parameters which have the biggest influence on the quality of convex copies, nor is it established how these parameters should be measured and which values should be allowed.

1 Introduction

The development of new hardware aiding blind and weak sighted people is observed together with the progress of new technologies, e.g. computers and electronic pocket notebooks are equipped with speech synthesisers. In spite of the current trends of artificial reading of Braille texts, traditional reading (developed in 1829) is still used. Braille's alphabet is read with the fingertips, from left to right (every finger covers one braille sign). Both hands are used in the process of reading (the right hand analyses while the left hand synthesizes signs).

Independent of count of points, each braille sign occupies the same area. Braille's alphabet is based on an array of six points in the area of a rectangle. The points are numbered:

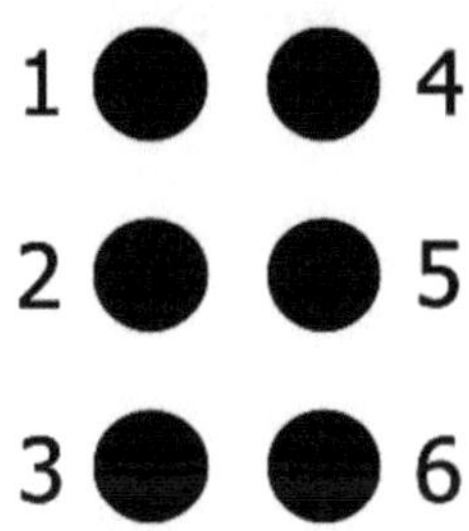

Fig. 1. Braille's six-point

Using different combinations of points, it is possible to get 64 characters: 1 without points (space), 6 one-point, 15 two-point, 20 three-point, 15 four-point, 6 five-point and 1 six-point. Each sign in Braille's alphabet represents one letter in the standard alphabet. Capital letters consist of two signs - the sign for the capital letter and the sign of the letter itself. Digits are similarly made up of two signs – the sign for the digit and the first letters of alphabet). Most letters are the same in all languages. The remaining signs represent language specific letters.

Braille's alphabet is used in special hardware for the blind, e.g. the Braille writing machine, Braille monitors.. The alphabet is also used to publish. books, newspapers, magazines, didactic help and dictionaries.

There are several technologies of making printouts with convex text and graphics. Technology is one of the factors which influence the quality of the copy, and the parameters describing the quality.

2 Parameters

Values of quality parameters of convex printouts are very different [1,2]. They are greatly influenced by: the technology of making a printout, the construction of hardware, even the kind of.

The blind say that the most important factor which influences the quality of convex printouts is the height of the points (this parameter determines, whether the text can be read or not. Even a well-convexed printout (with the right geometric parameters) might be read unwillingly. because of excessive roughness of the copy's surface. This causes induration of fingertips and reducing tactile sensitivity.

The parameters describing the quality of convex printouts are not normalised, therefore such copies have very different quality. The best estimate of quality is the opinion of a blind person. But what is also important, is the correlation of this estimate with technically measurable parameters.

The estimation of the parameters' range would force producers to make good quality printouts.

There are many parameters [3], which can describe quality of convex printouts, e.g.:

- height of points
- diameter of points
- distance between points
- distance between signs
- distance between lines
- roughness of the surface
- shape of point top
- technology of making a printout
- durability
- type of ground
- type of printout: one – or two-sided

The realised studies proved, that statistically important parameters are only height of points, diameter of points and roughness of surface.

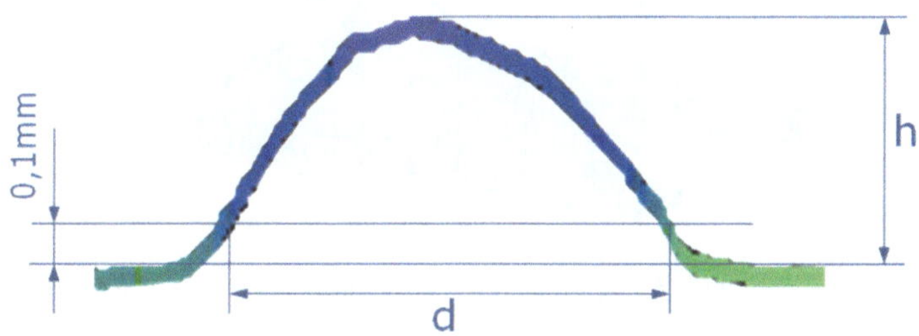

Fig. 2. Height and diameter of a point (d – diameter of point, h – height of point)

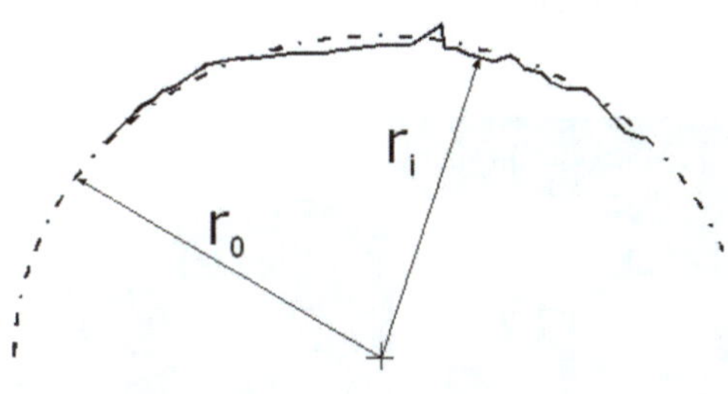

Fig. 3. Roughness of the surface

Roughness of a point's surface is estimated with parameter R_B

$$R_B = \sqrt{\frac{\sum_{i=1}^{n} (r_i - r_0)^2}{n-1}} \tag{1}$$

3 Measurements

The non-contact optic methods are the best methods to measure quality parameters. The most useful one is the fringe projection method, which allows for 3D scanning of an item, and describing its surface by a cloud of points.

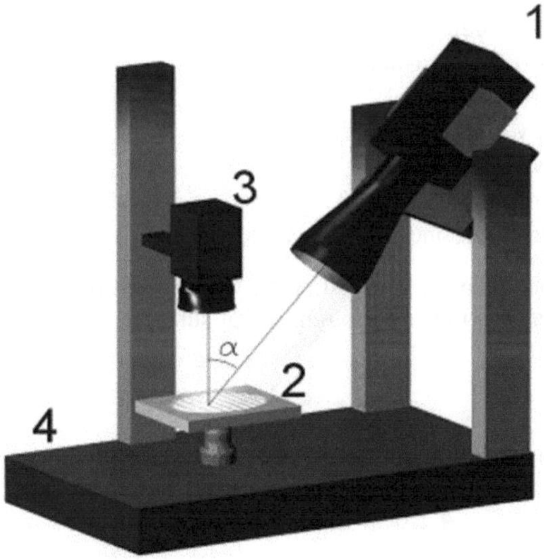

Fig. 4. Measurement stand (1- fringe projector, 2 – rotation table, 3- CCD camera, 4-base) [4]

While scanning fringes are projected on the measured object and they change systematically from dark to bright according to Gray's code. (figure 5) Such a projection is made at several angles.

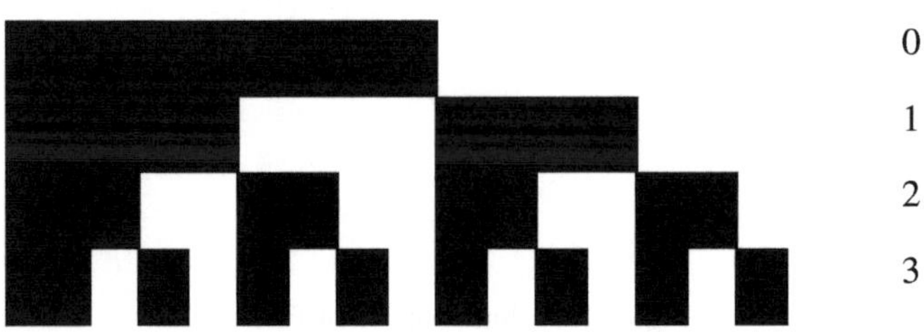

Fig. 5. Sequence of fringes (Gray's code)

In areas where the height is changed, the view of fringes is deformed (Fig. 6). The software determines the cloud of points analysing the sequence of views. (Fig. 7).

Fig. 6. Measured sample in the process of scanning

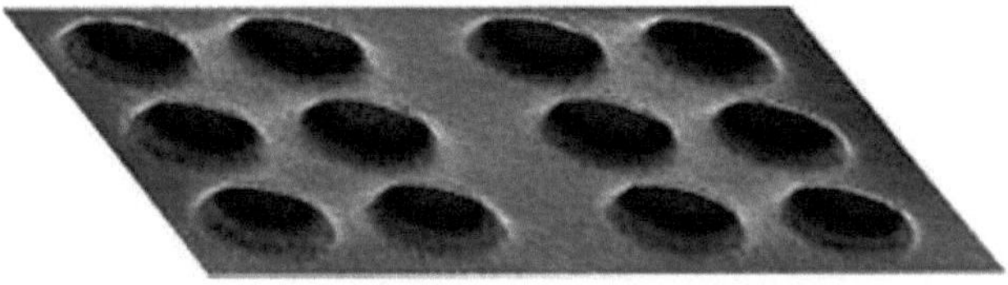

Fig. 7. Cloud of points of the measured sample

The next stage of measuring is the analysis of the received cloud of points using different software, which contains algorithms for estimating each parameter.

Measurements were made on the fringe-projection stand in the Optical Engineering Division of the Institute of Micromechanics and Photonics of the Warsaw University of Technology.

4 Studies

The preliminary studies allowed to determine the values of this these three parameters. Nowadays the representative group of visually impaired people in Poland evaluate specially prepared fragments of Braille's text with different quality parameters. Results of these studies will be published soon.

Acknowledgments. This work was supported by Polish Ministry of Science and Higher Education, Grant N N505 1919 33.

References

[1] Barczyk, R., Buczynski, L.: Estimating of Quality Parameters of Convex Copies for the Blind. In: IS&T NIP20: International Conference on Digital Printing Technologies, Salt Lake City, pp. 1011–1014 (2004)

[2] Barczyk, R.: Analysis of Qualitative Parameters of the 3D Convex Printing for Visually Impaired People, Diploma Work in Institute of Micromechanics and Photonics, Warsaw University of Technology (2004) (in polish)
[3] Jasińska, B.R., Choromańska, D.: The Most Important Factors Influencing Quality of Convex Printouts for the Blind. In: 3rd European Medical and Biological Engineering Conference EMBEC 2005, Prague, Czech Republic, November 20-25 (2005)
[4] Węgiel, M.: Analysis of the time dislocations and changes of object's shapes using virtual reality stand, Diploma Work in Institute of Micromechanics and Photonics, Warsaw University of Technology (2000) (in polish)

Early Detection of the Cardiac Insufficiency

M. Jamroży[1], T. Leyko[2], and K. Lewenstein[3]

[1] University of Technology in Warsaw, Faculty of Mechatronics, Metrology and Biomedical Engineering Institute, Warsaw, Poland
m.jamrozy@mchtr.pw.edu.pl
[2] Cardiology Rehabilitation Hospital in Konstancin Jeziorna, "Uzdrowisko Konstancin Zdrój" Sp. Z o.o., Poland
leyko55@gmail.com
[3] University of Technology in Warsaw, Faculty of Mechatronics, Metrology and Biomedical Engineering Institute, Warsaw, Poland
k.lewenstein@mchtr.pw.edu.pl

Abstract. Despite the huge technical progress in the cardiological diagnosis, heart auscultation is still an essential element of the physical study in cardiology. The possibility of hearing the acoustic phenomena produced by the contracting heart and the blood flowing through, enables the initial diagnosis of cardio-vascular pathology, especially cardiac valve defects. Replacing the traditional stethoscope with the electronic device (phonocardiograph) enables filtering noises, strengthening selected frequencies and referring them to the heart cycle. The development of phonocardiography had been suddenly stopped by introducing ultrasonic investigation into the cardiological diagnosis. The application of an innovative transducer (accelerometer) and the possibility of the digital processing of the acoustic signal, allowed to expand the acoustic analysis for the new frequencies (infrasounds). In the signal frequency analysis of the healthy heart, the largest spectrum strength is recorded among the frequencies between 6 to 12 Hz. In the case of damages in the left heart ventricle, the recorded spectrum is changed (the decrease of amplitude and higher frequencies appear).

1 Structure and Activity of the Cardio-Vascular System

The role of the circulatory system involves transferring blood, together with energetic components and products of metabolism which are transported to tissues and organs, thanks to which they can be used or removed from the organism. Heart is the organ which is responsible for the energy essential to activate movement, whereas blood vessels decide about the direction of the blood flow.

Heart has the weight of about 250-300 g. Muscle cells constitute about 70% of its mass, however from the point of view of the amount they are less than 1/3 of all the heart cells.

Heart muscle cells connect with each other by the use of highly specialized proteins and create a strictly defined structure. [3][4] Connections between heart muscle cells, apart from taking part in the contraction, also allow the efficient flow of the depolarizing current, which releases the process of contracting of the contractible elements , namely sarcomeres. It happens due to the cyclic increase of the calcium concentration in the environment of regulative proteins, belonging to sarcomeres. Spatial changes in the regulative proteins allow producing connections between contracting proteins- actin and myosin. Actin and myosin , due to the energy released from the hydrolysis of the high energetic bond called ATP, move towards each other, what lead to the contraction of the sarcomere. Synchronized contraction of all the sarcomeres included in the particular muscle cells causes contracting of all the heart muscle fibers.[1][4] Contraction of the sarcomeres, which is connected with the contraction of the muscle cells without changing ventricle's capacity, is possible due to the simultaneous stretch of the elastic elements included inside and outside the muscle cells.[4][5]

The same phenomena occurs in the right ventricle and pulmonary arteries, however the pressure is about 3-4 times lower. The blood transferred during the contraction stretches the elastic elements of the aorta and the pulmonary artery. During the relaxation phase, when the artery valves are closed, the energy gathered in the aorta and pulmonary artery causes farther blood transfer. With the frequency of about 70/' each of the heart ventricles pumps about 5-5,5 l. of blood per minute, what gives 7000 l. of blood per 24 hours. Only healthy, properly loaded and well nourished muscle can manage to do that. When any of the conditions mentioned is not fulfilled we can speak of the cardiac insufficiency.[4]

2 Cardiac Insufficiency

It is not possible to give one, universal definition of cardiac insufficiency. Most often we define it in terms of its symptoms which are: dysponea after effort, cough, attacks of night dysponea, weakness, cyanosis, swelling, palpitation etc. These symptoms are the result of the pumping function of the heart.[1][3][4] However, not only damages in the mechanical function of the heart determine the course of the disease. The so-called "compensational mechanisms" are here of crucial importance, among which we can point: activation of the sympathetic system and stimulation of the rennin-angiotensin-aldosterone axis. Short term influence of the above mentioned mechanisms causes keeping up the heart projection on the satisfactory level. However, when the cause of the damage is not removed, then the same mechanisms cause the sequence of reactions which increase the heart damage. As a result of the excessive sympathetic stimulation there is an increase in heart activity, and as a consequence there is an increased need of energetic substrates (especially oxygen). Activating the rennin-angiotensin- aldosterone system increases the volume of the plasma (higher strain on the heart), what is more, it changes the structure of the non-muscle heart elements, what leads to its fibrosis. The outcome of this process is a loss in elasticity, which is extremely important while transferring energy. Moreover, growth of the non-muscle fibers causes changes in the spatial arrangement of the muscle fibers, what lead to a change in the hemodynamic function of the heart ventricle. [2][3][4][5] In order to assess the level of cardiac insufficiency we use the functional classification according to the New York Heart Association (NYHA). We can distinguish four classes of the NYHA classification, which are as follows:

- NYHA I – patients with heart disease but with full physical activity
- NYHA II – patients with heart disease and a slightly limited physical activity
- NYHA III – patients with heart disease and considerably limited physical activity
- NYHA IV – patients who are unable to any physical effort[4]

The higher class of the cardiac insufficiency, the more serious prognosis and the shorter estimated period of the patient's survival. Consequently, the absolute mortality (of five years) among patients with cardiac insufficiency in the USA is about 50%, one year mortality in the case of the patients from NYHA IV is about 75%. Every year in the USA there is recorded about 260 000 deaths caused by the cardiac insufficiency. Cardiac insufficiency occurrence among patients between 35 and 65 year of age is about 1%, but among patients over 65 years old it grows to almost 10%.[4]

3 Characteristics of the Diagnostic Methods

Cardiac insufficiency is a dynamic process. In the first phase it takes the hidden course and the symptoms such as cough, difficulties in breathing or weakness are wrongly associated with general overtiredness, overwork or cold, they are hardly ever connected with heart disease. Whereas diagnosing the disease in the very first stage has crucial meaning for the patients, as eliminating the cause of the disease leading to the cardiac insufficiency can stop or even cause the regression of the disease. Unfortunately, diagnostic tools are limited because of the methods applied or difficult access to the highly specialized diagnostic laboratories. All these lead to the decrease of the chances to detect early disease symptoms. The most popular examination – electrocardiogram, usually doesn't show any pathological changes during this period. Similarly, the common chest radiograms are then appropriate in most cases.

Introducing examination with the use of ultrasounds can be marked as a turning point in the image diagnosis of the cardiac insufficiency. However, despite applying very advanced techniques including the use of tissue doppler, the very early stage of cardiac insufficiency cannot be detected, because it doesn't give any crucial pathological changes. Isotope and PET examination, as well as the magnetic resonance are still very expensive and hardly accessible and because of that they do not have a wide clinical application in detecting early cardiac insufficiency.[2][3][4] Taking into consideration difficulties with detecting early cardiac insufficiency and all the profits resulting from detecting it as early as possible, we made an attempt to use heart acoustic emission analysis to evaluate and detect early heart damage. This is the main goal of our research.

During heart beat cycle we can observe vibrations of the structures with frequencies characteristic for the particular histological structures. Depending on the degree of sarkomere stretch we can describe the tissue by the use of different reological methods. There are numerous factors influencing such model, e.g. relative stretch of the muscle, pathological changes, tissue's structure, the level of blood supply and oxidation. Adaptive mechanisms released by hemodynamic changes initiate and strengthen reconstruction of the heart muscle. Changing the expression of the genes responsible for the synthesis of the connective tissue lead to excessive heart fibrosis.

Infrasound signals generated during heart activity give information about the condition of the tissues. Methodology of gaining such information is an essential matter on the basis of which we can assess the condition of the heart using the analysis of vibrations [6].

The most crucial element in such vibration measurements is a converter. For this purpose we can use accelerometer or a funnel with the infrasound microphone. While choosing accelerometer we have to take into consideration its weight and measurement range. Accelerometer which can be used to make measurements from the surface of the patient's chest should be less inert, because the vibrations are small. Soft tissues are made of water in about 70% ; water is incompressible and that is why vibrations can act well in such an environment Another approach to measurement is the use of funnel with the infrasound microphone allowing measuring vibrations from the frequency of about 0 Hz. It enables the non-invasive examination and lowers safety demands. In this type of examination the patient's skin is the membrane. The measurement system prepared in such a way requires calibration, which can turn out to be expensive in the infrasound band. A great advantage of this solution is its low cost in mass production. The measurement trajectory used in the test examinations allows canvassing the signal, and after spectrum analysis – determining the cardiac insufficiency in the advanced stage. Imperfection of the measurement trajectory prevents precise determination of the signal's characteristics, which reflect particular changes. On the basis of the identical conditions of the examination, while comparing signals of the patient with a sound and an insufficient heart we are able to detect all the pathological changes. Signals obtained from the healthy person, measured in the same conditions are very similar to each other. Whereas signals taken from patients with cardiac insufficiency are considerably different in the area of frequencies measured in healthy persons. The measurement trajectory which is currently used by us doesn't allow defining either the period of disease or the causes of obtaining such a signal. The aim of our work is to design such a measurement trajectory so as to allow recording even the slightest changes in the signal, which could indicate the early stage of heart damage.

4 Short Summary

There had been conducted certain clinical examinations on the basis of the measurement trajectory, which in fact do not allow the quantitative evaluation. Tests we've made were assessed in terms of comparison and allow us to distinguish healthy persons from those suffering from the cardiac insufficiency. In order to conduct the quantitative analysis of the signal, there are professional measurement tools to be applied. The outcomes of our work are supposed to enable designing such a tool, which will be able to diagnose patients in different stages of the disease.

References

[1] Braunwald, E.: Atlas chorób wewnętrznych, Tom I, str. 22–30 (2003)
[2] Colucci, W.S., Braunwald, E.: Atlas niewydolności serca, str. 3–90 (2001)
[3] Braunwald, E.: Atlas chorób serca, str. 113–154 (2006)
[4] Braunwald, E.: Choroby serca. Tom 1, str. 21–48; 443--552 (2007)
[5] Szczelik A.: Choroby wewnętrzne, str. 5–13, 311--316 (2006)
[6] Nałęcz, M.: Biopomiary, str. 287–291 (2000)

[illegible]

System for Gaining Polarimetric Images of Pathologically Changed Tissues and Testing Optical Characteristics of Tissue Samples

N. Golnik, T. Pałko, and E. Żebrowska

Warsaw University of Technology, Faculty of Mechatronics,
Institute of Metrology and Biomedical Engineering, Warsaw, Poland

Abstract. Early detection of pathological changes in tissue is significant especially for cancer diagnosis. In many cases it decreases tumor development or completely cures patient. The most reliable method of cancer diagnosis currently used is microscopic cytological or histology analysis, which requires wide knowledge and experience form doctors. Polarimetric method presented in this article allows observations of cells or tissue sectors in light of different polarization degrees, under the microscope. Genesis of pathological changes entails in remodeling of tissue structure. Due to structure changes and rebuilding of collagen molecules, tissue reveals different optical properties than physiologically normal tissue. One of those is birefringence. Described work was focused on elaborating polarimetric method for testing optical characteristics of tissue samples, choosing the proper method for polarimetric image analysis and finding such optical characteristics, which will show correlation with cancer occurrence. The scientist aim was to extend knowledge of relation between arrangement of birefrigent elements and physiological state of tissue.

1 Introduction

Photon transmission through biologic materials depends on reflection, dispersion and absorption in this material. Dispersion is a dominant effect, which decides about tissue optical properties. That's why, main problem in gathering substantial biomedical information using absorption spectroscopy or optic imaging, is elimination or proper reflection of dispersed light. Using polarized beam allows for discrimination reflected and polarized light. Reflected light keeps polarization, while dispersed is being depolarized due to multiple dissipation or interaction with birefringent elements (for ex. collagen fibers). Visible radiation, especially around infrared range, has quite large ability for tissue penetration. It allows observing skin surface and hypodermic layers, from which depolarized light comes [7]. In a case of skin cancer, traditional initial diagnosis is based on evaluation of asymmetry, edge, color and diameter of change. Shifty changes are checked after biopsy which is an invasive, painful and time consuming procedure. Using polarized light assures noninvasive or less invasive diagnosis. Birefringence of cell structures may be a result of naturally present double refracting elements, proper staining or regular arrangement of fine parts with refractive index unlike the surrounding environment. Images gained in polarized light may contain substantial information about optical properties of examined tissues and fulfill data gathered after common luminosity analysis. Works, on using polarized light to biomedical examination, are being carried for many years. In 1949 it has been certified that nerve cells activity is correlated with the change of it's optic properties [1]. It is also known, that changes of tissue optical parameters are connected to it's metabolic or/and electric activity [2-5]. Therefore analysis of polarization degree of dispersed light may provide structural and functional information on examined tissue. One of the main implementation of polarized light in cancer diagnosis is detection of early cancerous skin changes or states before it. So far, many useful diagnostic methods has been elaborated. Jaques in his work [6] points that registration of two images in polarized light, with parallel and crossed polarizer, and calculation the state of linear polarization, enables receiving image with slightly higher contrast, which can reveal structures invisible in normal imaging.The main describing tool in polarimetric methods are Mueller matrix and Stockes vectors. Works on possibility of implementation of polarimetric microscopy in tissue structures imaging are intensively developed in Yuriy Fedkovych Chernivtsi National University. They proposed usage of amorphous substance model with uniaxial crystals. According to this model, change of tissue state (for example tumor occurrence) results in remodeling of spatial orientation and frequency of chain. Statistic and correlative methods, used for image analysis shown, that they allow for earlier medical diagnosis than presently used methods [10,11]. Based on results scientists from Chernivtsi National University proposed method for early cancer diagnosis.

2 Background and Purpose

The main purpose of described project was to test method for gained images analysis, for it's potential usability in cancerous changes diagnosis. The primary condition was receiving images in adequate quality, without interference from measuring system. As a result, polarimetric images gained from system, has much better quality that those described in literature concerning correlative methods for image analysis. Furthermore this images has quality sufficient for histopathologic tissue examination. They were sent to Pathomorphology Laboratory in Warsaw, where it's potential usability in telehistopathology was confirmed.

3 Materials and Methods

3.1 Preparation of Tissue Samples

Five basic tissues may be distinguish in human body: epithelial tissue, connective tissue, nervous tissue, muscle tissue and blood. Connective tissue consists two basic components: cells and intracellular fluid. Fluid contains many fibers keeping the proper state of tissue, and basal substance (gel) holds cells and fibers. Many fibers are built from collagen which reveals birefringence.Birefringence elements can also be found in muscle tissue. Micro-fibroblasts are built from actin (amorphous) and myosin (crystalline). In opposition to connective tissue, muscle tissue consist birefringent elements in cellular part. In connective tissue birefringent elements forms dead framework supporting cell structures. That's why pathological changes appearing in different tissues may differently influence observed tissue birefringence.

3.2 Algorithm

Full information about polarizing properties of object put under beam of light is contained in Mueller matrix [8]. This matrixes interact with 4-element Stockes vectors [9], which describes partly-polarized light, so they include 16 elements. According to definition, for defining 4-element Stockes vector, 6 light intensity measurements should be done for suitable states of polarization (4 linear and 2 circular). That's why, in the easiest version, for calculating 16-element Mueller matrix, it's necessary to do 24 measurements. For every (x,y) point of tissue image, Mueller matrix in this point is a product of diagonal amorphous part (which describes absorption) and crystal part (which describes birefringence of single element): $M_{ij}(x,y) = A_{ij}(x,y) \cdot C_{ij}(x,y)$. In this model Mueller matrix $M_{ij}(x,y)$:

$$M = A\begin{Vmatrix} 1 & 0 & 0 & 0 \\ 0 & \cos^2 2\alpha + \cos\delta \sin^2 2\alpha & \cos 2\alpha \sin 2\alpha(1-\cos\delta) & -\sin\delta \sin 2\alpha \\ 0 & \cos 2\alpha \sin 2\alpha(1-\cos\delta) & \sin^2 2\alpha + \cos\delta \cos^2 2\alpha & \sin\delta \cos 2\alpha \\ 0 & \sin\delta \sin 2\alpha & -\sin\delta \cos 2\alpha & \cos\delta \end{Vmatrix} \tag{1}$$

where:

$A=\exp(-\mu d)$ - describes light absorption in anisotropic environment ($-\mu$ - extinction factor, dthickness); δ - phase displacement made by birefringent element; α - angle between birefringent element axes and laboratory system axes.

In our simplified system, measurements time was shortened and only 3 components of Stockes vector and 3x3 Mueller matrix were calculated. At first, only images in the line of M_{22}, M_{23} and M_{33} elements (or their normalized equivalents) where calculated. Presently, we think, that essential information about spatial distribution of birefringent elements can be calculated from expression:

$$P=(2\times M_{11}-M_{22}-M_{33}) = A(1-\cos\delta) \tag{2}$$

or for normalized matrix:

$$p=(2-m_{22}-m_{33})=1-\cos\delta \tag{3}$$

Information about angle distribution can be found after analyzing:

$$\frac{M_{23}}{(2M_{11}-M_{22}-M_{33})} = \frac{1}{2}\sin 4\alpha \tag{4}$$

4 Results of Optical Characteristics of Tissue Samples

The first tested image analysis method was simple statistic analysis of opacity degrees of images responding to Mueller matrix elements, Stockes vector or distribution of opacity degrees on images gained for crossed and parallel polarizer and analyzer. Moments distribution of 4th order was calculated. Received results proved that statistic analysis of image opacity is too weak and doesn't allow for differencing regular and pathologically changed tissues. Exemplary results for mamma sample are shown in Fig.1.

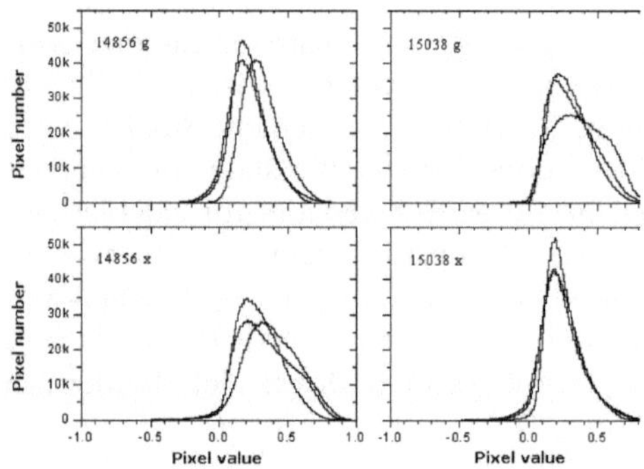

Fig. 1. Histograms of M22 Mueller matrix element image points values distribution for two mamma sample sets (x for regular and g for pathologic) calculated in 3 different places in sample.

The second method was autocorrelation, which measures signal correlation with it's copy moved with special value. Autocorrelation analysis was calculated for images gained with crossed polarizer and analyzer because in this position birefrigent elements are clearly visible. Images of autocorrelation for mamma tissue sample are shown in Fig.2 (sample 14856).

It has been observed during research, that important information, about structures visible in autocorrelation image, is lost. Using this methods requires then presenting full autocorrelation maps. Results received from another methods of image analysis (such as Fourier and multirepresentation entropy analysis) are too time-consuming to be clinical application. Analysis of curettage uterus shown that stained samples does not reveal measurable birefringence. Even images with parallel polarizers, images had very good quality and potentially could be used in telemedicine, images for crossed polarizers are completely dark. Specially prepared not-stained samples reveals birefringence, but interpretation is strongly hampered due to paraffin birefringence.

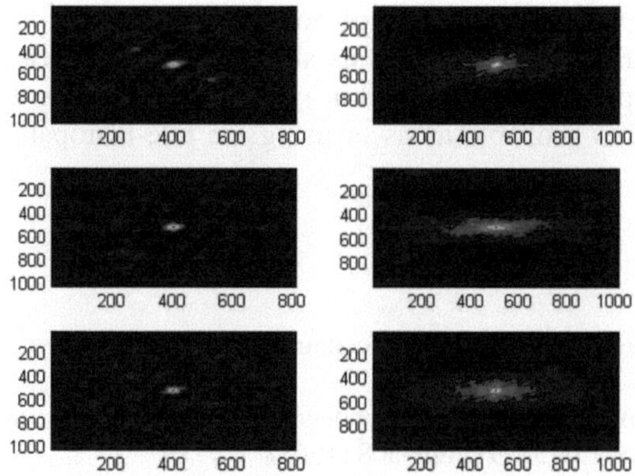

Fig. 2. Autocorrelation images for mamma tissue gained with crossed polarizer and analyzer (regular tissue – left, pathological tissue – right).

5 Results and Discussion

The essence of polarimetric method is visualization of elements in tissue which reveals birefringence (for example collagen fibers) and searching for correlation of distribution changes with pathological changes. The primary method is to calculate Mueller matrix and analyze it's elements. In this project the number of necessary measurements were shortened, and only 9 of Mueller matrix elements where calculated. Also, it has been pointed out that there is such combination of this matrix elements, which gives: information about distribution of phase displacement made by birefringence elements, and about distribution of angles made between axis of birefringence element and axis of laboratory system. Analysis of other Mueller elements does not give a clear results.Modified and tested polarimetric system used within this work was built as an attachment for optic microscope. It includes CCD camera which allows loading images with 1024x1344 resolution and 12 bit dynamics. For correction of images movements with different polarizer and analyzer position, special algorithm was written in MatLab. Thanks to this, gained precision was better than in earlier publications. For tissue image analysis a special program has been written in MatLab environment, which calculated Stockes vector and Muller matrix. Built system and proposed software assures archiving images with quality adequate for clinic diagnoses and may be used in telehistopathology.During research couple series of normal and pathologically changed tissue samples prepared in standard way was tested. It occurred that image analysis results strongly depends on sample preparation. Best samples would be unstained, but it's hard to get such samples not covered with paraffin. Information of using special stains [13-15] or special methods for preparing samples [12] can be found in literature.Results shows that polarimetric method may be useful in breast cancer diagnosis (in addition to histopathologic tests) and in brittle bone disease diagnosis (in both cases if biopsy is done). Analysis of matrix p element gave very good results for cervix of the uterus tissue. Confirming this results in medical research, may enable usage of polarimetric method in colposcopy.Wide series of endometrium samples analysis shown that in this case birefringence is poor. That's why polarimetric method usage is technically hard and limits practical application.Other probable areas of using are tele-histopathology and genetic engineering.

References

[1] Hill, D.K., Keynes, R.D.: Opacity changes in stimulated nerve. Journal of Physiology 108(3), 278–281 (1949)

[2] Chance, B., Cohen, F., Jobsis, F., Schoener, B.: Intracellular oxidation reduction state in-vivo. Science 137(3529), 499–508 (1962)

[3] Grinvald, A., Manker, A., Segal, M.: Visualisation of the spread of electrical activity in rat hippocampal slices by voltage sensitive optic probes. Journal of Physiology 333, 269–291 (1982)

[4] Lipton, P.: Effects of membrane depolarization on light scattering by cerebral cortical slices. Journal of Physiology 231(2), 365–383 (1973)

[5] Villringer, A., Chance, B.: Non-invasive optical spectroscopy and imaging of human brain function. Trends in Neurosciences 20(10), 435–442 (1997)

[6] Demos, S.G., Radousky, H.B., Alfano, R.R.: Deep subsurface imaging in tissues using spectral and polarization filtering. Optics Express 7(1), 23–28 (2000)

[7] Jacques, S.L., Ramella-Roman, J.C., Lee, K.: Imaging skin pathology with polarized light. Journal of Biomedical Optics 7(3), 329–340 (2002)

[8] Mueller, H.: The foundations of optics. Journal of Optical Society of America A38, 661 (1948)

[9] Stokes, G.G.: On the composition and resolution of streams of polarized light from different sources. Transactions of the Cambridge Philosophical Society 9, 399–416 (1852)

[10] Angelsky, O.V., et al.: Polarization-phase reconstruction of biotissues structure in their pathological changes diagnostics. In: Proc. International Conference Mechatronics, Warszawa, September 21-23, 2000, pp. 462–465 (2000)

[11] Pishak, V.P., Ushenko, O.G. (red): Laserowo-polarymetryczna diagnostyka w biologii i medycynie. Czerniowce (2000) (po ukraińsku)

[12] Sardar, D.K., Yow, R.M., Tsin, A.T.C., Sardar, R.: Optical scattering, absorption, and polarization of healthy and neovascularized human retinal tissues. Journal of Biomedical Optics 10(5), 51501–51501 (2005)

[13] Dziedzic-Gocławska, et al.: Polarizing Microscopy of Picrosirius Stained Bone Sections as a Method for Analysis of Spatial Distribution of Collagen Fibers by Optical Diffractometry. Bas. Appl. Histochemistry 26, 227–239 (1982)

[14] Allona, I., Vereda, M., Buchnera, A., Dayan, D.: Stromal differences in salivary gland tumors of a common histopathogenesis but with different biological behavior: A study with picrosirius red and polarizing microscopy. Acta Histochemica 108(4), 259–264 (2006)

[15] Shidham, V., Chivukula, M., Basir, Z., Shidham, G.: Evaluation of crystals in formalin-fixed, paraffin-embedded tissue sections for the differential diagnosis of pseudogout, gout, and tumoral calcinosis. Mod. Pathol. 14(8), 806–810 (2001)

Long-Term Monitoring of Transtibial Prosthesis Deformation

D. Paloušek[1], P. Krejčí[2], and J. Rosický[3]

[1] Institute of Machine and Industrial Design, Faculty of Mechanical Engineering, Brno University of Technology, Technická 2896/2, 616 69 Brno, Czech Republic
Tel.: +420541143261
palousek@fme.vutbr.cz

[2] Institute of Solid Mechanics, Mechatronics and Biomechanics, Faculty of Mechanical Engineering, Brno University of Technology, Technická 2896/2, 616 69, Brno, Czech Republic
Tel.: +420541142888
krejci.p@fme.vutbr.cz

[3] Department of Physiotherapy, Medico-Social Faculty, Ostrava University, Syllabova 19, 703 00 Ostrava, Czech Republic
Tel.: +420558663215
jiri.rosicky@osu.cz

Abstract. The paper deals with process of realization of one channel strain gauge amplifier. The amplifier supposed to be used for deformation measuring of tube adaptor of transtibial prosthesis. The aim of measuring is focused on long time data recording of deformation in rear-front direction during real load conditions. Results of measuring will be used for transtibial prosthesis adjustment for concrete patient. This paper describes measuring chain, development of hardware and software for long term data acquisition.

1 Introduction

There is an force and torque load between a socket of the transtibial prosthesis and the residual limb. This load varies during patient motion and depends among other on patient weight, speed of walking, type of movement and terrain. The load between the socket and the residual limb is also influenced by adjustment of prosthesis so-called Alignment [1]. Inappropriate prosthetic alignment can lead to excessive load of the residual limb, instability of prosthesis during movement (loss of safety) and can also leads to gait deviation. Therefore prosthetic alignment is performed in three steps - bench alignment, static alignment and dynamic alignment [2]. The optimal alignment can be assessed subjectively based on patient feeling during walk with prosthesis or objectively based on dynamic or kinematic analysis of gait [3]. The dynamic analysis is based on measuring of force interaction between prosthetic foot and ground while kinematic analysis is based on video analysis.

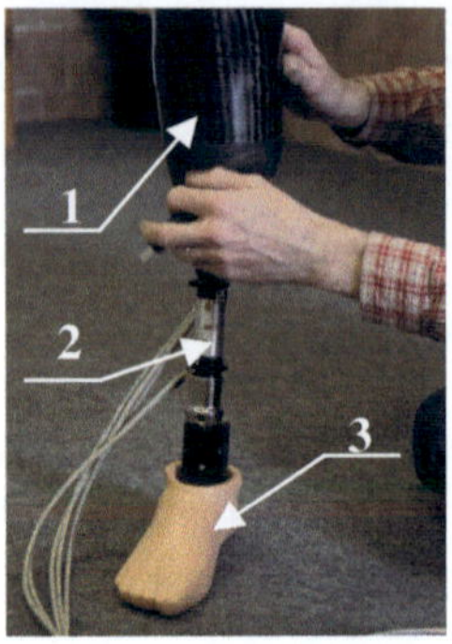

Fig. 1. Transtibial prosthesis (strain gauges on right leg) 1. Socket, 2. Tube adaptor, 3. Foot

Torque load of prosthesis can be calculated by combination of kinematic and dynamic analysis. Unfortunately, these approaches are linked to measuring and analysis in biomechanical laboratory and do not reflect everyday activities of patient. This fact leads to design of methodology that involves measuring of force and torque loads of prosthesis in real time during everyday activities. The methodology will be used for optimal alignment of transtibial prosthesis. The project is realized in cooperation with ING corporation, s.r.o. Ortopedická protetika Frýdek-Místek and Department of Physiotherapy, Medico-Social Faculty, University of Ostrava.

2 Objectives

The aim of the project is to realize long term monitoring of deformation of tube adaptor of transtibial prosthesis. It is required to develop monitoring system with simple operation, light-weight and small size. The measuring system will be attached to patient, therefore no constrains of movement is required. The connection between measuring system a driving computer (carried by patient in bag) will be realized by wireless network.

Goals of project are following:

1. Design of measuring chain based on required application.
2. Development of simple measuring amplifier.
3. Development of measuring software.
4. Testing of system and verification of functionality.

3 Realized Measuring

The need of long term deformation monitoring arises out of previous measuring performed in biomechanical laboratory [4]. Those measurements were oriented on deformation measurement of tube adaptor of transtibial prosthesis via strain gauges, reaction force measurement via force plates and kinematic and dynamic analysis via motion capture. The laboratory measurements were realized during straight walk, stairs walk and ramp sloping walk of patient. The decision of real time monitoring of patient walk was done based on problem detection in laboratory environment.

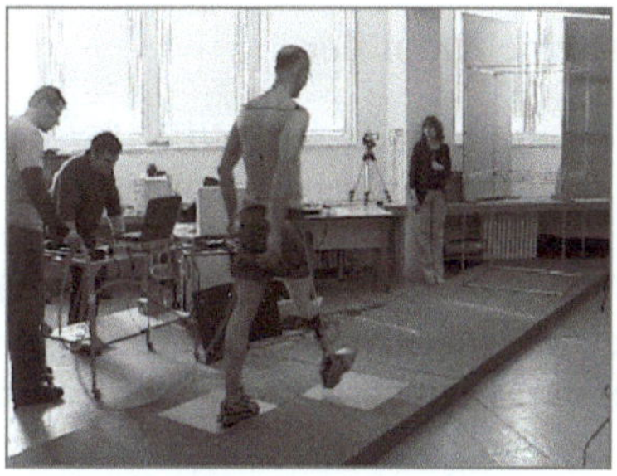

Fig. 2. Gait analysis in the laboratory

4 Measuring Chain

The measuring chain consists of strain gauges (attached to the tube adaptor of the transtibial prosthesis), transtibial prosthesis, wires, measuring amplifier, measuring pc card and measuring software. Strain gauges are used for measuring of deformation during bending. Signals form strain gauges are measured by measuring amplifier and processed by measuring card NI USB-6009.

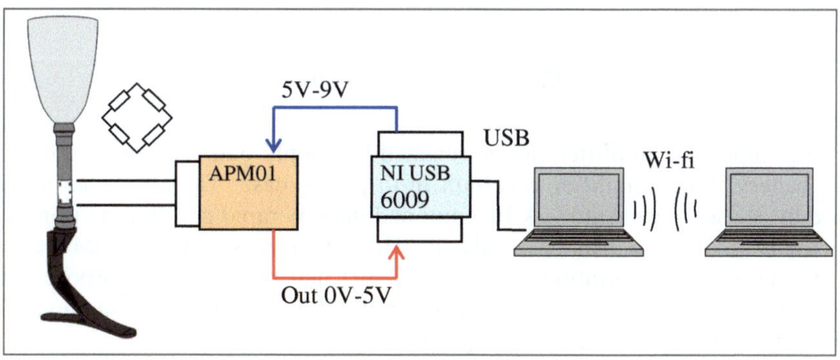

Fig. 3. Measuring chain

The special software was developed to manage measuring by subnotebook. This software is controlled by second notebook via wireless (wi-fi) connections. The measuring amplifier, measuring card and subnotebook is attached to and carried by patient during prosthesis deformation monitoring.

4.1 Transtibial Prosthesis

Prosthesis (Fig. 1) consists of socket, tube adaptor, connection adaptors and foot. The socked is designed and manufactured individually for each patient for normal life use. The tube adaptor is made from AlCu4Mg CSN 424201 (EN: AW-2017A equivalent) and is adjusted for patient height. The foot is made by Ossur company (type of foot is Sure-Flex).

4.2 Strain Gauge Measuring Amplifier

It was necessary to developed measuring amplifier for strain gauge measurement. The amplifier should be small and light in reason of its carriage by patient during long time measurement. The scheme of amplifier is shown on Fig. 4. and comes from INA125 datasheet. The INA 125 is a low power, high accuracy instrumentation amplifier with a precision voltage reference. It provides complete bridge excitation and precision differential input amplification on a single integrated circuit.

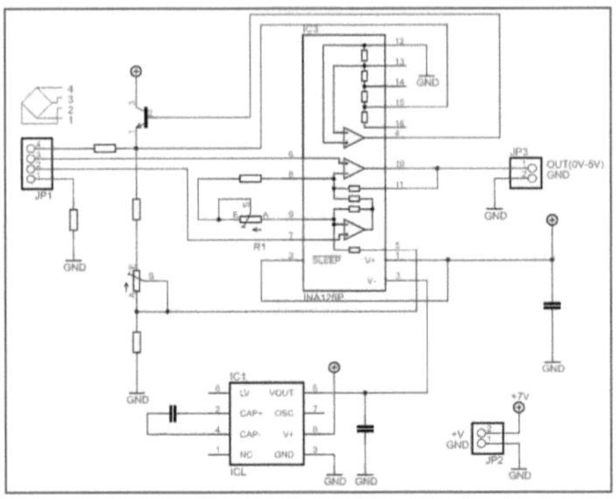

Fig. 4. Scheme of amplifier

Power supply of amplifier can be realized by supply output of measuring card or by battery. Final amplifier was built in to plastic case. The case was first designed by 3D software Autodesk Inventor and than by rapid prototyping printed to real case (Fig. 5) made from ABS plastic. The Stratasys SST 1200 FDM technology was used for 3D printing. The case is stiff and light for our purpose sufficiently. The design involves also connection to HBM Spider measuring amplifier by 15- pins connector for calibration purpose.

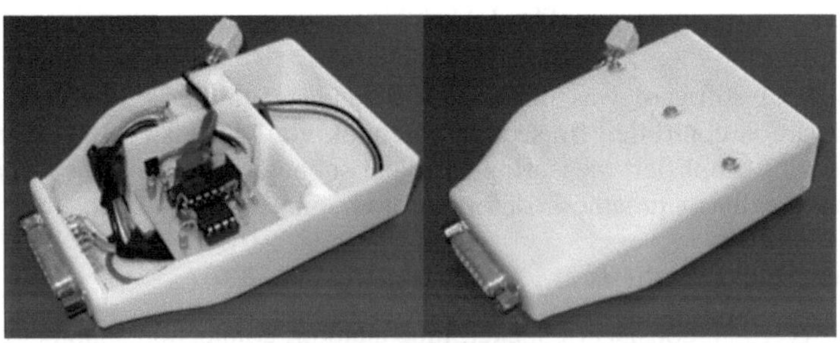

Fig. 5. Inside and outside view of amplifier

4.3 LabView

The measuring software was created in NI LabView that supports measuring card NI USB-6009 by drivers and tools for card control. Developed software (called SAM-1) also involves processing of measured data of tube adaptor deformation. The measuring card transforms analog amplified signal to digital form and sends it via USB interface to computer. The measured data are proceeded by low pass filter for further real time visualization. The proceeded data are also recorded to the file. The block diagram of developed software is shown on Fig. 6.

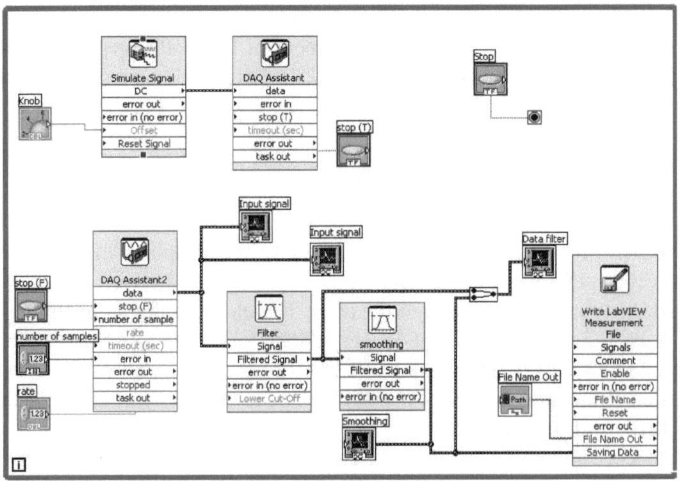

Fig. 6. Block diagram of measurement

The SAM-1 software can be used independently of LabView on each PC. Installation contains all necessary drivers and libraries for NI USB-6009 card. The software user interface (Fig. 7) is simple to use and can by operate also by patient during measurement. The software shows real time natural and also filtered data.

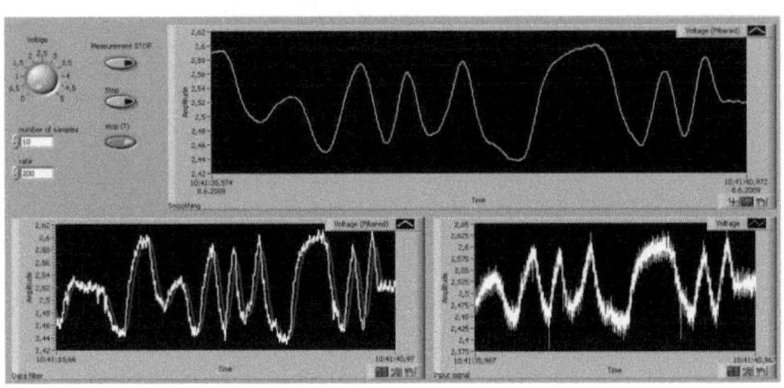

Fig. 7. Front panel of SAM-1

5 Conclusion

This paper presented results of measuring chain design and development for measuring of the prosthesis tube adaptor bending. The measuring chain for real time measurement was developed and its functionality was proofed by initial measurement. The research will continue by calibration of amplifier by mechanical testing machine ZWICK Z020. Then the measuring system will be used to patient measurement and retrieved data will be statistically elaborated by Matlab and Statistica software.

Acknowledgments. Project was realized thanks to support of ING corporation, s.r.o. – Ortopedická protetika Frýdek-Místek and Department of Physiotherapy, Medico-Social Faculty, University of Ostrava and MSM 0021630518 "Simulation modeling of mechatronic systems" project.

References

[1] Seymour, R.: Prosthetics and orthotics-Lower limb and Spinal. Lippincott Williams & Wilkins (2002) ISBN 0-7817-2854-1

[2] Reginald, M.: Amputation through the ages: The oldest major surgical operation. Australian & New Zealand Journal of Surgery 68(9), 675, 4, 6bw (AN 5549225) (1998)

[3] Bowker, J.H., Michael, J.W.: Atlas of Limb Prosthetics: Surgical, Prosthetic, and Rehabilitation Principles, 2nd edn., p. 930. American Academy of Orthopaedic Surgeons (1992) ISBN: 0-8016-0209-2

[4] Paloušek, D., Návrat, T., Rosický, J., Houfek, M., Krejčí, P.: Experimental Recognition of Loading Character of Transtibial Prosthesis. Engineering Mechanics 15(5), 355–364 (2008)

Tensile Stress Analysis of the Ceramic Head with Micro and Macro Shape Deviations of the Contact Areas

V. Fuis

Centre of Mechatronics – Institute of Thermomechanics AS CR and Institute of Solid Mechanics, Mechatronics and Biomechanics, Faculty on Mechanical Engineering, Brno University of Technology, Technicka 2, Brno, Czech Republic
fuis@fme.vutbr.cz

Abstract. The paper deals with the problems of ceramic head of hip joint endoprosthesis destructions in vivo, and with assessing the impact of shape deflections of conical surfaces on the tensile stress in the head. There are assumed two material characteristics of the stem – the first is linear elastic and the second is bilinear elasto-plastic modelled austenitic steel used in the bio-implants. Concerned are shape deviations from the ideal conical surfaces of the stem and the head of the endoprosthesis. The shape deviations may be modelled at the macro-level - this concerns model shape inaccuracies such as deviation from the nominal degree of taper, at the micro-level - when the stochastic distribution of unevenness on the contact areas is respected. The problem of stress in ceramic heads was solved using the finite element method – system ANSYS under ISO 7206-5 loading. In the paper are presented and analysed the results of solution of the macro shape deviations and micro shape deviations, obtained from measurements made on the cones of stems and heads. There are analysed three variants of the sizes of the measured micro shape deviations (measured, doubled and halved).

1 Introduction

The failure of cohesion of ceramic heads of total hip joint prostheses has been stated in a not negligible number of patients in the Czech Republic. The implant's failure of the ceramic head has always traumatic consequences for the patient, since a part of or even the whole endoprosthesis has to be re-operated. Hence, it is desired to reduce the number of implant re-operations to the minimum. Therefore the computational modelling (using FEM) of the stress and the failure probability (based on Weibull weakest link theory [1]) of the ceramics head was realised. In this case the influence of the material of the stem (elastic and elasto-plastic) and value of the micro shape deviations of the stem and the head contact cone were analysed.

2 Methods

The computational modelling of stress has been made on a system consisting of a testing steel stem and a ceramic head. The head has been put on the cone of the testing stem and the load of this system has been in compliance with ISO 7206-5 – Figure 1, which is the standard for determining static strength of ceramic heads for hip joint endoprostheses. With view to geometrical inputs, the deviations from ideal shapes can be divided into two groups - global (macro) deviations, i.e. the deviations from the stem's or head's nominal cone-shape - angle α in Fig. 1. Maximum allowed difference of the head's and stem's cones is α = 10' and this value is assumed in the computational modelling.

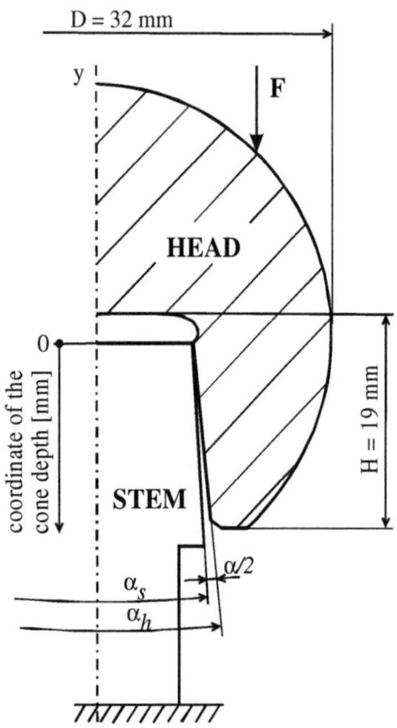

Fig. 1. System under ISO 7206-5 loading – macro shape deviation

The second group is local (micro) deviations which were measured using IMS UMPIRE device. These deviations are represented by micro shape deviations from the ideal cone are shown in Fig. 2. The deviations from the ideal cone shape are represented in the developed section of the cone in Figure 2a – measured head cone micro deviations (from -3.9 μm to +4.12 μm), Figure 2b – stem cone micro deviations (from -2.27 μm to +2.28 μm). The computational modeling was realised for three values of the micro shape deviations – the measured head and stem deviations – VAR. 1, the double of the measured deviations – VAR. 2 and the half of the measured deviations – VAR. ½. The micro shape deviations shown in Fig. 2 is the same for all three variants, only the scale is changed (doubled or halved).

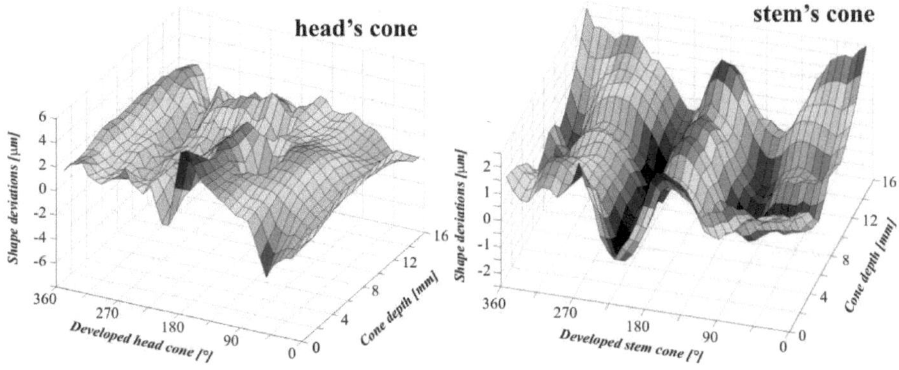

Fig. 2. Measured micro shape deviation of the head's and stem's cone areas – VAR. 1

The state of stress of the ceramic head of endoprosthesis can be strongly affected by the process of endoprosthesis implantation, when the surgeon fits the head on the cone of the stem. The fitting of the head on the cone of the stem is a random process, in view of the mutual position of the head to the stem (in sense of the head's slight turning around the axis y, defined by angle β - Fig. 3). Therefore a series of computations will be made for each pair and for various positions of the head towards the stem.

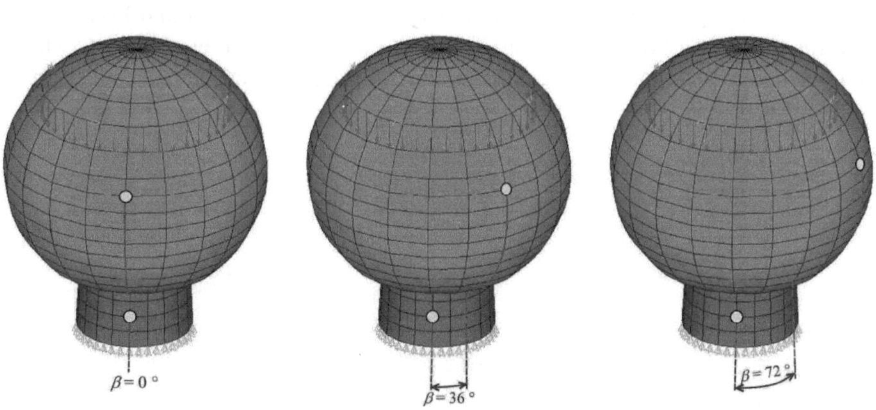

Fig. 3. Representation of various head to stem positions defined by angle β

The head is modelled as a linear isotropic continuum with the following constants – E_h = 390 GPa and μ_h = 0.23. The stem is modelled as a linear isotropic continuum (elastic) with E_s = 210 GPa and μ_h = 0.3 and as bilinear isotropic continuum (elasto-plastic) with E_s = 210 GPa and μ_h = 0.3 and $R_{p0.2}$ = 196 MPa and $R_{p1.0}$ = 235 MPa (according to the norm of the austenitic steel). The coefficient of friction between the head and the stem is f = 0.15 [2]. FEM ANSYS system has been used for modelling the stress in the system. From the viewpoint of the bonds of the system, it may be stated that besides the head - stem contact also the shift of the stem's lower part has been prevented. To ensure convergence, the head's rotation on the stem has also to be prevented. The load of the system corresponds to ISO 7206 - 5 - the force acts as linear load on a on a circle with a radius of 10 mm – Fig. 3.

3 Results

In a ceramic head a 3D state of stress sets in under loading. In view of the reliability of the head, which is made of brittle material, the most important are the tensile stresses [3]. For this reason only extreme tensile stresses in the ceramic head will further be analysed.

The distribution of maximum values of tensile stresses in the ceramic head (σ_{max}), in dependence on the value of the head' s loading, is shown in Figs. 4 and 5. The values σ_{max} for various positions of the head to the stem (various value of angle β) and for two types of the stem material (solid lines – elastic steel, dotted lines – elasto-plastic steel (austenitic)) form in Fig. 4 a belt of curves. The σ_{max} values for the elastic stem are significantly higher than for austenitic steel stem due to the plastification of the micro shape deviations of the elasto-plastic stem. The plastification caused the reduction of the contact pressure (increasing the contact area) on the contact cone and therefore it brings down the tensile stress in the head. The dispersion of the σ_{max} for austenitic steel are significantly lower (about 40% reduction) than for elastic steel due to the plastification of the stem. The elasto-plastic stem is more suitable for bio-implants due to the reduction of the tensile stress in the ceramic head.

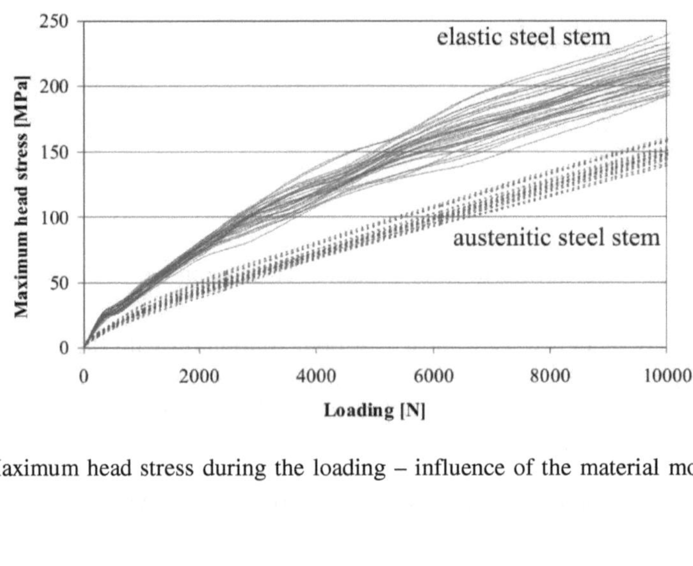

Fig. 4. Maximum head stress during the loading – influence of the material model of the stem

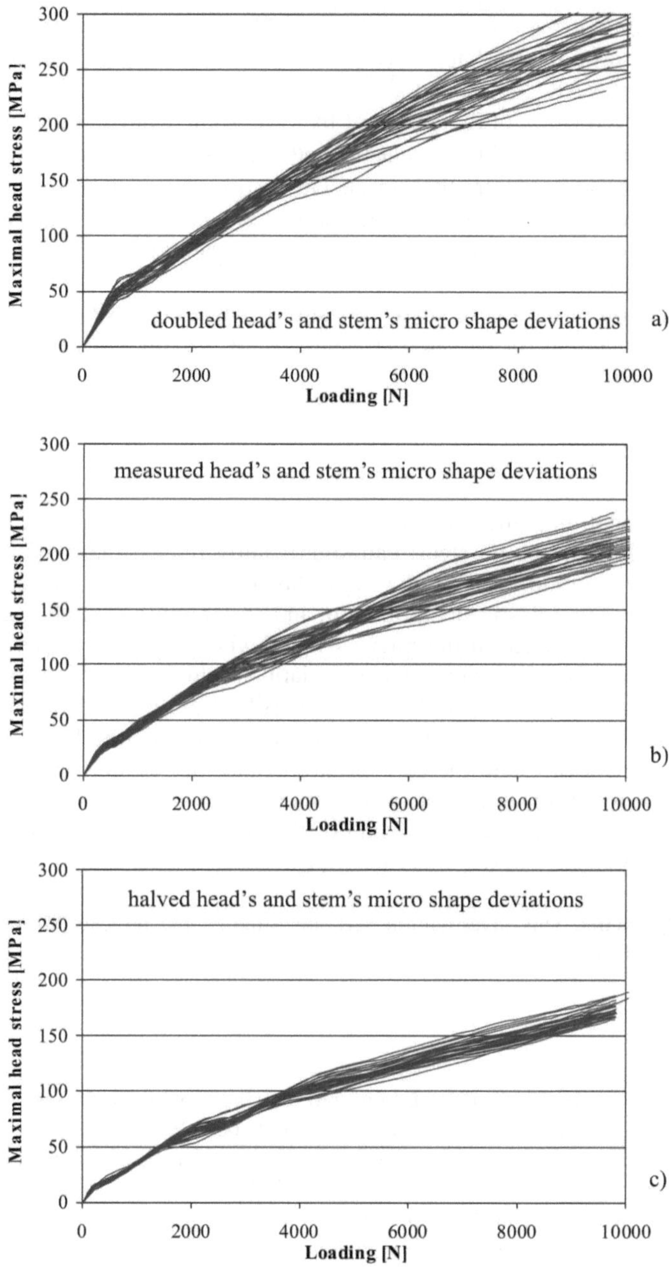

Fig. 5. Maximum head stress during the loading (elastic material of the stem) – influence of the value of the micro shape deviations: a) the micro shape deviations were doubled; (VAR. 2), b) the micro shape deviations without the changes (VAR. 1); c) the micro shape deviations were halved (VAR. ½)

The Fig. 5 shows the influence of the sizes of the head's and stem's micro shape deviations of the contact cone surface. There are assumed three variants – VAR.1 – measured micro shape deviations, VAR. 2 – doubled sizes of the measured values and VAR. ½ - halved sizes of the measured values. The Fig. 5 shows the significant reduction of the values of the σ_{max} in the head and dispersion of the σ_{max} curves during the loading for elastic material of the stem (curves in the Fig. 5b and solid curves in Fig. 4 are the same). The dispersions of the σ_{max} curves of the VAR. 1 and VAR. 2 are nearly same, only the σ_{max} values of the VAR. 2 are higher. Micro shape deviations for VAR. ½ caused the reduction of the σ_{max} values and dispersion too.

The reduction of the micro shape deviations (halved – VAR. ½) and using the elasto-plastic material for the stem are suitable for the reliability ceramic head of the total hip joint endoprosthesis.

4 Conclusion

By computational modelling it has been proved that the elasto-plastic stem material (for example austenitic steel) can significantly reduce the maximum tensile stresses in the ceramic head by the high plastification of the stem cone. Elastic material of the stem (elastic steel, titanium and chromium-cobalt alloys, ...) cannot reduce negative influence of the micro shape deviation of the contact cones. The value of the sizes of the micro shape deviations which are added to the macro shape deviation, significantly influence the tensile stress in the head and the value of the dispersion the of σ_{max} curves (different position of the head on the stem come). The tensile stress in the ceramic head cause the brittle fracture and therefore the reduction of the tensile stress (austenitic stem material or halved the micro shape deviations) causes the decreasing of the head's failure probability which is based on the Weibull weakest link theory [1].

Acknowledgements. This investigation was supported by the projects MSM0021630518 and AV0Z20760514.

References

[1] McLean, A.F., Hartsock, D.L.: Engineered materials handbook, Volume 4, Ceramics and Glasses. In: ASM International 1991, pp. 676–689 (1991)

[2] Fuis, V., Janíček, P.: Stress and reliability analyses of damaged ceramic femoral heads. In: Damage & Fracture Mechanics VII, pp. 475–486. WIT Press (2002)

[3] Fuis, V., Návrat, T., Hlavoň, P., Koukal, M., Houfek, M.: Analysis of contact pressure between the parts of total hip joint endoprosthesis with shape deviations. Journal of Biomechanics 40(suppl. 2) (2007)

Estimation of Sympathetic and Parasympathetic Level during Orthostatic Stress Using Artificial Neural Networks

M. Kaňa[1], M. Jiřina[1], and J. Holčík[2]

[1] Czech Technical University Prague, Faculty of Biomedical Engineering, Department of Biomedical Informatics, Nám.Sítná 3105, 27201 Kladno, Czech Republic
kana@fbmi.cvut.cz, jirina@fbmi.cvut.cz
[2] Masaryk University Brno, Institute of Biostatistics and Analyses, Kamenice 3, 62500 Brno, Czech Republic
holcik@iba.muni.cz

Abstract. This study deals with the development of a new method to quantify the effect of orthostatic stress on the cardiovascular system. Orthostatic hypotension in healthy subjects triggers the baroreflex, which induces increased sympathetic activity and decreased parasympathetic activity. We performed a tilt-table test on 19 healthy subjects while measuring electrocardiogram, galvanic skin resistance and blood pressure signals. We developed a method for inverse parameters identification using artificial neural networks to fit the experimental data and identify physiological parameters (sympathetic and parasympathetic level). We implemented a supervised controller in the form of mathematical model of the baroreflex which was used to estimate the sympathetic and parasympathetic levels for a selected set of experimental data. Obtained result was used as training set for our artificial neural network. The network was able to estimate the levels of sympathetic and parasympathetic discharge. The estimated values were successfully validated against the measured heart rate signal with low least-square error. Additionally we proposed a classifier, which was able to predict the sex, age, weight and health status of the patient based on estimated model parameters.

1 Introduction

Postural change from a lying position to an upright posture causes decrease of arterial blood pressure and triggers the baroreflex. Some methods to assess the baroreflex sensitivity are presented in [1]. While such simple methods provide interesting information, the contribution of neural pathways involved is not covered. The use of mathematical models provides a better insight into the non-linear regulation processes during the orthostatic stress. Artificial neural networks have shown good capability for system identification. They can be trained by adaptive learning on chosen measured data. In this work we propose artificial networks which are able to predict the sympathetic and parasympathetic tone change during the orthostatic stress, as well as predict the age, sex, weight and health status.

2 Materials and Methods

2.1 Measurements

11 women and 8 men underwent the Tilt-Table Test. Their age varies between 15 and 46 years. A Biopac MP35 system [2] was used for 3 leads electrocardiogram (ECG) and galvanic skin resistance (GSR) signals measurement. A cuff arterial blood pressure (BP) signal was measured using an OMRON M10-IT device. The ECG and GSR signals were sampled at 500 Hz and the BP signal at 10 Hz. All three signals were processed in Matlab and Statistica Neural Networks. After instrumentation, in a lying position on the tilt table, a 60 seconds baseline recording was performed on the subjects. The table was then tilt up to 70°, and the recording was resumed for additional 60 seconds.

2.2 Mathematical Model

Sympathetic and parasympathetic nerve discharge cannot be measured non-invasively. Therefore the use of mathematical models is required in order to estimate this information from the measurable heart rate, blood pressure and galvanic skin resistance. We modeled the change in arterial blood pressure BP during orthostatic stress with an adapted guassian function as follow, where BP_0 is the mean blood pressure, dP_{max} is the maximum decline in pressure, dP_{time} is the time-stamp when the maximum decline occurs, $dP_{duration}$ is the duration of pressure change till it returns to normal level, $dP_{beforerate}$ and $dP_{afterrate}$ are the rate of change of pressure before and after the baroreflex is activated:

$$BT(t) = BP_0 - dP_{\max} * e^{-\frac{(t-dP_{time}-dP_{duration})^2}{2*dP_{duration}}} + \max(t-dP_{time},0)*dP_{afterrate} + \max(dP_{time}-t,0)*dP_{beforerate} \quad (1)$$

We adapted and extended a non-linear mathematical model for the baroreflex regulation proposed in [3] with galvanic skin resistance response to the orthostatic stress as follow. N is the firing rate of the baroreceptors. N_0 is the baseline firing rate; M is the maximum firing rate. Ns, N_I, N_L represent the short, intermediate and long term threshold of receptors with corresponding time parameters τ_S, τ_I and τ_L. α, k_L, k_I and k_L are weighting factors.

$$N = N_S + N_I + N_L + N_0 \quad (2)$$

$$\dot{N}_S = k_S * \alpha * (BP - BP_0) * N * \frac{(M-N)}{(\frac{M}{2})^2} - \frac{N_S}{\tau_S} \quad (3)$$

$$\dot{N}_I = k_I * \alpha * (BP - BP_0) * N * \frac{(M-N)}{(\frac{M}{2})^2} - \frac{N_I}{\tau_I} \quad (4)$$

$$\dot{N}_L = k_L * \alpha * (BP - BP_0) * N * \frac{(M-N)}{(\frac{M}{2})^2} - \frac{N_L}{\tau_L} \quad (5)$$

The level of parasympathetic tone is given by T_{par}. The sympathetic response is modeled by T_{sym} using β as damping factor.

$$T_{par} = \frac{N}{M} \tag{6}$$

$$T_{sym} = \frac{1-\frac{N}{M}}{1+\beta * T_{par}} \tag{7}$$

The concentrations of norepinephrine C_{nor} and acetylcholine C_{ach} on the sinoatrial node depend on Tsym, Tpar and the time constants τ_{nor} and τ_{ach}.

$$C_{nor} = \frac{-\dot{C}_{nor} + T_{sym}}{\tau_{nor}} \tag{8}$$

$$C_{ach} = \frac{-\dot{C}_{ach} + T_{sym}}{\tau_{ach}} \tag{9}$$

The heart rate HR deviation from its mean value HR_0 depends on the scaling factors M_s and M_p on the concentrations of norepinephrine and of acetylcholine.

$$HR = HR_0 * (1 + M_S * C_{nor} - M_p * C_{ach}) \tag{10}$$

The sympathetic tone influences the galvanic skin resistance GSR using the scaling factor k_{pr}. GSRo is the baseline resistance.

$$GSR = GSR_0 * (1 + k_{pr} * T_{sym}) \tag{11}$$

The mathematical model contains 23 unknown parameters which were estimated from the experimental data using the Systems Biology Toolbox 2 [4].

Table 1. Model parameters 1

Ks	Ki	Kl	τ_S	τ_I	τ_L	M	No	α	BPo	dPduration	dPmax
9.61 +/ - 3	0.21 +/- 0.17	0.11 +/- 0.13	0.97 +/- 0.39	11.72 +/- 3.57	739.48 +/- 218.07	60.67 +/- 73.38	43.33 +/- 12.54	3.42 +/- 0.85	113.63 +/- 4.04	30.32 +/- 5.78	8.62 +/- 2.41

Table 2. Model parameters 2

dPtime	dPafterrate	dPbeforerate	B	τ_{nor}	τ_{ach}	HRo	Ms	Mp	Kpr	GSRo
79.12 +/- 25	-0.01 +/- 0.04	-0.06 +/- 0.04	6.77 +/- 3.74	0.61 +/- 0.89	0.87 +/- 0.77	67.78 +/- 4.07	2.88 +/- 1.25	16.94 +/- 4.12	0.77 +/- 0.16	8.66 +/- 1.97

2.3 Artificial Neural Networks

We used Statistica Neural Networks to investigate multi-layer perceptrons (MLP). The selected MLP neural network (see Fig. 1) has two hidden layers with 27 and 9 nodes, three inputs and three outputs. The inputs to the network are measured heart rate in beats/min., measured blood pressure in mmHg and measured galvanic skin resistance in μMho. The outputs from the MLP are simulated heart rate in beats/min., simulated parasympathetic response and simulated sympathetic response (see Fig. 2). Both the inputs and outputs have been normalized to the interval 0-1. Product of inputs and weights has been used as a postsynaptic function in both hidden layers. Hyperbolic tangent has been used an activation function in both layers as well. The MLP has been trained using training, validation and testing sets. The original data has been split into these sets randomly in the ratio 2:1:1. Two-phase training has been used. In the first run well-know backpropagation has been employed. In the consequent run the conjugate gradient training algorithm has been applied to precisely tune the network.

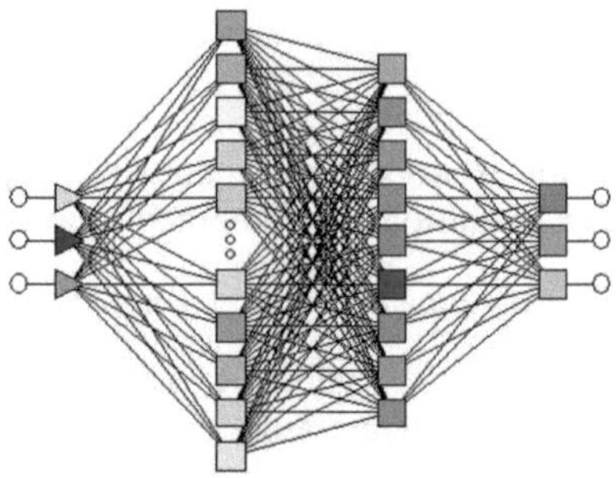

Fig. 1. Multilayer perceptron neural network architecture

The second task we were facing was a problem of classification patients according to the following characteristics: their sex (male or female), their age (15 - 46 years), their weight (50-110 kg) and the health status (1 for young healthy, 2 for elderly healthy, 3 for hypertensive). We have designed a similar network using MLP as well. The inputs to the classifier were the model parameters mentioned in Table 1 and 2; and the outputs were the corresponding patient characteristics.

3 Results

The 19 subjects were divided into two groups, who underwent the tilt table test in two distinct sessions. The first experiment session produced 22 datasets, which were used to fit the mathematical model using the downhill simplex method. The time course of arterial blood pressure and galvanic skin resistance as response to the orthostatic stress are depicted in Fig. 3. The obtained simulated sympathetic and parasympathetic tones, as well as heart rate were used as target outputs for the artificial neural network during training. The classifier used the age, weight, sex and health status information gathered during experiments as target outputs.

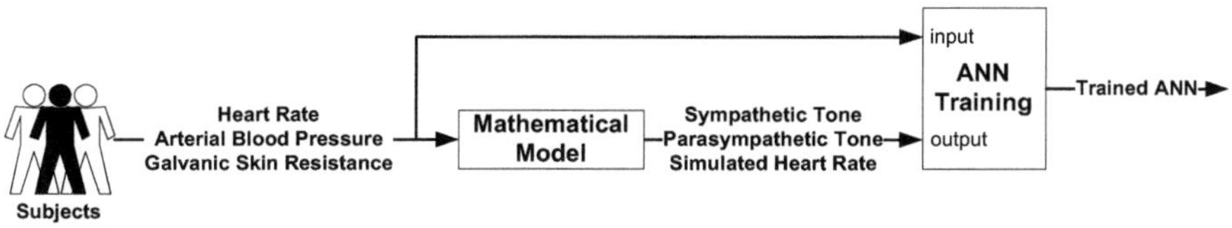

Fig. 2. Multilayer perceptron neural network training scheme

After the ANN and the classifier were trained, their efficiency was tested on data obtained during the second measurement session. 21 new data sets were recorded on 6 subjects. These data were applied as input to the ANN on one side and to the mathematical model on the other side. Both models output a simulated heart rate signal showing a good fit to the measured data, see Fig. 4. However the ANN appeared to include more non-linearity than the mathematical model. The ANN estimates sympathetic and parasympathetic tones as well. This output was compared to the one generated by the mathematical model. Fig. 5 shows the corresponding decreasing parasympathetic and increasing sympathetic tones with a good fit to the mathematical model.

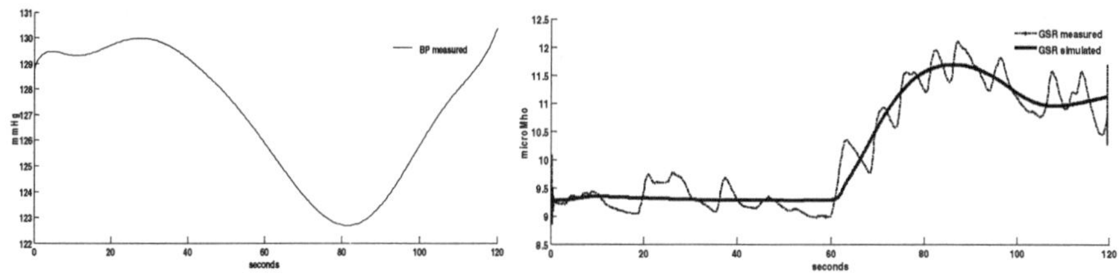

Fig. 3. BP and GSR response during orthostatic stress

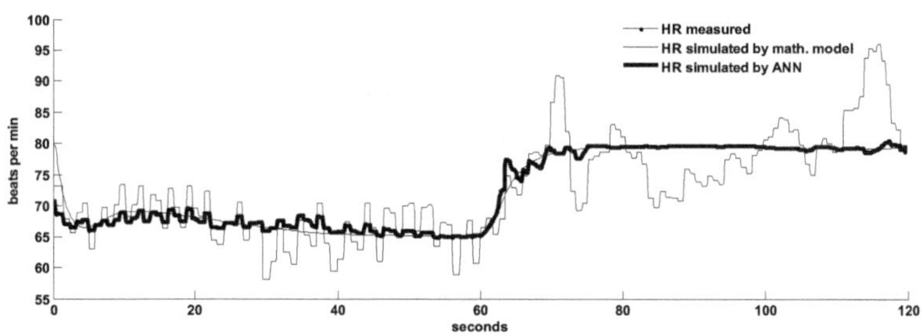

Fig. 4. Mean heart rate response during orthostatic stress

The estimated values of the mathematical model parameters (see Table 1 and 2) were applied as input to the classifier. We found out that the parameters contain useful information for predicting the health status and sex. The classifier was able to assign the right health status to unknown patients based on the corresponding estimated values of model parameters. More difficult task was the prediction of age and weight. In both cases the MLP has been employed as a regression model. After training the ANN responded correctly from the view of an order of values of the output values but incorrectly from the viewpoint of absolute values. Even so the network proved importance to give correct trends and interpretable results.

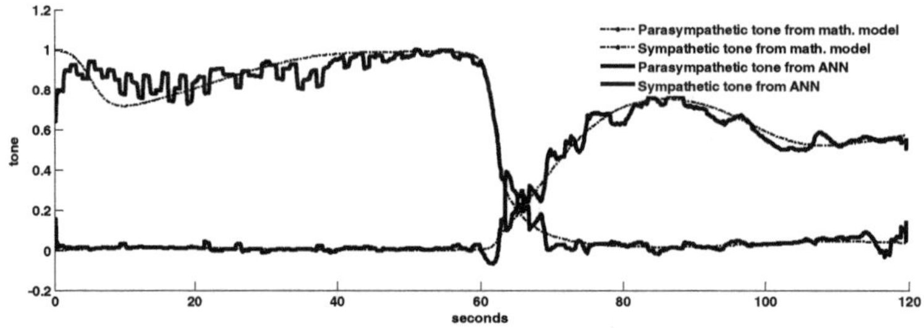

Fig. 5. Mean parasympathetic and sympathetic tone

The least-square error for the ANN and the classifier varies within range 0.11 - 0.06. More important is that the error signal was approximately the same for training, validation and testing set, what is a good measure of ANNs quality training. The mathematical model produced an error within range 0.0022- 0.088.

4 Conclusion

We could use the non-linear nature of artificial neural networks to describe some mechanism of cardiovascular control. The level of parasympathetic and sympathetic discharge on the sinoatrial node of the heart could be predicted, as well as the sympathetic tone on the vasculature and the patient health status. The network has been validated against measured data obtained during the tilt table test and the results agree with those presented in [3]. The success of the proposed network depends on supervised learning. This requires an accurate model of the cardiovascular control; otherwise it is not possible to train the ANN. A future research could be on unsupervised control during learning, where the controller tries diverse actions in order to reduce the error signal. Such a network might be more suitable for modeling autonomic cardiovascular control.

References

[1] Hilz, M.J., Dütsch, M.: Quantifitative Studies of Autonomic Function. Muscle Nerves 33, 6–20 (2006)

[2] Biopac Systems, Inc. (2009), http://www.biopac.com

[3] Olufsen, M.S., Tran, H.T., Ottesen, J.T.: Modeling baroreflex regulation of heart rate during orthostatic stress. Am. J. Physiol. Regul. Integr. Comp. Physiol. 291, R1355--R1368 (2006)

[4] Schmidt, H., Jirstrand, M.: Systems Biology Toolbox for MATLAB: A computational platform for research in Systems Biology. Bioinformatics 22(4), 514–515 (2006)

Human Downfall Simulation

J. Čulík, Z. Szabó, and R. Krupička

Czech Technical University of Prague, Faculty of Biomechanical Engineering,
Sitna 3105, 272 01 Kladno, Czech Republic
Tel.: +420-312608208; Fax: +420-312608204
culik@fbmi.cvut.cz, szabo@fbmi.cvut.cz

Abstract. Total knee and hip implants are usually designed for the static load of standing man, but the static load is not extreme. The top of research is to search extreme values forces and moments at leg and arm joints. The implants can be then designed for these extreme dynamics loads. Human fall was observed by camera system to record the position data of main points on human body. The 3D coordinates of the end of foot, ankle, knee, hip, pelvis, shoulder and wrist were stored. The simulation program of human fall was compiled at PC in language C++ using simulation system CDCSIS. The program has an input data - length, mass and inertia moments for each part of human body. The data are transformed according to actual coordinates of points on the body. Then the moving and turning accelerations are calculated and joint forces and moments are determined according to the d'Alembert's principle. At the conclusion the noise is canceled.

1 Introduce

If any implant is put to human joint we must know loading of this implant. Very often the load of standing human is used but it is not very predicative. If the implant is designed according to the strength then a downfall search is suitable. If the fatigue occurs then it is possible to find the force of human gait in time and if the implant is designed on strength then a downfall search is suitable.. The object of this article is to determine the joint forces and moments in the course of fall.

The human body is divided to elements: foot, calf, femur, trunk and arm (hume-rus and radius part). The input data for computing algorithm has two types: the constant data for all body elements and data for specific movement. The constant data are:

- Mass and inertia moments of parts of the human body.
- Lengths and widths of human body part.

Measured data for actual movement:

- Coordinates of observed points.

The figurant motion was observed by Lukotronic MCU 200 camera system. The observed system and model of the body with the positions of the active markers is demonstrated at fig. 1. The observed points are: end of the foot, ankle, knee, hip pelvis, shoulder and wrist. The accelerations of observed points are calculated numerically. The accelerations of gravity centers and rotation acceleration of human body parts are calculated from accelerations of individual parts.

The moments of inertia are determined for femur and tibia plane and/or humerus and radius plane; respective perpendicular planes. The equilibrium equations according to the d'Alembert's principle are written for each body parts. Three forces and three moments at each joint have to be calculated. The calculation of forces and moments starts from leg and/or arm ends. If these points are not in contact with floor the forces and moments are zero but they are not unknown. From the equilibrium conditions are determined forces and moments on the other side of body. If the end of the foot has a contact with floor then the forces and moments are not zero and the equilibrium condition for the whole human body must be considered.

2 Materials and Methods

The Lukotronic MCU 200 camera capturing system was used for motion capturing. Each camera unit consists of 3 single infrared cameras, that measure special move-ments of active infrared-markers in real-time. By means of the three single cameras the motions of the markers are determined in three dimensions [2].

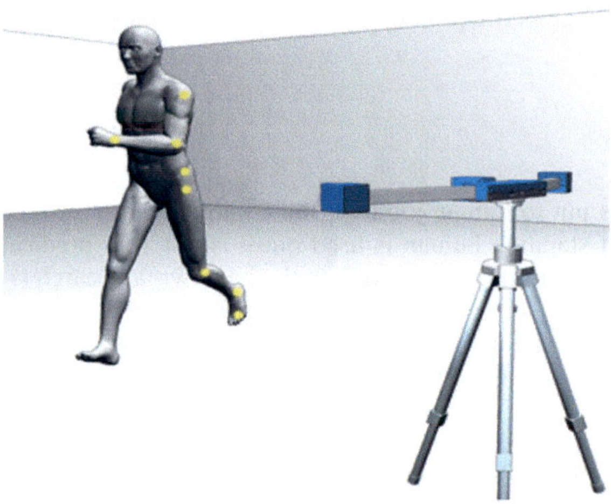

Fig. 1. Lukotronic MCU 200 camera system and model of the body with the positions of the active markers.

We placed eight markers on body anatomical landmarks. The system can capture eight markers at five meter distance from the camera. Capturing frequency was set to 25 Hz. The system returns 3 dimensional coordinate of each marker. The origin of the coordination system is in the middle camera (see Fig. 1).

3 Results

The input data are 3D coordinates of the points on the body measured with camera system at regular time intervals. The average length of body elements and their mass and moments of inertia come into the program as constants. The length of the element is calculated from the measured coordinates. The constant values are corrected according to scale of average and measured length values.

The accelerations of measured points are calculated numerically from

$$\ddot{x}_i = \frac{x_{i-1} - 2x_i + x_{i+1}}{h^2} \tag{1}$$

where h is time step, x_{i-1}, x_i, x_{i+1} are values of coordinates in consecutive time points. The acceleration of gravity center is calculated from accelerations of the end of body elements.

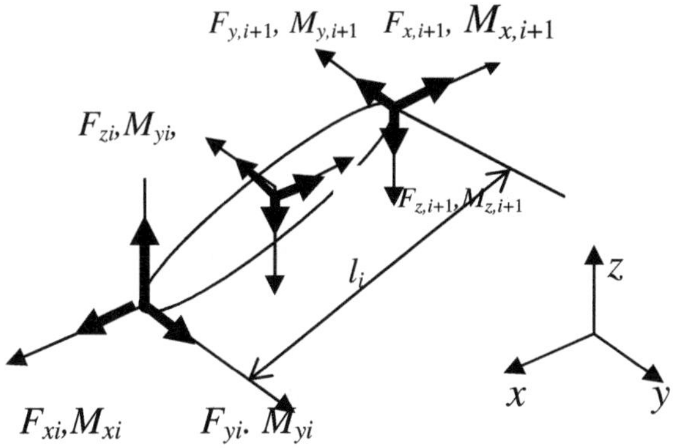

Fig. 2. Part of the human body with end forces and moments.

The human body element, the end forces and the moments are illustrated on Fig.2. The distance of the gravity center from the left side is denoted as a_i and from the right side is denoted as b_i.

The equilibrium condition of external and inertia forces is

$$\vec{F}_i - \vec{F}_{i+1} - \vec{a}_i m_i = \vec{0} \tag{2}$$

The moment equilibrium conditions are written for femur and tibia plane of the leg and for humerus and radius plane of the arm. The unit vector $\vec{r}_1$ is perpendicular to this plane and it is result of vector multiplication of vectors in direction of femur and tibie and/or humerus and radius. The $\vec{r}_2$ is a unit vector at the axis direction of the body element.

The $\vec{r}_3$ is a unit vector perpendicular to the femur – tibia plane and/or humerus – radius plane

$$\vec{r}_3 = \vec{r}_1 \times \vec{r}_2 . \tag{3}$$

The rotate acceleration vector has the following coordinates

$$\vec{\varepsilon} = \begin{Bmatrix} -\dfrac{a_{2y} - a_{1y}}{l} + \dfrac{a_{2z} - a_{1z}}{l} \\ \dfrac{a_{2x} - a_{1x}}{l} - \dfrac{a_{2z} - a_{1z}}{l} \\ -\dfrac{a_{2x} - a_{1x}}{l} + \dfrac{a_{2y} - a_{1y}}{l} \end{Bmatrix} . \tag{4}$$

The rotate acceleration ε_1 according to vector $\vec{r}_1$ and the rotate acceleration ε_2 according to vector $\vec{r}_3$ are

$$\varepsilon_1 = \vec{\varepsilon}.\vec{r}_1 , \; \varepsilon_2 = \vec{\varepsilon}.\vec{r}_3 . \tag{5}$$

The inertia moments at the directions mentioned above are $I_1\varepsilon_1$, $I_2\varepsilon_2$. The moment of inertia is defined as

$$\vec{M}_i^* = I_1\varepsilon_1\vec{r}_1 + I_2\varepsilon_2\vec{r}_2 . \tag{6}$$

The moment equilibrium conditions are

$$\vec{F}_i \times \vec{\Delta}_1 - \vec{F}_{i+1} \times \vec{\Delta}_2 + \vec{M}_i - \vec{M}_{i+1} - \vec{M}_i^* = \vec{0} , \tag{7}$$

where $\vec{F}_i, \vec{F}_{i+1}, \vec{M}_i, \vec{M}_{i+1}$ are force and moment vectors at the start and the end of the element. Vectors $\vec{\Delta}_1, \vec{\Delta}_2$ have the following coordinates

$$\{\Delta_1\} = \frac{a}{l} \begin{Bmatrix} x_2 - x_1 \\ y_2 - y_1 \\ z_2 - z_1 \end{Bmatrix}, \; \{\Delta_2\} = \frac{b}{l} \begin{Bmatrix} x_2 - x_1 \\ y_2 - y_1 \\ z_2 - z_1 \end{Bmatrix} .$$

The distances between the center of gravity and the forepart (index 1) and/or the end of the element (index 2) are *a, b*.

The forces and moments are calculated from the end of the leg (point 3 – the end of foot) and/or the end of the arm (point 6 – wrist). In the case, when the arm is not in contact with the floor, then

$$\vec{F}_6 = \vec{0}, \; \vec{M}_6 = \vec{0} . \tag{10}$$

If the leg is not in contact with the floor (point 1 – end of fuss) then

$$\vec{F}_1 = \vec{0}, \; \vec{M}_1 = \vec{0} .$$

If the leg is in contact with floor (point 3 – fuss under the ancle) and the 2nd leg is not, then

$$\vec{F}_3 = -\sum m_i \vec{a}_{c,i} , \tag{8}$$

$$\vec{M}_3 = -\sum \vec{M}_i^* - \sum \vec{r}_{3,i} \times m_i \vec{a}_{c,i} , \tag{9}$$

where $\vec{a}_{c,i}$ is acceleration of center of gravity, $\vec{M}_i^*$ can be calculated from (6) and

$$\vec{r}_{3,i} = \vec{X}_i + \vec{\Delta}_{1,i} - \vec{X}_3 ,$$

where $\vec{X}_i$ are coordinates of point i.

If both of the legs and/or leg and wrist are in contact with floor (points 3 and 8) then it is supposed that $\vec{M}_3 = \vec{0}$ and $\vec{F}_3$ is nonzero; at point 8 only moment M_8 is non zero, on the axis connecting points 3 and 8 and F_{y8}, F_{z8}. The moment equilibrium condition for point 3 is

$$(\vec{X}_8 - \vec{X}_3) \times \vec{F}_8 - \sum \vec{r}_{c,8,i} \times m_i \vec{a}_{c,i} - \sum \vec{M}_i^* + M_8 \vec{r}_{38} = \vec{0} , \tag{10}$$

where $\vec{X}_3, \vec{X}_8$ are coordinates of the points 3, 8. The vector

$$\vec{r}_{c,8,i} = \vec{X}_i + \vec{\Delta}_i - \vec{X}_8$$

is coordinates of element centers of gravity, where $\vec{X}_i$ are coordinates of element starts and $\vec{\Delta}_i$ are local coordinates of gravity centers. The unit vector at direction from the point 3 to the point 8 is

$$\vec{r}_{38} = \frac{1}{l}\left(\vec{X}_8 - \vec{X}_3\right) ,$$

where l is distance between points 3 and 8. The vector equation (10) (3 scalar equations) has 4 unknowns – the vector $\vec{F}_8$ and the scalar M_8, but we set $F_{8x} = 0$.

The force $\vec{F}_3$ can be calculated from the force equilibrium condition

$$\vec{F}_3 + \vec{F}_8 - \sum m_i \vec{a}_{c,i} = 0 . \tag{11}$$

4 Conclusion

The computer simulation according to the previous algorithm was done. The program uses the simulation system CDCSIS having two parts: fall animation and joint forces and moments calculation. The outputs of the second part of simulation are the time course of signals of joint forces and moments. The examples of results: the graphs of vertical forces in time during the falls to knee are shown in the Fig. 3. The upper signals at the graphs are the forces at the ankle, the middle one at the knee and the bottom at the hip (the time in seconds is from zero to value displayed above, the forces are in Newton, tensile is positive and push is negative). The jump in the signal is in time moment when the knee has gone in contact or passed the contact with floor. The minimum and maximum of signals is written under the each graph.

The occurrence of results depended on the time step of measurement. The origin of inaccuracy is too large time step, numerical calculation of the 2nd derivation and signal noise. The high-speed camera using is planned to follow research program. The body point acceleration will be measure with accelerometers.

The used camera capturing system and the calculation algorithm will be used for human downfall simulation. The goal of next research will be determination of extremes of joint forces; the extremely forces will be used for implant design. The knowledge of the knee, hip and finger joints forces are very important in the mentioned cases.

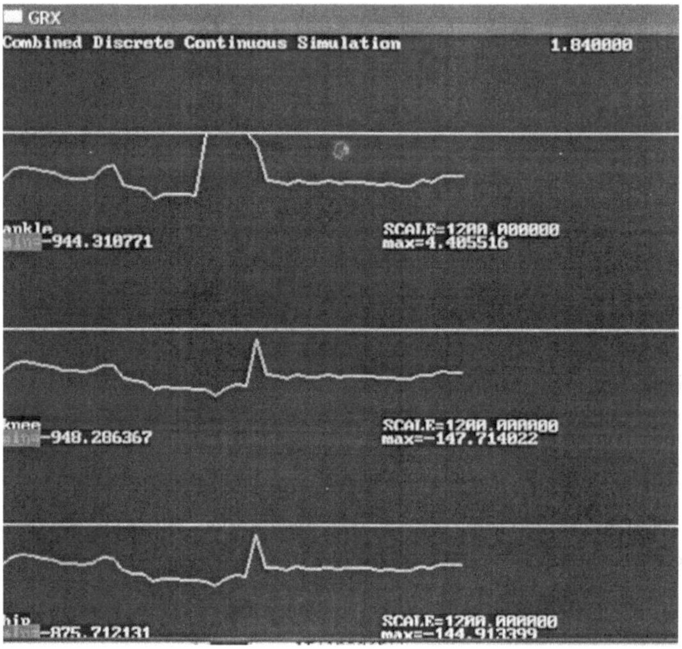

Fig. 3. Vertical forces of the ankle, knee and hip during the fall on knee.

Acknowledgment. The research was support by the Czech Technical University grant SM6840770012 "Trans-disciplinal Research at Biomedical Engineering Area".

References

[1] Culik, J.: CDCSIS C++. Manual of Czech Technical University in Prague (2008)

[2] Lukotronic, Motion measurement system (2009), http://www.lukotronic.com/

[3] Szabó, Z., Krupička, R., Rozinek, O.: Technical Background of 3D Motion Analysis of Patients with Neurological Diseases. In: IX. International Conference Symbiosis 2008, pp. 32–34. Slesian Technical University, Gliwice (2008)

Heuristic Methods in Gait Analysis of Disabled People

B. Kabziński and D. Jasińska-Choromańska

Warsaw University of Technology faculty of Mechatronics,
Instytut Mikromechaniki i Fotoniki, A. Boboli 8 Warszawa, Polska
danuta@mchtr.pw.edu.pl

1 Introduction

Movement is used to achieve the optimal position to acquire the desired impulse or perform an activity. The basic condition of gait development is a reliable nerve-muscle controlling system, which begins in the brain, traverses the spinal movement center, and ends in the motion receptors, in the muscles and hamstrings. This allows to preserve balance and perform complex movements.

Gait pattern may change depending on many factors as the patient's age, muscle weakening or pain. The ageing of a patient's organism results in degenerative changes, balance movement coordination disorders, which persistently change the patient's gait characteristics. Pain is also a very significant factor changing gait pattern because the patient adapts gait in order to minimize pain during movement. The problem is when pain withdraws while the old gait pattern remains [3].

Every abnormality in gait sequence results in larger energetic consumption which leads to the overloading of the cardiac and respiratory systems. Therefore the process of classification of the walking pattern has key meaning for the length of life In analysing pathological gait, normal function is the model against which disability is judged. Deviations from the normal pattern define the functional error.

These errors include all segments from toes to trunk and are applicable to all types of pathology.

The walking cycle (stride), described as successive motion phases, is divided into two basic periods: stance and swing, which are further divided depending on their functional tasks into eight phases (Fig. 1).

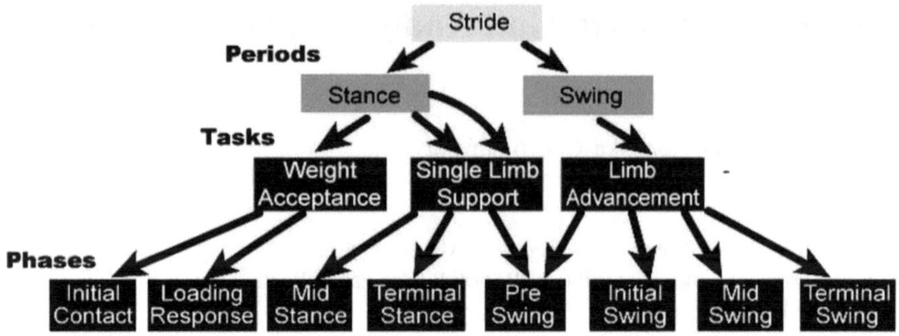

Fig. 1. Gait systematics

Analysing human walk consists not only of observing motion, but also of other relevant techniques which are involved in this process and make analysis complete.

2 Methods

Methods of gait analyzing:

- motion analysis
- electromyography
- ground reaction force
- energy expenditure

2.1 Motion Analysis

Motion is much easier to observe than to measure. While the major arcs of joint motion occur in the sagittal plane, there are also subtle actions occurring in the coronal and transverse planes. These deviations in the sagittal plane during movement are often much greater in the disabled walker and they may have great importance for drawing conclusions. Motion can be measured by 2 different means: electrogoniometers which are attached to limb joints and measure joint flexion.

2.2 Electromyography

Electrical signals, which accompany the chemical stimulation of muscle fibers, travel through the muscles and adjacent soft tissues. With appropriate instrumentation, these myoelectrical signals can be recorded and analysed to determine the timing and relative intensity of the muscular effort.[1]The purpose is to enable the clinician to sense the relative intensity of a muscle's action during the stride.

A sampling rate of 2500 Hz captures all the significant data. To simplify data storage, many investigators use sampling rates of 500Hz. There is a corresponding loss of data that may be significant, but in general purpose application of gait analysing it is sufficient. As a result information about the muscle stimulation pattern is gathered and correlated with the gait cycle to determine whether the muscles are activated in the appropriate moment.

2.3 Ground Reaction Force

As body weight drops onto and moves across the supporting foot, vertical, horizontal and rotatory forces are generated on the floor that can be measured with appropriate instrumentation. These ground reaction forces are equal in intensity and opposite in direction to those being experienced by the weight-bearing limb. The ground reaction forces can be represented as a single vector that combines the simultaneous vertical, sagittal and coronal forces. The graph is scaled in the percentage of total body weight and the percentage of ground contact time (GC) (Fig. 2).

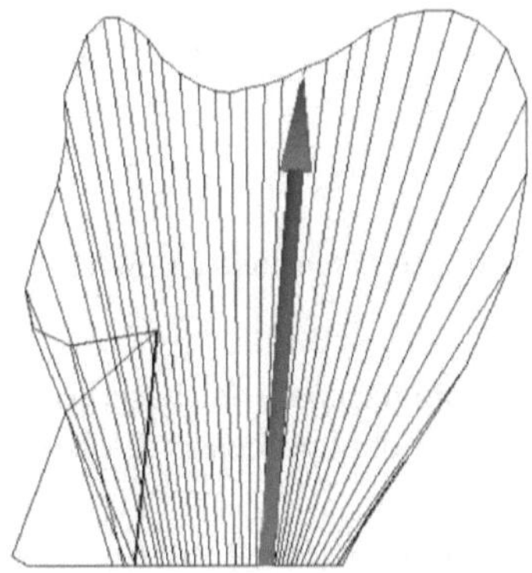

Fig. 2. Pattern of sagittal vectors during a stride

3 Discussion

While using all of those methods, a clinician gets significant amount of data. It considers sagittal, hip, knee and tarsal joint rotation angles, rotation moments, and generated power, ground reaction force, muscle group activation pattern. Each spatial parameter is described in 3D, as a result of analysis obtains set of graph (Fig. 3) which have to be interpreted correctly and as a result a diagnosis should be given. Correlation of such amount of data and influence of one to another, knowledge which is more significant than others requires experience and skills from scientific staff. Such abilities are to be gained only by long practice and still diagnosis can be influenced by various factors e.g. psychical condition, bad weather health disturbances, so diagnosis is highly subjective, It is common problem that the same patient has two different diagnosis.

Natural effect of such problems in diagnostics of gait disturbances is searching for more reliable and quantitative methods than subjective human assessment.

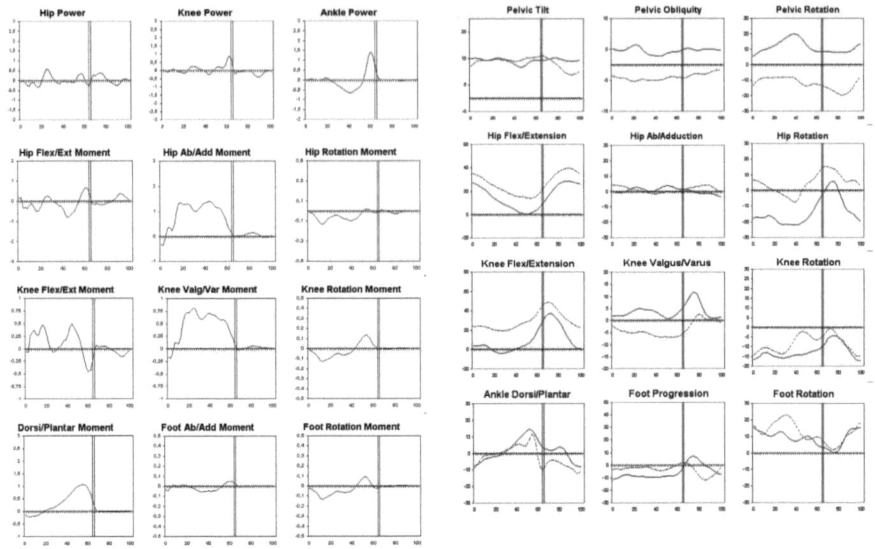

Fig. 3. Example of gait parameters graphs for left and right limb

A solution seems to be a computer analysis (based on heuristic methods) and limiting amount of input data using data processing. These methods are:

- Normalization in time – reduction of sample quantity, resampling – equal amount of samples for each parameter.
- Fast Fourier Transformation (FFT) reflects the frequency distribution of temporal signals and is another method of reducing amount of input data.
- Extracting parameters e.g. peak values of parameters (commonly used)

Different methods are used for automatic recognition of movement patterns. One of them is support vector machines based on finding optimal separating hyper planes of data sets [2]. Other is waveform data reducing technique to a statistical measures of distance which indicate whether a patient has a similar gait pattern to a normal one [4]. Fuzzy logic techniques were used based on determining to which data set certain gait pattern is assigned. What seems most promising for automatic gait pattern recognition is application of artificial neural network (ANN) combined with input data limiting techniques, but effect strongly depends on type of input data used during analysis. If only temporal data are used (time of double support and right and left single support phases) a result is limited to ability of assigning whether gait pattern is close to normal one or not, or determining seed of walking measured in statures s-1 [5]. However, if taking into consideration other gait parameters (hip and knee joint angles) this method could determine different types of gait characteristics [3] and even types of malfunctions causing specific gait pattern. The problem that seems to be relevant is proper selection of gait parameters that can describe gait characteristic and methods that limit the amount of input data to be analyzed by ANN. The problem of parameter selection is also relevant for reason of result comparison between different gait assessment centers.

4 Methods

To achieve such a set of representative parameters an artificial neural network would be used. First step is reducing number of input data, as previously mentioned, it is significant quantity and it has to be strongly reduced to be able to use it further research exploiting heuristic methods. Each dynamic parameter (shown on Fig. 3) consisting of 50 samples was reduced to 1 or 2 values e.g. mean, max value or local minimum, of a waveform. Such drastic data reduction was necessary for further analysis utilizing heuristic methods. Next step is to train and test ANN using all available data, and then follow the same procedure with different sets of input data. For this purpose a basic, back propagation neural network and self organising.

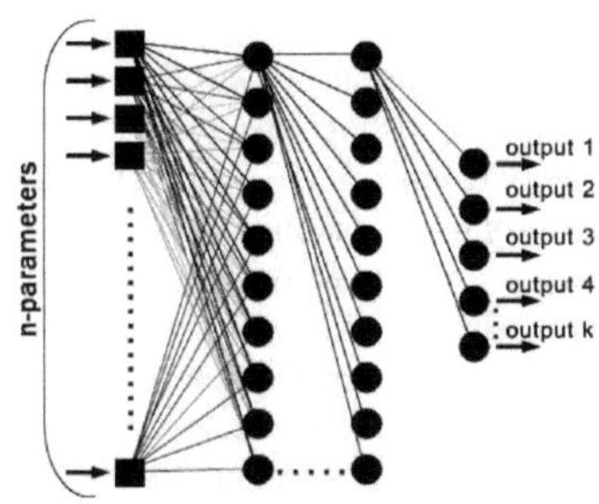

Fig. 4. Artificial Neural Network structure (not all connections shown)

A self-organizing map consists of components called nodes or neurons. Associated with each node is a weight vector of the same dimension as the input data vectors and a position in the map space. The usual arrangement of nodes is a regular spacing in a hexagonal or rectangular grid. The self-organizing map describes a mapping from a higher dimensional input space to a lower dimensional map space. The procedure for placing a vector from data space onto the map is to find the node with the closest weight vector to the vector taken from data space and to assign the map coordinates of this node to our vector.

As a result ANN would classify gait patterns into several categories. All results will be compared with clinician's diagnosis to verify how a certain set of data is reliable. As a conclusion would be a selection of best set of input data.

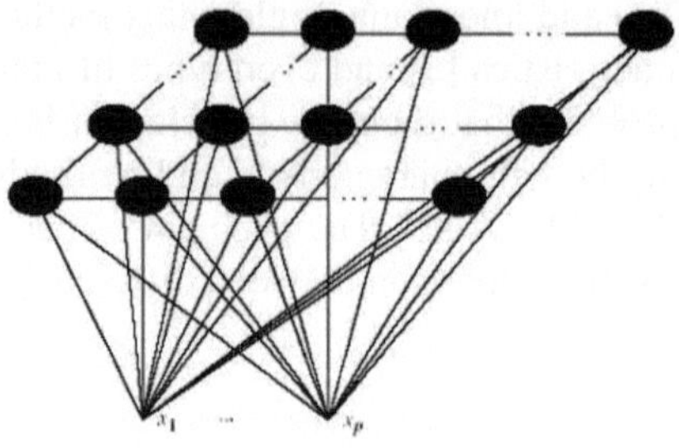

Fig. 5. Graphical representation of self organizing map

Acknowledgments. This work was supported in part by the Polish Ministry of Science and higher education, under Grant 1926/B/T02/2007/33.

References

[1] Perry, J.: Gait Analysis. Normal and Pathological Function, Slack Incorporated (1992)

[2] Begg, R., Kamruzzaman, J.: A machine learning approach for automated recognition of monement patterns using basic, kinetic and kinematic gait data. Journal of Biomechanics 38, 401–408 (2005)

[3] Barton, J.G., Lees, A.: An application of neural network for distinguishing gait patterns on the basis of hip-knee angle diagrams. Gait and Posture 5, 28–33 (1997)

[4] Deluzzio, K.J., Wyss, U.P., Costigan, P.A., Sorbie, C., Zee, B.: Gait Assesment in unicommpartmental knee arthroplastry patients: Principal component modelling of gait waveforms and clinical status. Human Movement Science 18, 701–711 (1999)

[5] Gioftsos, G., Grieve, D.W.: The use of neural networks to recognize patterns of human movement: gait patterns. Clinical Biomechanics 10, 179–183 (1995)

[6] Jasińska-Choromańska, D., Kabziński, B.: Methods for analyzing human motion functions. Elektronika, Nr 8-9, SIGMA, Warszawa (2004)

Author Index

MIX
Papier aus verantwortungsvollen Quellen
Paper from responsible sources
FSC® C105338

If you have any concerns about our products,
you can contact us on
ProductSafety@springernature.com

In case Publisher is established outside the EU,
the EU authorized representative is:
Springer Nature Customer Service Center GmbH
Europaplatz 3, 69115 Heidelberg, Germany

Printed by Libri Plureos GmbH
in Hamburg, Germany